Meanings of Pain

Simon van Rysewyk
Editor

Meanings of Pain

Volume 2: Common Types of Pain and Language

Springer

Editor
Simon van Rysewyk
School of Humanities, Department
of Philosophy and Gender Studies
University of Tasmania
Hobart, Tasmania, Australia

ISBN 978-3-030-24156-8 ISBN 978-3-030-24154-4 (eBook)
https://doi.org/10.1007/978-3-030-24154-4

This Springer imprint is published by the registered company Springer Nature Switzerland AG.
The registered company address is: Gewerbestrasse 11, 6330 Cham, Switzerland

Foreword

As a clinician/researcher I have spent more than 20 years listening to more than 23,000 patients with persistent pain, trying to understand who they are and what is their problem and how can we help them. It has been a humbling and human experience to share their road with them. They (the patients) have struggled to describe their pain, their distress and their losses. More often than not their experience of the medical system has been frustrating and unrewarding for them and in some cases accentuated the problem.

My experience as a physician is obviously influenced by my own cultural and personal take on pain, in my case a triple level spinal fracture as a medical student (and 4 years of back pain followed by recovery) and a C6 radiculopathy as a pain physician. I am one of the lucky ones—my nervous system seems hardwired to return to homeostasis over several months. My radiculopathy experience proved instructional to me on several levels—a complete failure of any medical treatment to make a difference to the agonizing pain, a hardening of my response to patients under my care (I more frequently ceased their opioids during this period) and ultimately a resolution of my pain when I took a medically prescribed holiday from my practice.

Why mention this? Because too often medicine takes a reductionist approach to the problem of pain. That most assuredly leaves the transfer of understanding of the experience hollow and superficial and any treatment of it necessarily incomplete and often unsatisfying. This book addresses a critical aspect of this problem, namely how to capture the lived experience and how to interpret the meaning of pain both for the experiencer and for the observer.

This book should provide the reader with not only a vocabulary of language about pain and a number of ways in which meaning can be explored and captured but above all it should engender in the reader a sense of enquiry and self-reflection on how to take this information and apply it in the clinic. Many of us could do with less of the doing and more of the thinking about what the doing is for and what the doing represents. This multi-author, multi-disciplinary and multi-country contribution is a welcome addition to the knowledge base of persistent pain.

It is a striking fact that we know more and can predict more on the behavioural responses of the rat to acute and persistent pain than we understand the reactions of our fellow human beings in a cultural/societal/interpersonal and family context to persistent pain. This book bridges that gap of understanding.

As an avid reader of primary foundational texts in medicine such as the textbooks and treatises of Sir William Osler, Silas Weir Mitchell, John J Bonica and Harvey Cushing, amongst other giants of medicine, I have been struck not only by their deep empathy for the people they treated but also by their extensive period of listening to and observing of the patient. Hours of observation was not unusual for these doctors and from their deep observational skills they assembled many first proper documentations of conditions, pathological hypotheses and attempts at treatments (some spectacularly successful). The long stay inpatient nature of care in the hospital at that time no doubt aided their observation as well as the serial observation of long periods of the natural history of the condition, but it was the acuteness of their attention that stood out in their published works. In our delivery of modern (minutes) medicine, we could do well to spend some time immersing ourselves in the lives of our patients if we want to understand what they are going through. To do that we will need a conceptual framework of that experience just as much as a conceptual framework of diseases and their pathology, and this book will hopefully give you, the reader, those concepts.

It is my opinion that this three-volume work will stand as the definitive reference work in this field. I believe it will enrich the professional and personal lives of health care providers, researchers and people who have persistent pain and their family members. The combination of framework chapters with chapters devoted to analysing the lived experience of pain conditions gives the requisite breadth and depth to the subject. I commend Simon van Rysewyk for the herculean task of taking on the birth of a tome so wide in scope and ambition.

Newcastle, NSW, Australia Marc A. Russo

Contents

Chapter 1
Exploring the Meanings of Pain: My Pain Story

Joletta Belton

Abstract First-person narratives of the lived experience of pain, and the meanings of that experience, are uncommon, especially from persons who are not also clinicians or researchers. Yet such narratives could be particularly useful in understanding pain. First-person accounts, stories of pain, can lend unique insights into the lived experience of pain, how individuals make meaning of it, how they come to those meanings, and how those meanings can change over time. Such narratives could lead to new areas of inquiry and explorations of new possible treatment paths. This chapter provides such a narrative, offering a glimpse into one person's lived experience of pain and its meanings. It demonstrates how our individual narratives, our stories, help us make sense of our experiences, including pain. It demonstrates how our narratives can change over time as new information and understandings lead to new meanings, and how such changing narratives and meanings can be a part of a therapeutic process that can lead to better outcomes for patients and clinicians alike.

Clinical Implications: This chapter provides a first-person account of the lived experience of pain and recovery. It explores the meanings of pain, how they came to be, and how those meanings change over the course of time, from early onset of pain through worsening, unexplained pain to recovery from pain.

Keywords Lived experience · Pain · Chronic pain · Narrative · First-person phenomenology · Meanings of pain · Patient experience · Recovery · Healing

1 Introduction

What are the meanings of pain? It is a question that has been asked by researchers and clinicians, philosophers and poets, for centuries. A question that is no doubt asked by many of the millions of people who live with pain. It is an incredibly

J. Belton (✉)
Endless Possibilities Initiative, Fraser, CO, USA
e-mail: Jo@epicolorado.org; http://www.mycuppajo.com/

S. van Rysewyk (ed.), *Meanings of Pain*,
https://doi.org/10.1007/978-3-030-24154-4_1

difficult question to answer. The meanings of pain are deeply personal, yet also shared. They change over time, within individuals and within cultures [1]. Just as the meanings of pain are ever-changing, the experience of pain is ever-changing, too. Perhaps that is one and the same thing. Pain changes from day-to-day, moment-to-moment, year-to-year. It has a different character in different contexts, yet also stays the same, one of pain's many paradoxes [2]. Pain is certainly not just a sensation, not just a symptom, not any one "thing" to be objectified at all [3]. Pain is *lived* rather than known, *experienced* rather than felt or thought about. We *have* pain, taking ownership of it in unique ways. We rarely, if ever, say we have joy, have love, have fear. Pain takes ownership of us as well. It disrupts our lives [4], destabilizes our worlds [2], threatens our identities, our futures...our very existence [5–7]. Pain is the unmaking of the world [8].

This chapter reflects what my pain has meant to me over the years, my own changing meanings. From the early days, when I first felt a twinge in my hip stepping off a fire engine on a routine call at work, to the ongoing worsening pain that unexpectedly ensued, to seeking care and failing treatments, and to an eventual path to recovery. It's a story of darkness and despair, of not being understood and not understanding, as well as one of light and hope, of connection and making sense of things.

To be clear, when I say "my pain" I mean very specifically the pain I felt in my right hip for many years, not the many pains that came before, the pains that came during, or the pains that have come since. I've never referred to the other pain I've experienced as "my pain." That in itself is interesting. I don't know exactly when it became "my pain." I suspect it did so when it didn't resolve as all the other pains have, when it became *chronic*. There was no distinct demarcation between acute and chronic pain, though. It was the same pain, just continuing on for longer than it *should* have [6]. This pain, *my pain*, was different. I suspect some of the difference lies within the meanings associated with it, the meanings I explore in this chapter.

I could not explore these meanings without help. There are references and quotations throughout this chapter from researchers, clinicians, writers, and fellow humans who've lived with pain whose words and work have helped me give voice to what has oftentimes been inexpressible. There are many more whose works and guidance have influenced me in immeasurable ways and have helped me to make sense of things but are not referenced here. To all of them I extend my deepest gratitude.

2 My Pain Story

My pain started during a routine call, a medical aid in the middle of the night. My crew and I had transported a patient to the hospital and were about to pull away when I realized I'd left my clipboard in the emergency room. I'd stepped off fire engines thousands of times over the course of my career as a firefighter paramedic without incident but, this one time, I missed the step and felt a twinge in my hip. Just a twinge. Little did I know that the missed step would land me on a path of ongoing, worsening pain that didn't make sense. That this small twinge would lead to a downward spiral into a darkening abyss of despair, hopelessness, and loss. Of pain and suffering, of isolation and withdrawal from the people, places, and experiences that mattered to me. That made me *me*.

I was lost. If you'd asked me to define myself before this pain set in, I would have answered: "I'm a firefighter." To me that encapsulated all that I was. Strong, fit, capable. A badass. A hero. A firefighter.

And then, one day, I wasn't.

The pain eventually led to my medical retirement. With the loss of my identity, I also lost my purpose and self-worth, my way of being in the world and of relating to the world. I lost my social roles, my hobbies, my meaningful activities. I was no longer a runner, a weightlifter, a medic, an athlete. No longer a firefighter. No longer *me*. The world disappeared, my horizon of meaningfulness diminished until nothing, but the pain remained [7, p. 116]. Pain became everything. I became pain.

That story, my story, is not found in my medical record. When I look through my years of paperwork, reports by numerous physicians, clinicians, and claims adjusters, my experience is nowhere to be found. In its place are lists of symptoms and limitations, iterations of a hip injury history, myriad (and sometimes conflicting) labels and diagnoses, and a litany of failed treatments. My pain is nowhere in there. My losses, my grief. My suffering glossed over or missing altogether.

My record says things like "pleasant thin female," "still has discomfort with sitting." It doesn't say "distressed and worried human being who hasn't been able to sit for over 2 years, which has wholly altered her life and her person." No sitting meant no driving a car, no meeting friends for coffee or dinners out, no socializing or visiting family, no trips to the grocery store, no watching a movie on the couch with my husband. No work. No play. No life.

It is noted in the pages of my medical record that I was forced to medically retire from my career, but there are no notes on how medical retirement meant losing the identity that had defined me. How with it I lost my sense of worth and purpose. How I lost my friends, financial security, and future. How I had lost a drastic amount of weight and was no longer recognizable to myself, my family, my friends. I was not "pleasantly thin," I was *terrifyingly* thin. Firefighters I had trained and worked alongside for years were shocked when they saw me, they thought I was dying. Their shock, the shock of all of this experience for me, is not in my paperwork. What pain meant to me, what it meant for my life, for my future, was not documented.

My medical history tells the story of a pained and troublesome hip that didn't play by the biomedical rules. A hip that didn't respond to treatments as expected. It only tells the story of a hip. I write this story to right that story.

3 Meanings When Nothing Made Sense

Pain as Assault on the Self [6] "The whole life and personal identity of the person in pain becomes strangely alien and foreign" [7, p. 121]. Pain changes everything. Especially pain that goes on beyond when it should have ended and for which there is no explanation, no reason. It changes us. We are not ourselves. By far the most profound aspect of my pain experience was losing my sense of who I was, my sense of worth and purpose, of *thereness*. I lost my identity, my place in the world. The person looking back at me in the mirror was unrecognizable, unknowable. Small, frail, weak. Broken. An imposter [4]. A stranger I could not accept as *me* [6].

The real me had been hijacked by a foreign invader named pain. The enemy. My body the traitor that had betrayed me [9]. My hip, my whole leg, no longer felt like mine. It felt *other*. Alien [7]. My body, my identity, my life all became so strange and foreign. A landscape I could not navigate, for which I had no map, no direction, no guiding light to show me the way forward. I withdrew from the world, becoming disconnected from all the things that mattered to me. The people, places, and experiences that had given me a sense of purpose and worth, that had made my life meaningful. I was like a hurt animal, withdrawing into the shell of my former self to seek some sort of protection and safety, to try and make sense of all that seemed so insensible.

David Biro [10, p. 18] wrote that "pain is an all-consuming internal experience that threatens to destroy everything but itself—family, friends, language, the world, one's thoughts, and ultimately even one's self." It is not the integrity of our tissues that is threatened by pain, it is our integrity as persons, our continued existence, our very being, that is threatened [5].

Searching for Answers When the twinge didn't go away, when it persisted, when it morphed into ongoing pain that never let up, the need for a diagnosis, for an answer, became more and more urgent. This pain that did not get better, this threat to my very being and way of life, needed an explanation. Perhaps more pressing, it needed a solution. I wanted the pain, which I vaguely conceived of as some sort of injury, some mechanistic problem, some parts of the machine not working right, to be found and fixed [11]. It seemed so simple. Just a matter of finding the fault and fixing it. Just a matter of getting the machine back up and running.

Turned out it wasn't so simple. Despite all sorts of faults being found over the years, the pain was never fixed. The faults found differed between professions and clinicians, and clinicians within professions, based on the particular lens of education, training, and biases through which they viewed my hip and my pain. All sorts of fixes were tried: medications, physical therapy, injections, surgery, chiropractic care, acupuncture. . .but I failed them all.

Those years were a roller coaster ride of relief and high hopes when I would be told *this* is the reason you have pain and *this* treatment will fix it, followed by the depths of despair when treatment after treatment didn't work. When my pain persisted, or worse, got worse. Despite my diligence, my adherence to every plan, despite my motivation to recover and get back to work, get back to my life, get back to being me, I failed to get better. And to be certain, it was *I* who failed. *We* fail treatments, they do not fail us [11].

It was devastating. All those years, all those treatments, all those failures. And I still I had no explanation for my pain. No solution. No way forward.

Shame After years of searching for answers and failing treatments, the only conclusion to be drawn was that there was nothing *really* wrong with my hip. The damage had been repaired by surgery, my anatomy was fixed. There were no more faults to be found and corrected, we'd run out of plausible possibilities. If there was nothing *really* wrong with my hip, then naturally there had to be something really wrong with *me.* In the absence of a disordered spine, it must be a disordered mind [4, p. 231].

My self-blame rose exponentially.

And I was ashamed. So ashamed. Ashamed of my pain. Ashamed of my failures. Ashamed I didn't get better when I *should* have. Ashamed I wasn't handling it all better. Ashamed of the person I'd become, this broken, useless, *weak* person. I was ashamed I had let my fellow firefighters down. Ashamed I'd let my family and friends down. Ashamed I'd let my health professionals down, too. People I liked very much, people I trusted. People I wanted to tell their treatments were working, even when they weren't.

I was especially ashamed I'd let my husband down. Ashamed I was no longer the person he fell in love with, no longer the person he married. Ashamed I was no longer even present, as I was wholly focused on the pain. Pain that demanded all of my attention, used up all of my resources, and consumed all of my thoughts until there was no room for anything else.

Pain Is in Everything [7] "The pain of a small and particular thing. . .can totalize itself. The specific pain diffuses like a malignant mist throughout the experienced world" [2, p. 255]. Pain is lived. Experienced. Not just felt or sensed or thought about. Lived. When it sticks around long enough, pain imbues every aspect of our being, coloring every thought, every emotion, every moment, every story. Pain is in everything, a total experience [7, p. 117]. It becomes the mist through which we see the world, the medium in which we live out our lives.

Every decision in my life became centered around pain. Believing that pain meant damage meant that every painful step, every painful moment, meant more damage, more harm. That meant nothing could be done without forethought, no matter how minute the task. What used to be thoughtless and easeful now required extensive planning and effort. Every movement, every posture, from sitting to standing to walking to getting out of bed. So much work to just lie down, sit, take a step. In anticipation of pain and making things worse, doing more harm, I was constantly bracing, tensing, guarding, protecting, which only increased my pain. When the pain increased, so, too, my vigilance and focus on the pain, which only served to increase my worry and anxiety, which of course only increased the bracing, tension, guarding, and protection that increased my pain. . .and so on and so on and so on.

Worry "Pain is an ideal habitat for worry to flourish" [12, p. 234]. My worry certainly flourished, fertilized by years of unclear diagnoses and failed treatments. It was also fertilized by the language of healthcare. Words like "dysfunctional," "weak," "unstable," "imbalanced," "torn," "out-of-alignment," "degenerating," "injured," "failed." Words repeated so often they became my own, a part of my belief system about myself, about what I could and could not do. I still remember all of the NOs in my workers' compensation paperwork: no running, no lifting, no squatting, no climbing, no awkward positions, no sitting for more than 20 min. . .no no no. Words that came to define me, not just my pain.

After failing all the treatments all I was left with were these labels, these limitations, and nothing to do about them. I was worried I was never going to get better, that things would never change. I was worried about the future, that all it held in store was pain and suffering. I was worried about my relationship with my husband, our financial security, our future together. I was worried I was never going to be able to fix this pain that had caused all this worry. Worried that this was as good as it was going to get.

Ending the Search [9] Three years after the twinge, two years after surgery, I was declared permanent and stationary. A medico-legal decree that my condition had plateaued and there was nothing more to be done. A decree that I was never going to get better. That this was, in fact, as good as it was going to get.

"There's nothing more we can do." A statement so freely said with no acknowledgement whatsoever of what that means to those of us receiving the sentence. And a sentence it is. We often hear those words when we're at our lowest point. When we've desperately searched for answers yet still don't understand our pain. When we've failed every treatment. When we are no longer ourselves and the future seems so dark and despairing. When we don't feel heard, seen, or understood. When it seems no one is listening, and we wonder if they ever were [13]. When we have lost so much and are suffering so greatly.

"At first it was a heavy sense of loss and sorrow, wherein I could distinguish little else. By imperceptible degrees, it became a hopeless consciousness of all that I had lost—love, friendship, interest; of all that had been shattered. . .the whole airy castle of my life; of all that I remained—a ruined blank and waste, lying wide around me, unbroken to the dark horizon." David Copperfield, Charles Dickens.

After years of searching for the biomedical fix to my pain and failing miserably, after hitting rock bottom, my life a ruined blank and waste, there was nothing more to be done. The only choice left to me was to withdraw from the healthcare system altogether [14]. As dismal as that seemed, it actually led me to a different path forward.

4 Changing Meanings, Making Sense

Reconceptualizing Pain "When will this pain go away and real life recommence?" [2, p. 6] During those worst years of my pain, those years of searching in vain for answers, of losing my identity, of shame and blame and worry, my life had been on hold [15]. The pained life I was living was somehow not my *real* life, just as I was not the *real* me. My real life would only recommence once my pain was sorted out, fixed, gone.

It never occurred to me during that time that I *was* living my real life. That it *was* the real me. What a terrible thing to think at the time. It being my real life, the real me, only became possible when my pain came to mean something different. Only then was it acceptable.

My way forward was through science and stories that helped me make sense of my experiences and enabled me to reconceptualize my pain and myself with pain. Three years to the day of the twinge I medically retired from the fire department, having worked in a civilian position after my failed surgery prevented me from going back to work as a firefighter. In the months prior to my retirement (much earlier than I'd ever expected or wanted), I started graduate school to earn my Master of Science in human movement, hoping to discover the structural or biomechanical explanation for my pain, and what to do about it, that everyone else had missed. I was still searching for *the* fix, so I chose pain science as my research focus. It should come as no surprise that my search did not turn out how I expected and instead challenged the notion that there was a singular cause, a singular fix, to be found for my pain at all.

Rather than the fix what I found was validation. Legitimacy. Understanding. When I learned that pain is a complex, emergent experience underpinned by biology and influenced by myriad psychological, social, and cultural factors, I finally felt like my pain was validated. Finally felt like my pain was *real.* With real, credible, scientific explanations for it. My pain was not just "all in my head," not exaggerated, not made-up, not my fault. I was not crazy, not a malinger. Not just weak of mind, body, or character. It was an immense relief. A burden was lifted. I wasn't to blame for my pain. At the same time I felt I could take responsibility for how I moved forward, that I had the capacity and ability to do so.

When I learned just how relevant our thoughts, emotions, beliefs, and expectations are to our experience of pain, how influential our relationships, cultures, environments, the language we use and hear, the stories we tell about ourselves to ourselves and to others [16], I finally felt like I could *do* something. We are ever-changing beings, after all. Adaptable, resilient. Bioplastic [17]. Our biology constantly changing in response to our environment, our internal and external contexts, our new and repeated experiences, until our very last breath [16–18]. I felt empowered for the first time in a long time. Not so helpless and hopeless. There was realistic hope for change.

Embracing Uncertainty I also came to recognize the inherent uncertainty in understanding pain in all its complexity. One might expect that to be disheartening, but instead it lifted some of the weight of blame and shame of all my failed treatments off my shoulders, and off the shoulders of others. It lightened the load. We don't know all there is to know, and that is ok. It is better to embrace uncertainty than give people false hope that their pain is X and Y will fix it. When the explanations do not explain, and the fixes do not fix, the meanings of our pain become ever more threatening, carry ever more weight, bear even more consequences.

Honesty can still be hopeful. There is always something that can be done. We can always tell a better story. A story that makes biographical and biological sense and tells a better account of reality than the current one [19]. A story that helps us reconcile our pained selves with our real selves, so we are whole once again. Ourselves again.

Other Ways of Knowing "Art and science are alike in their quest to reveal the world" [20, p. 41]. The seeds planted by the science of pain had to be fertilized with things outside the realm of the scientific literature, too. To rediscover my place in the world, to feel I belonged in that world, I needed writers, poets, and songwriters, people who've long been trying to tell the truth of the human experience, to help me make sense of things. Many of my truths were found on the pages of nineteenth century literature and in the stories of Stephen King. I saw my pain, my suffering, myself, in song lyrics, poems, and movies. Countless stories helped me to see I was not alone, not aberrant, not an outlier. That I was connected to our collective human experience [11, 21]. Placing my pain, myself, into the larger context of humanity helped ease my suffering. Reading literature, something I had abandoned in my years of pain, opened the door to creative thinking again. To curiosity and reflection. To possibility.

Through science I began to make sense of my pain. Through stories I began to make sense of my life, of myself, with pain.

Mitigating Threat Understanding that pain was not a damage meter, that it was not a direct reflection of the state of my tissues [16], meant pain was no longer such a threat to my future, to my being, to my very existence. My life no longer had to be on hold until I was rid of it. When the meanings of pain became less threatening, less sinister, less despairing, the sensations I felt in in my hip—the thunks, clunks, and shifts, the tightness and weirdness—no longer meant there was *something really wrong in there*. They were just sensations in my hip. Sensations I could interpret differently. I could notice the sensations without fear, without so much worry, without having to attend to my hip so vigilantly. They were just weird, different. I also didn't have to pretend they weren't there, didn't have to pretend that pain wasn't there. Pain was no longer the enemy, my body no longer a traitor. I could respect pain without fearing it. I could acknowledge my pain and still get on with living.

Reconnecting No longer laden with a sense of impending destruction, with the threats abated, I was no longer at war with my pain and my body, which meant I finally had the capacity to focus on things other than the pain. When pain wasn't so threatening, I could make space for it, which meant there was room for everything else. For all of the things that mattered to me. The people, places, and experiences that gave my life meaning, that made me feel like me.

Reconnecting with nature was one of those things, and something else I had withdrawn from in my pain. While literature opened the door onto the world again, venturing back outside made it literal. No longer so worried that I was doing harm to myself with each painful step, I could more easefully exist in the world. I didn't have to plan every movement, every moment. I was free. No longer a prisoner in my home, no longer so wholly inside my hip and my head, no longer a prisoner of pain, I saw the world again, as though for the first time. I started taking pictures, literally seeing the world through a different lens and with a fresh perspective. It was a revelation. The world was still out there, in all its mystery and wonder, despite my absence from it for so long. I felt insignificant in the best possible way. A part of something much larger than myself, much larger than my pain.

"I came one evening before sunset, down into a valley, where I was to rest...I think some long-unwonted sense of beauty and tranquility, some softening influence awakened by its peace, moved faintly in my breast. I remember pausing once, with a kind of sorrow that was not all oppressive, not quite despairing. I remember almost hoping that some better change was possible within me." David Copperfield, Charles Dickens.

5 New Meanings, Sense Made

Myself Again Some better change was possible within me. Things did change. My life changed, my pain changed. I changed. I found myself again. Perhaps I found myself for the first time. I regained my sense of worth, of value. I became whole once again, no longer broken, frail, weak. No longer disregarded. I was whole. Strong, adaptable, resilient. Kind and caring, loved and loving. I was *me*.

It took a long time for me to realize I hadn't been who I was because I had been a firefighter. That I had been a firefighter because of who I was. The more I (re) discovered and (re)engaged with my values, the less worry and attention I paid to my pain. The more I engaged with the things that gave my life purpose and meaning, the bigger my life became, and the smaller the pain relative to that bigger life. The pain wasn't gone, it was just no longer the center of everything. And that was everything.

Acceptance I could finally accept pain as a part of my story, a part of my human experience, a part of me. I could finally accept what had happened, accept my pain, accept all the changes that had taken place, all the suffering that was endured, all the darkness and the chaos, because it didn't mean accepting it as my future, too. There was so much more to my life, to me, than pain. So much more to my story. And there was so much yet to be written.

A Better Story We are the stories we tell ourselves, and our selves are also "being *formed* in what is told" [22, p. 55]. Our stories shape our understanding of our experiences. They reflect what we've gone through and help us create what is yet to come. For a long time, I told a dark and dismal story. A story of pain, loss, and despair. Of brokenness. And that was the story I lived. But stories can change. The story I tell of my pain now is much different than the story I told seven years ago, or four, or just last year. It is a changing narrative, which affords the possibility of telling a better story in each retelling [21].

Removed from the chaos, removed from the constant threat of pain, I now see my experience as a quest [22]. One that led me down a path of pain and suffering, but also down a path of learning, discovery, and growth, of light, hope, and possibility. A path that led to a better understanding of myself and of others. Of being kinder and more compassionate. Of being more grateful for what I have rather than focusing so much on what I've lost. It was also a path that forced me to be more creative, to find the ways of engaging with my values that work within my limitations, while also gently challenging those boundaries from time to time to be sure I am not limiting myself unnecessarily. I did not know I was on this quest along the way, I can only see it now that I've gotten to this point along the path [22].

Patience and Persistence None of this happened overnight. It was a gradual process over the course of years. It took time for cognitive shifts to become behavioral shifts to become biological shifts. It wasn't linear. There was no step-by-step process. And there were setbacks. As I came to new understandings, new ways of thinking and being, I came to experience moments of *not-being-in-pain* [2]. When you are in pain for many years, not-being-in-pain, even for just a few moments, is a revelation (if a bit of an untrusted one). What was surprising was that no matter how brief the not-being-in-pain lasted, when the pain came back it would often derail me. It would bring back the old fears, old worries, and the oh-too-familiar despair. It was as though I utterly forgot what the years of being in ongoing, debilitating pain had been like. And once back in pain, or during a flare-up, it would be as though I'd never had any moments of not-being-in-pain. I'd once again feel I was sentenced to a lifetime of pain and suffering and misery.

This happened repeatedly, even with all that I had come to know, come to understand, come to believe. I had to keep coming back to what I knew, keep telling myself a new and better story about my pain, keep repeating a more therapeutic narrative rather than reverting back to the old, familiar, confining story of pain, disability, and lost identity I lived for so many years [3]. Over time, with a bit of courage and a whole lot of patience and persistence, the new and better story won out.

Clinicians and Meaning Making "One need not attain full physical recovery in order to heal" [2, p. 15]. When I was in the depths of my pain, nothing made sense, there was no meaning, no end to my suffering in sight. It was chaos, there was no cohesive story to be told [22]. We seek care to make sense of the things, to bring order to the chaos, to end our suffering. We seek care because we are not equipped to handle what we are going through on our own. Howard Brody [21, p. 85] wrote that when we seek treatment we are really saying:

"Something is happening to me that seems abnormal, and either I cannot think of a story that will explain it, or the only story I can think of is very frightening. Can you help me to tell a better story, one that will cause me less distress, about this experience?" Or, if you prefer, "my story is broken; can you help me fix it?"

Tell Me Your Story A couple years ago I started having pain in my left hip, my "good" hip. Without realizing it, I'd gradually begun to limit my life again, not doing things some of the things I loved doing, planning my movements, holding rigid postures. I started reverting back to the pained me, but it was a slow reversion, so I could readily rationalize it all. It so happened that about seven months into this new pain I was a patient demo for physical therapist and researcher, Peter O'Sullivan, at the 2017 San Diego Pain Summit. At the start of the demonstration he asked me to tell him my story, to start wherever I'd like. Surprisingly, to myself at least, I didn't start with what was going on in my left hip. Instead I went all the way back to seven years before when I'd stepped off the fire engine and felt the twinge, despite my right hip not even bothering at the time.

I teared up, realizing in the telling of my story how afraid I was of going back to that time. Back to the beginning of this long and often treacherous path through ongoing, worsening pain that had ended the career that had defined me and had taken so much. I had rebuilt my life. I was volunteering as an adaptive snowboard coach for people with disabilities, running a non-profit I co-founded to empower people living with pain to live well, writing my blog, traveling, spending time with family and friends. I could drive again, sit on furniture, go out for dinner or a coffee, the simple things I vowed never to take for granted ever again. And I was terrified, deep in my biology if not my consciousness, that I was going to lose it all again. That pain would become everything once more.

It was a profound moment. I hadn't given voice to those worries, those fears, before I was asked to tell my story. Not even to myself. Even with all that I knew, even with years of having lived with pain and studied pain and worked through pain, I still needed help making sense of this new pain, this new challenge. I still needed help connecting the dots and putting this experience into a more comprehensive context, so I could come to my own conclusions and new understandings and tell a better story about this pain. One that made biographical and biological sense and put things in perspective.

Stories in the Clinical Encounter Stories are always told in the clinical encounter, whether we think they are or not. The stories told are often of a biomedical sort, the health care professional seen as the expert and the "only voice that counts" [11, p. 227]. The story tends to be one of faults that have been found with their litany of labels, accompanied by myriad diagnoses or, perhaps worse, ruled-out diagnoses. We are told everything that is wrong with us, or that there is nothing *really* wrong with us at all. We're told what's to be done, or that nothing can be done, and then the story ends.

It's often not a very good story. Not a very helpful or healing story. These biomedical stories are told to us about us, but we are nowhere to be found in their telling. Too often, the character most egregiously missing in these stories is the person in pain. It is the story of an elbow [23] or of a hip. We are depersonalized and thus dehumanized, our stories too often treated as superfluous, a waste of precious time, rather than as a valuable form of knowledge that could, and should, inform the treatment plan and path forward [21, 24].

We are not machines to be fixed, not problems to be solved. We are human beings in need of compassion and understanding, guidance and support. We need help to make sense of what is happening and to determine how best to move forward with what we learn from one another. Our stories can lead to greater understanding of not just *our* pain, but to a greater understanding of pain. Within the shared humanity between patient and clinician, with all the inherent uncertainty and all of our human fallibility and vulnerability, there can be crafted therapeutic narratives that provide credible explanations for the patient's pain as well as realistic paths forward for them. It is their story, after all. The story they will live outside the clinic walls.

Love and Be Loved The other profound moment along my journey occurred the year after my medical retirement. I interviewed neuroscientist and physical therapist Lorimer Moseley while in graduate school for a project. I grilled him for 45 minutes on pain science, still believing my answer, *the* answer, would be in there somewhere. That I would discover just what I needed to do to be rid of this pain that had so altered myself and my life. At the end of the interview I asked him what the one thing he'd want people in chronic pain to know or to do. His response was to love and be loved.

To love and be loved.

I was floored. There are few moments in our lives that truly change the course of things, and that was one of those moments for me. It was in that moment I saw a different possible future. It was what helped me to reframe the problem to be solved [12]. It was the light I needed to guide the way, the nudge I needed to reconceptualize my pain and see a different path forward. A path not so focused on being rid of pain and was instead focused on getting back to the things that mattered to me, that meant something, that made my life worth living. It was what helped shift my perspective, what opened the door onto better possible futures and future selves, what made my recovery possible.

The two most profound things said to me by clinicians were "tell me your story" and "love and be loved." Not what most would expect and not what I knew I needed to hear until I heard them. Even more important for this discussion, though, is that in both of these instances they listened to my story and we had a conversation. I felt heard, believed, and validated, which confirmed my worth as a human being and that what I said was of value. That I was of value. That cannot be overstated. When we are in the depths of our pain, when we feel invisible and disregarded, when we do not like who we've become, it can be really hard to see our worth, our value. Our own humanity.

6 Conclusion

I started this chapter saying my pain story started with a step off a fire engine. Truth is, my pain story started long before that. We each come to every experience of pain with a lifetime of learning, memories, and beliefs. A lifetime of knowledge, societal and cultural influences, and experiences. A lifetime of illnesses, injuries and recoveries, thoughts, emotions, and expectations. We each come to every episode of pain with biological and philosophical foundations uniquely our own. With pasts and plans for the future that affect our present. With a sense of who we are and who we will continue to be.

Pain disrupts all of that. Disrupts all that seems so certain and true. All that we thought we knew.

My medical record tells the story of a hip with pain, but my hip did not feel pain. My hip did not make meaning of pain. My tissues, while certainly playing a role in my experience, did not have any experience at all. I did. Just me. Yet while it was only I who felt the pain, it wasn't just I who felt its effects. Pain affected those around me, too. My employers and co-workers, my friends and family, even strangers on the street I may have been curt and unkind to when my pain was overwhelming. Pain most affected my husband and dog as our lives together were completely upended. Pain changed everything. It changed me. The changes were unasked for and unwanted, the person I'd become unwelcome. It was not the present my past was supposed to lead up to [22, p. 55] and none of it made any sense.

Gratefully, it was in making sense of my experience, through science and stories, through the help of many guides within and without the health care system, that the threat was abated, that my suffering ended, that life resumed. That I was *me* again. A changed me but still me. A self I could accept, a self I wanted to be. My experience of pain and all that it wrought can now be placed in a wider context, a broader narrative. I can now tell a better story. A more cohesive story, with a cohesive self [25] at its center.

I've found my place in the world again.

As much as I've written about my pain over the years, I've learned so much in writing this chapter. It forced me to explore the meanings of my pain over time, which meant having to go back to those dark years and remember as best I could. It forced me to highlight meanings I thought were most salient, and the changing meanings that were most relevant. It made me think about my recovery as well. What got me to where I am today, what might have gotten me here sooner. Lous Hesusius [26] wrote in her book, *Experiencing Chronic Pain in Society,* that all she ever wanted was someone to listen. It seems a sensible place to start to revolutionize our understanding of pain and our approaches to pain research and treatment. Listen. Hear us. Hear our stories.

When we feel listened to, heard, validated, and understood, it makes all the rest of it possible. It allows us to explore what our experience means and discover new possibilities. When we are in the depths of pain and our personhood is sustained by the personhood of others, we have a chance to recover from our suffering [5]. When we hear words that promote healing [21], that emphasize our strengths, courage, and persistence, our resilience and adaptability, we have a better chance at better outcomes. When we share power with our health care team, when we jointly construct narratives that make biological and biographical sense, we can tell, we can live, better stories. Stories where we are an agent in our own rescue [27]. Stories where we are the heroes of our tale.

When we can fix our broken stories we can heal, even if we are not cured.

References

1. Bourke J. The story of pain: from prayer to painkillers. Oxford: Oxford University Press; 2014. 396 p.
2. Leder D. The experiential paradoxes of pain. J Med Philos. 2016;41(5):444–60.
3. Morris DB. Narrative and pain: toward an integrative approach. In: Moore RJ, editor. Handbook of pain and palliative care. New York: Springer; 2013. p. 733–51.
4. Bunzli S, Smith A, Schütze R, O'Sullivan P. The lived experience of pain-related fear in people with chronic low back pain. In: van Rysewyk S, editor. Meanings of pain. Cham: Springer; 2016. p. 227–50.
5. Cassel EJ. The nature of suffering and the goals of medicine. N Engl J Med. 1982;306 (11):639–45.
6. Smith JA, Osborn M. Pain as an assault on the self: an interpretative phenomenological analysis of the psychological impact of chronic benign low back pain. Psychol Health. 2007;22 (5):517–34.
7. Svenaeus F. The phenomenology of chronic pain: embodiment and alienation. Cont Philos Rev. 2015;48:107–22.
8. Scarry E. The body in pain: the making and the unmaking of the world. New York: Oxford University Press; 1985. 400 p.
9. Toye F, Seers K, Hannink E, Barker K. A mega-ethnography of eleven qualitative evidence syntheses exploring the experience of living with chronic non-malignant pain. BMC Med Res Methodol. 2017;17:116.
10. Biro D. Listening to pain: finding words, compassion, and relief. New York: WW Norton; 2011. 256 p.
11. Loftus S. Pain and its metaphors: a dialogical approach. J Med Humanit. 2011;32:213–30.
12. Eccleston C, Crombez G. Worry and chronic pain: a misdirected problem solving model. Pain. 2007;132(3):233–6.
13. Thacker MA, Moseley GL. First person neuroscience and the understanding of pain. Med J Aust. 2012;196(6):410–1.
14. McGowan L, Luker K, Creed F, Chew-Graham CA. How do you explain a pain that can't be seen? The narratives of women with chronic pelvic pain and their disengagement with the diagnostic cycle. Br J Health Psychol. 2007;12(2):261–74.
15. Bunzli S, Watkins R, Smith A, Schütze R, O'Sullivan P. Lives on hold: a qualitative synthesis exploring the experience of chronic low-back pain. Clin J Pain. 2013;29(10):907–16.
16. Butler DS, Moseley GL. Explain pain. 2nd ed. Adelaide: NOI Publications; 2013. 133 p.
17. Moseley GL, Butler DS. The explain pain handbook: protectometer. Adelaide: NOI Publications; 2015. 49 p.
18. Doidge N. The brain that changes itself: stories of personal triumph from the frontiers of brain science. New York: Penguin Books; 2007. 427 p.
19. Launer J. Narrative-based practice in health and social care: conversations inviting change. 2nd ed. Abingdon: Routledge; 2018. 164 p.

20. Banville J. Beauty, charm, and strangeness: science as metaphor. Science. 1998;281 (5373):40–1.
21. Brody H. "My story is broken; can you help me fix it?" Medical ethics and the joint construction of narrative. Lit Med. 1994;13(1):79–92.
22. Frank F. The wounded storyteller: body, illness, and ethics. 2nd ed. Chicago: The University of Chicago Press; 2013. 279 p.
23. van Rysewyk S. A call for study on the meanings of pain. In: van Rysewyk S, editor. Meanings of pain. Cham: Springer; 2016. p. 1–22.
24. Charon R. Narrative and medicine. N Engl J Med. 2004;350(9):862–4.
25. Thompson BL. Making sense: regaining self-coherence. In: van Rysewyk S, editor. Meanings of pain. Cham: Springer; 2016. p. 309–24.
26. Heshusius L. Experiencing chronic pain in society. Arroyo Grande: CreateSpace Independent Publishing Platform; 2017. 272 p.
27. van der Kolk B. The body keeps the score: mind, brain and body in the transformation of trauma. New York: Penguin Books; 2014. 445 p.

Chapter 2
After the Tango in the Doorway: An Autoethnography of Living with Persistent Pain

Bronwyn Lennox Thompson

Abstract Persistent pain is a common health problem and increasingly, qualitative research is being used to explore the impact on daily lived experience. Stigmatisation and "othering" is reported in these studies, and health professionals indicate they struggle to know how best to help this group of people. In this autoethnography, I provide an account of my life as a clinician, educator, researcher and social media commentator who lives with fibromyalgia. Through this narrative I consider the social factors influencing my attitudes towards my own pain over time, and the position I hold as a member of many different social groups discussing persistent pain management.

Clinical Implications

- People seeking help for their pain are influenced by family habits and attitudes towards healthcare and accepting clinical opinion.
- Although persistent pain is common, "othering" or the tendency to see those with persistent pain as different from clinicians (in a negative way) is frequently encountered. This may be inadvertent but results in stigma.
- Information about pain mechanisms is insufficient for supporting a new way of living life for people with persistent pain.
- People with persistent pain may reject the narratives of people who have learned to live well with their pain, and this can be isolating and difficult to deal with.
- There is little research investigating how clinicians advise people of their persistent pain diagnosis, leaving a gap in our understanding of the best way to convey this information.
- Researchers who themselves live with persistent pain provide a unique insight into what it is like and may offer new ways for clinicians to carry out their work.

Keywords Othering · Health professional attitudes · Persistent pain · Autoethnography

---

B. Lennox Thompson (✉)
Orthopaedic Surgery and Musculoskeletal Medicine, University of Otago, Christchurch, New Zealand
e-mail: Bronwyn.thompson@otago.ac.nz

S. van Rysewyk (ed.), *Meanings of Pain*,
https://doi.org/10.1007/978-3-030-24154-4_2

1 Introduction

I remember the first time I felt unexplained pain. I was 12 years old, cycling home from school riding straight into a headwind. The pain in the middle of my back, just under my right shoulder-blade had been there all day, all week. I couldn't recall anything I'd done, it just appeared there. I couldn't get comfortable: there was no position that eased it, no wriggling, stretching, twisting that changed it. "Growing pains" said my mother, and that was that.

2 Writing an Autoethnography

An autoethnography is a personal story written in first person, and used to problematise, challenge, reflect and probe common cultural perspectives relevant to that story [1, 2]. Autoethnography is usually accompanied by orienting information about the author [3]. I am a 54-year-old woman living with persistent pain, diagnosed as fibromyalgia. I work as an Academic Coordinator, responsible for the postgraduate programmes in pain and pain management at University of Otago, New Zealand. I research the ways people live well with persistent pain, and how I, as an educator, can improve the way health professionals interact with those who experience pain and seek help. I am an active social media participant, author of the blog Healthskills.co.nz (established 2007), and administrator of the Facebook group Exploring Pain: Research and Meaning, with a membership of over 21,000 from around the world. I offer these details to provide readers with more information about my world, helping to establish my perspective and, in a sense, authority, to write about what it is like to live with pain, teach about pain, research aspects of pain, and converse with people from around the world with a common interest in pain and how we treat people experiencing pain.

Autoethnography is intended to shed light on both the personal and the broader sociocultural context [4]. Adams, Ellis and Holman-Jones [4] write that "Autoethnographers speak against, or provide alternatives to, dominant, taken-for-granted, and harmful cultural scripts, stories, and stereotypes" [4, p. 3]. While pain is a common thread across the groups of which I am a member, differing areas of interest, focus, attitudes and beliefs, and considerations colour the discussions. Reflecting these various facets of pain and its management requires shifting language and perspective: these shifts are part of my life and the way I present myself in the world. What can I learn from my experiences, and how can my experiences shed light on the ways persistent pain is conceptualised? Can these insights prompt reflection amongst clinicians, researchers and educators?

The materials used in autoethnography include research literature but extend beyond this to personal material drawn from the author's life [4]. I have been an avid diarist for most of my life, and I have reviewed my assorted writings as I have reflected on my experiences over time. Naturally I present a very personal view and one that cannot be truly verified. Subjective experiences can't be validated, and it is ironic that researchers, clinicians, lawyers, insurers, family and friends experience the same challenge when listening to a person describe their pain: is any account of pain "true"? Can it be believed? As a community we struggle with the idea that when a person indicates they are sore, we have no measure against which we can substantiate that claim [5, 6]. We rely, therefore, on the judged authenticity of the person; on their tendency to be believable. Similarly, this paper can only be judged on how authentic and believable my narrative and conclusions are. By being transparent about my stance and sources of information, readers are invited to judge the trustworthiness of this work [7].

Clinicians and researchers develop an understanding of their clinical or research culture as they pursue their training, and as part of induction into professional behaviour [8]. For us all there was a time when the assumptions we hold today were new and foreign. Over time, our understanding of "what it is like to be" a clinician develops into a familiar persona and it becomes difficult to recognise conventions for what they are. Difficulty recognizing attitudes and practices as a distinctive lens through which both an individual and the group in which we belong is one of the characteristics of culture, whether related to the local community culture, or in this instance, to the culture of a profession, or even a workplace [9]. I observe that we seamlessly transition between our home life, work life, leisure life—often without recognising the different ways we inhabit these lives. Autoethnography provides both the author, and the reader, with an opportunity to become aware of commonplace attitudes and practices of which we may otherwise be oblivious [1].

I have organised this account with brief accounts of my experiences, as I reconstruct them from memory, my diaries, photographs, books and discussions with family and friends. Accompanying each account, I provide an analysis of the experiences. The questions I posed myself throughout my analysis were "Why did I perceive, think, behave and evaluate the way I did, and how do I relate this to others in thoughts and actions?" [1, p. 212] Finally, I relate these findings to the various contexts I inhabit today to shed light on implicit attitudes to stimulate discussion and self-reflection [10].

3 Going to the Doctor

Not Something to Talk About

I was a bookworm as a child. I preferred this to playing sports, though I adored dance. I'd wanted to dance since I was 5 years old, and finally enrolled after my mother took my younger sister to an orthopaedic surgeon to examine her hip. She'd been born with a hip displacement, and we'd taken the trip to Auckland from Gisborne for the last appointment. While at the appointment the idea of dance was discussed, and my mother was reassured that "all little girls should learn ballet." At 7 years old I was suddenly permitted! I danced my heart out and was good at it, passing all my exams with honours. No problems with pain when I danced.

Going to the doctors was a regular practice in our family, at least for coughs, colds and sinus troubles. Musculoskeletal pain—not so much. Medical practitioners could say The Words, Write The Prescription. This was the 1970's. Era of Sinequin (I recall looking at the puppet showing two faces—one sad, one happy—in the waiting room at the family doctors). My mother had postnatal depression, believed wholeheartedly that her mood problems were a result of an imbalance of neurotransmitters and swallowed the pills, never thinking there was anything else she could do except follow the doctor's orders. It was the same for any ailment. "Go and see the doctor and get something for it" was our family motto.

But pain? Not something to talk about. Unless there was blood, or visible injury. You can't have pain without being able to see what's wrong. Except for migraine. For this, my mother took to bed.

The 1970s were a time of change in the pain world. In the mid-1960s Melzack and Wall had published their seminal paper "Pain mechanisms: A new theory" [11]. Gate control theory not only incorporated modulation at the dorsal horn, but also included a "central control system." Psychologists rapidly became interested in pain and pain management. Fordyce and colleagues started conceptualising chronic pain as a multidimensional problem, with experience, biology, judgements and behaviours, including behavioural reinforcement from those in the individual's social world, all contributing to the way pain is experienced and expressed [12]. Cognitive therapists started examining contributions from the "top down" (Melzack and Wall's "central control"). In 1974 the International Association for the Study of Pain was formed in Seattle by Professor John Bonica. "IASP was a different organization from its inception. John mandated that it be truly interdisciplinary. Not just a group of physician providers, it was to consist of nurses, dentists, physical and occupational therapists, psychologists, social workers, pharmacists and basic scientists. All professionals interested in reducing pain and suffering were invited to participate on an equal basis" [13].

Change, however, did not happen quickly in small towns in New Zealand in the mid-1970s. While Melzack, Bonica and Fordyce were working tirelessly to shift perceptions on an international stage, back home in Gisborne, on the East Coast of the North Island of New Zealand, pain in children and adolescents was either "growing pains" or "you really want to get out of Phys.Ed. today, don't you." I learned to live around my pain. It was just there and there was nothing to be done about it.

4 The Tango in the Doorway

In the early 1980s, I started working as an occupational therapist in what was then called adult psychopaedics, or intellectually disabled adults; these people often had other psychiatric and physical health problems and lived in a supported residence. It was there that a resident and I collided in a doorway. He was over six feet tall, and "well-nourished." He used a gutter frame walker, and his official diagnosis was "spastic diplegia" as a result of being assaulted as a baby. He was now in his late twenties and we had worked together for 6 months on gaining his independence. He and I tangoed in the doorway, and he overbalanced, landing on top of me, walking frame and all.

The whole event was shocking. I remember the fear of wondering if I'd caused the fall, the shame of letting a person land on the ground. I had paperwork to complete, but I couldn't recall the details. Did I go through the door first? Was I rushing? Did he stumble? Did the walking frame get caught on the joint between the doorstep and the concrete path? I felt responsible. I worried that he would lose trust in me. He had a minor graze on his elbow. I had bruises on my ribs, elbows and knees. And a terribly sore back. I remember sitting in the tea room trying to catch my breath and fill out the incident report. No-one asked how I was, whether I was OK. I was only months out from graduation, and this was the first time I'd ever had anything like this happen. But, being young and generally fit, I thought my injuries would settle down, so I drew a deep and shaky breath and went on with my day.

Several weeks later, my bruises had faded, and the incident investigation was complete with no-one pointing the finger at me—but my back pain was still there. If anything, it was worse. It nagged at me all through the day, at night as I tried to sleep, when I sat, stood, walked, cycled, bent over, and got dressed. It was present all the time. I decided to see my doctor. This time, for the first time I can recall, the doctor gave me a prescription for anti-inflammatories, yes, but also wrote a referral for physiotherapy. The diagnosis was "lumbago."

At the time in New Zealand, the McKenzie Institute [14] *had just been established. The physiotherapists I saw had embraced this treatment with enthusiasm. Three times a week I was given hot packs for 10 min, traction, then manipulations, and sent off to "do exercises." I don't recall any information being given to me for this approach, but I was dutifully "doing my exercises" every day. Nothing much happened. Once I saw a different physiotherapist: she didn't use manipulation, but instead hooked me up to an interferential machine. It sucked and throbbed and pulsed—nothing much happened. Another time I was given ultrasound. Nothing much happened. There were no reasons given for why different therapists carried out different treatments. My body was the problem, so my need for explanations was irrelevant.*

After about 18 months of treatment I was trying to be active, but I was still sore. I didn't know why, and I didn't care much. I had changed jobs, moved into an assessment and rehabilitation ward for elderly people. There was a lot to love about this job: strong support for occupational therapists from the geriatricians; a busy but supportive and integrated team environment where physiotherapists and occupational therapists worked closely together. I enjoyed the work immensely.

I was involved in a lot of self-care assessments and treatment. This mean donning a waterproof coat, gumboots, and standing in a bathroom helping people undress, shower, and dress again, all the while working on spatial neglect, sitting balance, observing for cognitive and perceptual difficulties and working to reduce the impact of these on the person's capabilities. It also meant physically helping people transfer to and from bed to wheelchair to shower stool and back. Bending over, straightening up, twisting—all the things that really hurt. After about a year, I went to see the occupational physician for the hospital who suggested an x-ray, and 2 weeks off work. I dutifully went and got the x-ray, but thought I'd be able to carry on working provided I didn't have to bend and stretch too much. My manager, on the other hand, thought I should stay off work because if the doctor says I should, then I should. I was banned from coming in to work for the next 2 weeks.

When the x-ray results came in, I read the radiologists report—I had a "transitional vertebra." In my naiveté I thought "At last! There's something that can be done! This vertebra must be causing the problem, so can I get surgery to fix it?"

5 The Good Doctor Said...

My family doctor nodded when she read the report, and said "well, perhaps it has something to do with the problem, but seeing as your pain hasn't gone away, I'm referring you to The Auckland Regional Pain Service." I had no idea what this was all about, but she was my doctor, so I obediently went off to the appointment.

What an eye-opener! This doctor listened to me. Took his time examining me. Asked me questions like "Have you found any medications that help?" I said I hadn't. He said "Well, that's not surprising, because lots of people with your kind of pain don't find medications very helpful" Wow. That floored me! Medications, the prescription, the ritual, wouldn't always help for pain?

He then went on to say the words that totally blew my mind: "I'm afraid there's nothing more that medicine can do for you, you will need to learn how to live with this."

I was devastated. I don't remember much else of our conversation—except he left me with the advice to read a book. That book was "The Challenge of Pain" by Melzack and Wall [15]. *Going home that day my world was shattered. "I'm meant to go away and live like this forever?!" I thought.*

This is the point at which I usually begin my story of my persistent pain problem. It's not entirely true, as you can see, because I'd already had encounters with both pain and healthcare since childhood. And it's interesting that I often omit reference to those earlier encounters, attitudes, and assumptions. And often I leave my narrative devoid of emotion, yet the emotional impact of the pain I experienced, the nature of its onset, lack of treatment response, referrals, differing opinions, and that feeling of being abandoned to deal with this ongoing, unrelenting and nauseating pain on my own, in my early twenties, still echoes in my mind.

I suppose I refer to this consultation as my introduction to living with pain because of two things: (1) I was told that there is no quick fix, no pill, no surgery, nothing that would take the pain away. By a doctor. A specialist. And (2) this doctor pointed me towards something for me to get my head around. An explanation. In hindsight, and certainly not in the immediate months after that consultation, I am grateful for his wisdom. I wasn't sent for more imaging, possibly orthopaedic assessment and surgery. I wasn't told to go do certain exercises, nor given an explanation to make me feel better. As a generally healthy and active health professional, this doctor expected that I would use my head and work through my situation having been given a book reference that provided accurate and thoughtful information. After a few months of shock and horror, I got the book and started reading.

What an amazing book! The key messages I got from Melzack and Wall [15] in the mid 1980s, was that my pain was real. That pain is complex. That what happens in my body isn't necessarily the way my brain interprets it—and vice versa. That I was not doing myself any harm by keeping active even though my pain would regularly do what it wanted. There was a science to understanding pain, and I could understand it.

As I reflect on this encounter today, I realise how much my childhood socialisation into "going to the doctor," along with trust in medical treatments, influenced my reaction. I wonder if I was to be in the same position, same age, but today: would I be as willing to go along with this diagnosis? People seeking help for a health problem have, at their fingertips, Google and all of social media to consult [16]. While use of social media has posed many ethical questions [17], the proliferation of evidence-based websites and discussion forums providing health information has meant a more informed general public. In turn, health practice has needed to respond to a more critical stance from people seeking help [16].

Tensions are present when people are informed by internet-based knowledge, encounter clinicians with different views about the condition and/or its management, or who are unused to a collaborative approach to health management [18]. It is arguable whether information availability has initiated this shift in attitude, or whether it is part of a larger shift in public awareness of variations in clinical practice and/or trust in authority figures. Brown, Elston and Gabe [19] point to a "functional-democratisation of patient-doctor relations via civilising processes" [19, p. 164]. In other words, the power asymmetry between those who know, and those who do not has reduced, while there has been greater emphasis on caring and empathy by clinicians [20]. These authors also argue that changes in society such as an increased emphasis on articulating consumer choice and rights in health policy have contributed to greater awareness of choice amongst patients. Concurrent developments in "patient-centredness" [21] and overt endorsements of partnerships between health professionals and the people they help (for example, World Health Organization) also serve to reduce the "doctor knows best" attitude inculcated in me as a child.

Managing disparate views about treatment today is thus something clinicians must navigate, and particularly when the person seeking help will need to adopt new behaviours. The shift from patients as treatment recipients to people needing to engage in daily behaviour change is perhaps one of the more common transitions healthcare systems and communities have had to adopt over the past 20 years [22]. It is this shift towards helping people to live well alongside a long-standing illness that remains one of the major challenges for healthcare systems today [23]. In my case, I was given treatment without explanation because my back was the "problem;" I was given a diagnosis and I accepted this; I was referred for treatment and I went. I had little opportunity, or desire, to question what was offered. Would it be different today?

It's also interesting to note—and perhaps I made little reference to it at the time—that the emotional and social impact on me as a person was completely overlooked. At no point was the impact of my back pain on my life and my "self" explored; even when being given a diagnosis of chronic pain I had to work out how to do the ordinary things of everyday life *by myself.* Today I ask myself: has our approach to persistent pain progressed any since 1985? Much qualitative research shows that it has not [24].

6 Is Explanation Enough?

It would be easy to stop there, and think my story is over. To assume that because I had an explanation that was enough, and to think that I am, today, working as an educator, researcher, clinician and social media commentator on persistent pain *because* of my persistent pain. To assume, I've heard said, makes an "ass" out of "u" and "me."

Certainly, the idea of "explaining pain" has caught on like wildfire since Moseley's "intensive neurophysiology education" trial for low back pain [25]. It has, in some instances, become an isolated treatment, prevention strategy [26, 27], perhaps a "fad" as one blogger puts it [28].

Providing people with persistent pain with relevant information about how the nervous system works has been incorporated into pain management programmes since at least the 1970s [29]. The initial concepts were drawn from Melzack and Wall's gate control theory [11], and later extended with cognitive behavioural models [30, 31]. The key principles of this information were to "educate the patient about human anatomy, physiology, and psychology" [29, p. 506] but—and this is an essential point—programmes also incorporated Fordyce's behavioural approaches, and cognitive therapies [32]. Programmes also included physical reactivation, return to work or vocational planning, and the facility in which I worked, exposure to activities in the community such as going to the shopping mall, supermarket, catching the bus and going to the park.

Wilbert Fordyce said, "information is to behaviour change as spaghetti is to a brick" [33]. While I found The Challenge of Pain intriguing, fascinating, the key messages were that my pain didn't represent ongoing damage, and that I could use psychological principles to learn to modulate my pain. I could learn to live well. The book did not, on its own, change *how* I lived my life. That, it seemed, was up to me. Alone.

I am not unique in feeling confused, bereft and disabled at that time, even though equipped with new knowledge. Stories of what it is like to live with a persistent pain label abound in qualitative research (although my label was, at that time, simply "chronic pain"). My feelings at the time are well-depicted in Lempp, Hatch, Carville and Choy [34] "I was a complete mess;" "I was feeling suicidal;" "very, very weepy;" "mentally and physically shattered;" "feeling very, very down;" "I am in a mental cocoon, in a cage, in a bubble;" "I was at a low point, in a black hole" [35, p. 6]. Arnold, Crofford, Mease, Burgess, Palmer, Abetz and Martin [36] reported that women "expressed feelings of embarrassment, frustration, guilt, isolation, and shame"; "feeling guilt or shame when rearranging priorities due to fibromyalgia, such as taking better care of themselves..." [36, p. 118].

As above, I ask myself how well do we support those who finally get a diagnosis of persistent pain? Are people told that their pain is unlikely to ever go completely? I searched in vain for research papers examining the process of giving a diagnosis of chronic pain—whilst there are many papers on giving a "life limiting diagnosis" (in other words, you have a terminal illness), and there are papers investigating what it is like to *treat* people with chronic pain—breaking the news that your life may be long yet painful does not appear to be a popular research topic. This leaves clinicians with little guidance as to how to have the conversation, when to have the discussion, and how to give what person might need most.

7 Tensions Between a Culture of Medicine and a Person's Need

At the heart of this ambivalence towards persistent pain may be the clinician's unease at being unable to "solve the problem." Just as I had thought that there must be something that can help me, something to "take it away," so clinicians may, even today. Toye, Seers and Barker [37] identified themes after synthesising studies of healthcare professionals' experiences in this field:

1. There is skepticism within medicine for illness where diagnosis is difficult, and in turn there is "a siren song" of diagnosis to which many are unable to resist. The skepticism within medicine arises from a culture in which subjectivity is shunned in deference to objective findings, but pain is always subjective.
2. Clinicians must then "navigate juxtaposed models of medicine" as they attempt to reconcile the evidence in front of them of a person who is distressed and disabled yet for whom there is little in the way of objective clinical findings.
3. This in turn leads to a need for "navigating the patient-clinician borderland"—a place where clinician's knowledge is either absent or unhelpful, while the person

seeking help continues to hope and call for something that cannot be readily provided.

4. Clinicians, particularly primary care general practitioners, are often placed in a position of "dual advocacy" where the person in front of them desires help, but providing that help may involve what the *clinician* feels is unnecessary, or unhelpful, investigation or treatment.
5. The personal cost of these encounters may be one of the primary obstacles to positive attitude change towards those who ask for support.
6. Toye, Seers and Barker [37] complete their analysis with a theme of the "craft of pain management," alluding to the ongoing delicate balancing act with which clinicians listen to their patient, and respond to their needs in a way that defies categorisation or algorithms. They conclude with a model depicting the various tensions that form around clinical management for people with persistent pain: between a "dualistic biomedical model" and "an embodied psychosocial model;" "professional distance" and "proximity;" "professional expertise" and "patient empowerment;" "making concessions for therapeutic alliance" and "evidence-based utility;" and finally, "patient advocacy" and "health-care system advocacy" [37].

These tensions are rarely spoken of in health professional training, particularly undergraduate training, where technical expertise, certainty, and objectivity are prized.

Contrasting with this reticence to indicate that long-term pain may be a reality, in a grounded theory study of people living well with persistent pain I found that people only began considering a shift towards living with pain once they had been told their problem needed *management* rather than *cure* [38]. Other investigators have also found this: In people with neuropathic pain associated with spinal cord injury, one participant said, "And then I finally said to myself, nothing's going to work. I might as well try to live with it, and learn to live with it, and since then I haven't tried pursuing any type of pain relief" [35]. "All the previous treatments dealt with taking away the pain. This is the first time one gets a treatment that focuses on acceptance of the pain, and you really understand that this is chronic pain that will never disappear; it's the first time one has received the message from this angle" [39]. Perhaps, as Mark Sullivan argues, health professionals need to reconceptualise their contribution to health: is the job of a health professional to treat disease, or to "advance the person's capacity for personally meaningful action"? My occupational therapy heart beats faster at a view of health in which doing, being and becoming is the goal of treatment [40]. Occupational therapy is based upon the belief that by engaging in occupation (personally meaningful activities), people achieve health, and that occupational engagement is a process of both developing and expressing a sense of self [41]. One of the most pervasive effects of persistent pain is to interfere with what people can do, how they do things, and ultimately their self-concept.

Resolving the dilemma of *when* to begin talking about pain as a long-term problem is not easy. Ultimately, the decision to pursue or stop pain reduction is one made by the person experiencing pain, albeit with variability in terms of how much self-determination each person accepts (or wants). Informed decision-making between the person seeking help and his or her health professional requires accurate information about the available options. If clinicians struggle to talk about pain as a long-term problem, and perhaps fail to see learning to live well with persistent pain as a positive option it is unlikely a pain *acceptance* rather than pain *reduction* approach will be included in clinical conversations. Pain intensity is, as Ballantyne and Sullivan point out, not the best measure of the success of chronic pain treatment [42]. Persistent pain is a multidimensional problem, affecting people in ways unique to each individual depending on the meaning given to the experience, so too the relevant outcomes should reflect the priorities unique to each individual. Achieving an understanding of what these priorities are requires a therapeutic stance that moves beyond assuming that "reduced pain = improved life."

8 What Am I Today and How Did I Get There? And Why Does that Matter in a Book About the Meanings of Pain?

Today I live well with persistent pain. Although I have pain I can, and do, most activities I choose. It is impossible to determine how my life is different as a result of my experience, and I am hard-pressed to say what my pain means to me. I notice where I am sore, and because pain shifts in location, quality and intensity, I am aware of different parts of my body at different times, while other parts rarely reach my awareness because they are seldom sore. My body does all the things I want and need it to, but I undertake activities with attention and awareness. Pain may be present for some time in a body area before I imbue that experience with the added interpretation "Oh, time I did something about that" and seek a clinical opinion. Perhaps my ongoing lived experience of pain has reduced my empathy for others seeking to reduce theirs? Maybe I have become more sensitive to pain as a problem and over-emphasise the need for pain to be addressed with more compassion? I will never truly know for, like the experience of pain, these qualities cannot be measured or weighed.

Ultimately my widespread and persistent pain and associated symptoms is probably part of a nociplastic process. Nociplastic pain implies changes in the way somato-sensory information is processed and is defined as "pain that arises from altered nociception despite no clear evidence of actual or threatened tissue damage causing the activation of peripheral nociceptors or evidence for disease or lesion of the somatosensory system causing the pain" [43]. I usually describe myself as living well with fibromyalgia [44] because this is the best fit so far for the set of symptoms I experience. When I am thinking of myself as "the person with pain," this is the label I use most: a person living well with fibromyalgia.

But I have other labels and roles also. These roles place me in a curious predicament when I simultaneously remember I am a person living with persistent pain, an educator, a researcher and a social media engager.

9 "Us" and "Them": Othering

In conversation with health professionals, for me there is an uncomfortable tendency to hear people living with pain referred to as "them." I've had people tell me that people living with pain "lack living skills," are "non-copers." I've heard about "unmotivated" patients. Patients who are "non-compliant." Patients who "don't want to get better." Patients who are "motivated by secondary gain."

I ask myself "Am I unmotivated? Am I motivated by secondary gain? Am I non-compliant? Am I a difficult patient?" If I'm not, am I a "motivated" patient? One not influenced by "secondary gain"? Am I compliant? Am I easy? What would turn me into "one of those patients" that clinicians dread? Would I change—or is it something about "them"?

I've heard health professionals talk about the need to "keep a distance," to "be professional," to not express sadness or distress at hearing a person's story—no matter how traumatic it is. I've listened to health professionals judge the lifestyles and priorities of those people living with pain, and describe weight gain, and use of mind-altering substances as "laziness," and "not really wanting to work." The lenses through which clinicians view people are coloured by both professional training and their own socio-cultural background. To health professionals, health is a priority—while to people seeking help, the problem may not be "health" but circumstance, leading to talking past one another, and missed opportunities to collaborate [45–47].

Othering is "a process that identifies those that are thought to be different from oneself or the mainstream, and it can reinforce and reproduce positions of domination and subordination" [48, p. 253]. Othering can reflect differences in gender, ethnicity, age, education, socioeconomic status, and cultural differences. When a person seeks help for a painful condition there are expectations about how the person ought to behave: they should defer to the expertise of the clinician; they should provide information when requested; they should follow the treatments; and they should recover. Leht, Fjellman-Wiklund, Stalnacke, Hammarstrom and Wiklund [49] describe the way women attending a persistent pain management centre presented themselves as "a proper patient, ready to change," often deferring to gendered notions of how women ought to behave, while men emphasised the importance of valuing themselves and presenting themselves as assertive and active. Lehti and colleagues also found that health professionals identify with those patients who are articulate and educated, yet one participant noted that she had to be careful to "fit into the norm of a proper patient by not looking too good (i.e., healthy) or too bad (i.e., ill)" [49, p. 49].

Health professionals undertake training over years to develop skills to treat people who seek their help. Dedication to such training suggests a strong underlying motivation, and a common theme for choosing medicine as a career is to help people [50], often based on personal or family experiences with ill health and encounters with health systems [51]. While this motivation may remain, neuroimaging experiments show low-regulation in areas usually associated with empathy when physicians observe others undergoing painful procedures, and that this occurs early in neurobiological processing [52]. Decety and colleagues consider this ability to reduce negative emotions subsequently allows greater recruitment of cognitive capabilities when going to the person's aid. In Decety's words, physicians "must try to resonate and understand the patient without becoming emotionally over-involved in a way that can preclude effective medical management" [52, p. 1682]. Morris, however, points out the heavy reliance on animal models in medical research but scant attention to writing about animal pain and asserts that "dismissive attitudes toward animal pain reflect wider anthropocentric and biomedical difficulties in responding to *otherness*" [53]. He goes on to say, "What little we know about the pain of animals is invisibly entangled in routine assumptions about differences between humans and animals...The category animal exposes the anthropocentric twists and biases that underlie immemorial distinctions that separate animals from humans: *us* from *them*." Defining exactly how different an animal is from a human is a messy business, but Morris argues that humans have determined animal pain to have little or no meaning, and this same disregard for meaning has also been applied to pain experienced by "enemies, slaves, women, children, minorities" who are powerless, and therefore "other" [53].

Philosopher Lajos Brons [54] points out that the notion of recognising the "*not*-self" (others) is an integral part of recognising "self" (our felt, or lived sense of being a unique individual) and therefore inescapable in daily social interactions. The concept of learning more about ourselves is embodied when we travel to different countries and come across practices we are unfamiliar with such as ceremonies or celebrations, clothing choices, or food choices. For some of us, the aphorism "When in Rome, do as the Romans do" may allow us to assent to practices we may never have agreed to do in our own country, while for others this may represent a threat to propriety. In either event, we become aware of something about the "other" and ourselves.

Brons goes on to argue that one way to limit the negative effects of othering is by appealing to the concept of "charity." Charity is, according to Brons, a stance where the "other's" actions are considered reasonable. Reasonableness assumes that the other person has rational, coherent, and true reasons for doing and saying what they do—even if, at first, we may not discern the underlying reasons. By invoking a charitable interpretation on another's actions, we are in turn asked to question our own preconceptions, our assumptions about the reasons the person did what they did [54]. Perhaps this philosophy could be applied to our own actions, thus preserving a forgiving attitude towards decisions we make, both in the line of healthcare and in our personal lives.

As one of those "others" living with persistent pain, I can feel uncomfortable when I interact with clinicians on social media. It is confronting to be told that the label I accept for my widespread pain is "not a proper diagnosis." Technically I suppose this is correct: a diagnosis in medicine requires "a causative agent or process with a fairly high degree of certainty" [55] and there is no clarity about the cause—or indeed the processes—involved in the widespread pain that I experience. The impact on me, however, is a denial of my reality, and my confidence in even a partial understanding of fibromyalgia becomes reduced, and I feel vulnerable. And yet, when I converse with other clinicians online, I add a different dimension to their understanding, for if people living with persistent pain feel neglected and misunderstood then people who live well with persistent pain are almost invisible. It is this dimension of being a person who lives with persistent pain and being informed about pain concepts that allows me to notice the various sensitivities, as I bring some alternative, and hopefully humane, perspectives to light.

A further challenge occurs for me when I interact with people who are currently distressed and disabled by their pain. In one discussion I was told that my pain was "less intense because you can keep doing things and you don't take medications," yet in an earlier comment this person remarked that "but no-one can tell you how bad your pain is because they don't feel it, my pain is so bad I need to take opioids all the time." On the one hand in this group there was vocal support for subjectivity and validity of self-report; yet on the other, a contradictory narrative of living well and dealing with persistent pain without medication was rejected. In this instance, I was the "other."

10 Learning to Live with Pain

As I write this paper, I reflect on the way I've changed the way I deal with my pain. When it first began, I expected, as most people do, it would resolve if I followed the advice of the health professionals. The pain has never abated, and has waxed and waned over the years, being a constant reminder of my body and its limitations. Despite working in the field of chronic pain management for most of my clinical career, and now researching and teaching, I am still learning to live well with pain. And, like many, I have learned in isolation, without a coach. The impact of living with pain, even when living well, is still with me every day.

There is a hidden world when living with pain. To be explicit: underneath the façade of living well lies a bedrock of constant micro-choices. Choices about where to place priorities for the day, what to choose not to do, how to go about daily living. There are larger choices too: do I pursue a new therapy that's just come onto the market? Do I go on that trip, knowing the effects it will have? Do I say no to this opportunity, because I'm already just managing all the things I'm doing right now? Can I push this week, to get this project done, because it's part of my job—or do I maintain that venerable coping strategy of "pacing" and ultimately miss out on opportunities because the effect of pushing through will likely lay me low on energy for some weeks? What of the new pain I have? Do I ignore it, knowing that so many of my pains will be there for a month or so, then fade away? Or is it something I could treat? Should I seek a diagnosis, only to be told it's not visible on imaging, or do I hold off but later find I could have had surgery that would have obliterated the problem?

Knowing that I have choices, and having a range of options to choose from, may be one of the more liberating aspects of having developed a repertoire of coping skills. Being flexible in the way I approach my daily life allows me a sense of personal agency. Perhaps this flexibility and agency is what distinguishes those who consider they live well with pain from those who are distressed and disabled?

11 Physician, Heal Thyself

When I disclose my persistent pain, I am often asked "Oh is that why you work in this area?" It would be reasonably easy to think this. Somehow, I am uncomfortable with this belief. It may be some of the muttered judgements overheard in clinical tea-rooms after a person with pain has said "Oh I'd like to start counselling people like me" that this is self-interest, or it is somehow harmful because of limited insight into one's own biases, perhaps? Indeed, in some circumstances the complete opposite belief is held: in the early days of Alcoholics Anonymous, recovered alcoholics were thought to be the only people able to help other recovering alcoholics [56]. This is no longer the case.

Two key points emerge for me: firstly, the prevalence of chronic pain in the general population in New Zealand is approximately one in five adults [57]; thus, it is probable that a great many clinicians working in this area of healthcare also live with chronic pain. Secondly, the sense of discomfort I experience when I worry that others may judge me because I am open about my pain is only a tiny fraction of that experienced by people in a less well-informed context.

I "came out," much like someone disclosing their sexual orientation, when I felt that I wasn't being authentic about my situation and what I was also dealing with. I wanted to reduce the stigma attached to persistent pain. While I was telling people that it's OK to acknowledge persistent pain and not to be ashamed of it, I was simultaneously keeping my pain to myself.

Being open about my persistent pain has not been easy, in part because of the fear of being misinterpreted, but also perhaps because clinicians can themselves feel uncomfortable about how to relate to an "other" when she is also one of "us." One misinterpretation of "pain education" is that by being educated, pain is meant to reduce, perhaps be eliminated. I am educated, my pain has not gone. This is confusing!

Diagnosis, treatment, rehabilitation, education and, naturally, social media are social events, yet persistent pain is a personal experience, and cannot be directly shared. Traversing this territory, determining whether my pain is *this* or *that*; my right to be a person living with this reality *and* a clinician *and* an educator *and* researcher requires cognitive dexterity. Being a person straddling (or tiptoeing through maybe) these different worlds involves shifts in what I say and do. While I know certain things about pain neurobiology and psychology and sociology, I ask myself if this means I have earned the right to tell others what it is like to live with pain? Am I a "representative" of people who live well with pain?

I hope I offer clinicians one way to gain insight into the life of someone who lives with pain. To help challenge attitudes of "them" and "us." People living with pain are not stereotypes, we are as many and varied as are the people in the supermarket. And we are also "us," health professionals.

12 Alone But Not Alone, Yet Alone

Frequently in qualitative studies, people living with pain talk of feeling isolated and of being alone. I also feel alone. I cannot be the only person to live in these multiple worlds where I am "one of them" while also being "one of us." Clinicians' attitudes towards people living with pain are readily found in research literature, but it is difficult to locate work examining clinicians' experiences of themselves also being a person living with and working with people who have persistent pain. This is at odds with the stance adopted early in medical research where using oneself as the "subject" when trialling novel treatments was considered the most ethical way to understand human responses, a position Price and Aydede espoused more recently [58]. Perhaps one way of bridging the gap between "them" and "us" is to remember the adage from disability activists: Nothing about us without us [59].

References

1. Chang H. Autoethnography: raising cultural consciousness of self and others. In: Methodological developments in ethnography. Bingley: Emerald Group; 2007. p. 207–21.
2. Ellis C, Adams TE, Bochner AP. Autoethnography: an overview. Hist Soc Res. 2011:273–90.
3. Wall S. Easier said than done: writing an autoethnography. Int J Qual Methods. 2008;7 (1):38–53.
4. Adams TE, Ellis C, Jones SH. Autoethnography. In: Matthes J, editor. The international encyclopedia of communication research methods. Hoboken, NJ: Wiley-Blackwell; 2017.
5. Schafer G, Prkachin KM, Kaseweter KA, Williams AC. Health care providers' judgments in chronic pain: the influence of gender and trustworthiness. Pain. 2016;157(8):1618–25.
6. Ashton-James CE, Nicholas MK. Appearance of trustworthiness: an implicit source of bias in judgments of patients' pain. Pain. 2016;157(8):1583–5.
7. Williams EN, Morrow SL. Achieving trustworthiness in qualitative research: a pan-paradigmatic perspective. Psychother Res. 2009;19(4–5):576–82.
8. Strouse SM, Nickerson CJ. Professional culture brokers: nursing faculty perceptions of nursing culture and their role in student formation. Nurse Educ Pract. 2016;18:10–5.
9. Spencer-Oatey H. What is culture? A compilation of quotations. GlobalPAD Core Concepts. Available at GlobalPAD Open House. Retrieved; 2015.
10. Denshire S. On auto-ethnography. Curr Sociol. 2014;62(6):831–50.
11. Melzack R, Wall PD. Pain mechanisms: a new theory. Science. 1965;150(3699):971.
12. Fordyce WE, Fowler RS, DeLateur B. An application of behavior modification technique to a problem of chronic pain. Behav Res Ther. 1968;6(1):105–7.
13. Loeser JD. In Memoriam: John J. Bonica [Internet]. Seattle: International Association for the Study of Pain; 1994 [cited 2019 1/8/2019]. Available from: https://www.iasp-pain.org/JohnBonica.

14. McKenzie Institute New Zealand. About Robin McKenzie [Internet]; undated [cited 02/02/2019]. Available from: http://www.mckenzieinstitute.org/nz/en/about-us/undefined/.
15. Melzack R, Wall PD. The challenge of pain. 2nd ed. London: Penguin; 1996. 447 p.
16. Antheunis ML, Tates K, Nieboer TE. Patients' and health professionals' use of social media in health care: motives, barriers and expectations. Patient Educ Couns. 2013;92(3):426–31.
17. Maughan DL, Economou A. Social networking sites: a clinical dilemma? J Med Ethics. 2015;41(2):203–5.
18. Tan SSL, Goonawardene N. Internet health information seeking and the patient-physician relationship: a systematic review. J Med Int Res. 2017;19(1):e9.
19. Brown P, Elston MA, Gabe J. From patient deference towards negotiated and precarious informality: an Eliasian analysis of english general practitioners' understandings of changing patient relations. Soc Sci Med. 2015;146:164–72.
20. Dukhu S, Purcell C, Bulley C. Person-centred care in the physiotherapeutic management of long-term conditions: a critical review of components, barriers and facilitators. Int Pract Dev J. 2018;8(2).
21. Phillips RL, Short A, Kenning A, Dugdale P, Nugus P, McGowan R, et al. Achieving patient-centred care: the potential and challenge of the patient-as-professional role. Health Expect: Int J Pub Part Health Care Health Policy. 2015;18(6):2616–28.
22. Poulin P, Shergill Y, Romanow H, Busse JW, Chambers CT, Cooper L, et al. Researching what matters to improve chronic pain care in Canada: a priority-setting partnership process to support patient-oriented research. Can J Pain. 2018;2(1):191–204.
23. Zwar N, Harris M, Griffiths R, Roland M, Dennis S, Powell Davies G, et al. A systematic review of chronic disease management. New South Wales: Research Centre for Primary Health Care and Equity, Medicine SoPHaC; 2006.
24. Hibbard JH, Greene J, Sacks RM, Overton V, Parrotta C. Improving population health management strategies: identifying patients who are more likely to be users of avoidable costly care and those more likely to develop a new chronic disease. Health Serv Res. 2017;52(4):1297–309.
25. Moseley G, Nicholas MK, Hodges PW. A randomized controlled trial of intensive neurophysiology education in chronic low back pain. Clin J Pain. 2004;20(5):324–30.
26. Louw A, Diener I, Butler DS, Puentedura EJ. The effect of neuroscience education on pain, disability, anxiety, and stress in chronic musculoskeletal pain. Arch Phys Med Rehab. 2011;92(12):2041–56.
27. Louw A, Diener I, Butler DS, Puentedura EJ. Preoperative education addressing postoperative pain in total joint arthroplasty: review of content and educational delivery methods. Physiother Theory Pract. 2013;29(3):175–94.
28. Seedman J. The truth about pain science, exercise, and movement [Internet]. 2019 [cited 15/1/2019]. Available from: https://www.advancedhumanperformance.com/blog/pain-science-exercise.
29. Loeser JD. Multidisciplinary pain management. In: Merskey H, Loeser JD, Dubner R, editors. The paths of pain. Seattle: IASP Press; 2005. p. 503–11.
30. Flor H, Turk DC. Chronic pain: an integrated biobehavioral approach. Seattle: Lippincott Williams & Wilkins; 2015. 547 p.
31. Main CJ, Sullivan MJL, Watson PJ. Pain management: practical applications of the biopsychosocial perspective. 2nd ed. Edinburgh: Churchill Livingstone; 2008. 432 p.
32. Rosenstiel AK, Keefe FJ. The use of coping strategies in chronic low back pain patients: relationship to patient characteristics and current adjustment. Pain. 1983;17(1):33–44.
33. Butler S. The wit and wisdom of Wilbert (Bill) Fordyce (1923–2009). Scand J Pain. 2017;16(1):160–3.
34. Lempp HK, Hatch SL, Carville SF, Choy EH. Patients' experiences of living with and receiving treatment for fibromyalgia syndrome: a qualitative study. BMC Musculoskel Dis. 2009;10:124.
35. Henwood P, Ellis J, Logan J, Dubouloz C-J, D'Eon J. Acceptance of chronic neuropathic pain in spinal cord injured persons: a qualitative approach. Pain Manag Nurs. 2012;13(4):215–22.

36. Arnold LM, Crofford LJ, Mease PJ, Burgess SM, Palmer SC, Abetz L, et al. Patient perspectives on the impact of fibromyalgia. Patient Educ Couns. 2008;73(1):114–20.
37. Toye F, Seers K, Barker KL. Meta-ethnography to understand healthcare professionals' experience of treating adults with chronic non-malignant pain. BMJ Open. 2017;7(12):e018411.
38. Lennox Thompson B, Gage J, Kirk R. Living well with chronic pain: a classical grounded theory. Disabil Rehabil. 2019:1–12.
39. Pietilä Holmner E, Stålnacke B-M, Enthoven P, Stenberg G. The acceptance. J Rehabil Med. 2018;50(1):73–9.
40. Wilcock AA. Reflections on doing, being and becoming. Aust Occup Ther J. 1999;46(1):1–11.
41. Walder K, Molineux M. Occupational adaptation and identity reconstruction: a grounded theory synthesis of qualitative studies exploring adults' experiences of adjustment to chronic disease, major illness or injury. J Occup Sci. 2017;24(2):225–43.
42. Ballantyne JC, Sullivan MD. Intensity of chronic pain: the wrong metric? New Engl J Med. 2015;373(22):2098–9.
43. International Association for the Study of Pain (IASP). IASP council adopts task force recommendation for third mechanistic descriptor of pain [Internet]. Washington, DC: International association for the study of pain. 2017 [cited 10/01/2019]. Available from: https://www.iasp-pain.org/PublicationsNews/NewsDetail.aspx?ItemNumber=6862.
44. Jones GT, Atzeni F, Beasley M, Flüß E, Sarzi-Puttini P, Macfarlane GJ. The prevalence of fibromyalgia in the general population: a comparison of the American college of rheumatology 1990, 2010, and Modified 2010 Classification Criteria. Arthritis Rheum. 2015;67(2):568–75.
45. Allegretti A, Borkan J, Reis S, Griffiths F. Paired interviews of shared experiences around chronic low back pain: classic mismatch between patients and their doctors. Fam Pract. 2010;27 (6):676–83.
46. Henry SG, Bell RA, Fenton JJ, Kravitz RL. Goals of chronic pain management: do patients and primary care physicians agree and does it matter? Clin J Pain. 2017;33(11):955–61.
47. Hardy JK, Crofford LJ, Segerstrom SC. Goal conflict, distress, and pain in women with fibromyalgia: a daily diary study. J Psychosom Res. 2011;70(6):534–40.
48. Johnson JL, Bottorff JL, Browne AJ, Grewal S, Hilton BA, Clarke H. Othering and being othered in the context of health care services. Health Commun. 2004;16(2):255–71.
49. Lehti A, Fjellman-Wiklund A, Stalnacke BM, Hammarstrom A, Wiklund M. Walking down 'Via Dolorosa' from primary health care to the specialty pain clinic-patient and professional perceptions of inequity in rehabilitation of chronic pain. Scand J Caring Sci. 2017;31(1):45–53.
50. Woodward A, Thomas S, Jalloh MB, Rees J, Leather A. Reasons to pursue a career in medicine: a qualitative study in Sierra Leone. Global Health Res Policy. 2017;2:34.
51. Wouters A, Croiset G, Isik U, Kusurkar RA. Motivation of Dutch high school students from various backgrounds for applying to study medicine: a qualitative study. BMJ Open. 2017;7(5): e014779.
52. Decety J, Yang CY, Cheng Y. Physicians down-regulate their pain empathy response: an event-related brain potential study. Neuroimage. 2010;50(4):1676–82.
53. Morris DB. Animal pain: the limits of meaning. In: van Rysewyk S, editor. Meanings of pain. Cham: Springer; 2017. p. 389–99.
54. Brons L. Othering: an analysis. Transcience: J Global Stud. 2015;6(1):69–90.
55. Calvo FC, Karras BT, Phillips RL, Kimball AM, Wolf F, editors. Diagnoses, syndromes, and diseases: a knowledge representation problem. Washington: AMIA; 2003:802.
56. Leach B, Norris JL. Factors in the Development of Alcoholics Anonymous (A.A.). In: Kissin B, Begleiter H, editors. The biology of alcoholism: volume 5: treatment and rehabilitation of the chronic alcoholic. Boston: Springer; 1977. p. 441–543.
57. New Zealand Ministry of Health. Annual data explorer 2016/17: New Zealand health survey [Internet]. 2017 [cited 04/02/2019]. Available from: https://minhealthnz.shinyapps.io/nz-health-survey-2016-17-annual-update/.

58. Price DD, Aydede M. The experimental use of introspection in the scientific study of pain and its integration with third-person methodologies: the experiential-phenomenological approach. In: Aydede M, editor. Pain: new essays on its nature and the methodology of its study. Cambridge: The MIT Press; 2005. p. 243–74.
59. Charlton JI. Nothing about us without us: disability oppression and empowerment. Berkeley: University of California Press; 2000. 220 p.

Chapter 3
Diagnosing Human Suffering and Pain: Integrating Phenomenology in Science and Medicine

Smadar Bustan

Abstract This chapter shows that phenomenology may be critical to diagnose human suffering in science and medicine, significantly improving patients' well-being, while providing a more thorough understanding and adequate management of on-going suffering in the context of chronic pain and other major illnesses. Historical and contemporary references from philosophy, medicine, and science explain why, to date, there is no golden rule or consensual definition of suffering. As a solution, the paper calls for a paradigmatic conceptual shift, explaining the fundamental principles, and how they are translated into an experimentally and clinically tested "Pain-related Suffering assessment tool," as demonstrated by Bustan's Fan Models. The tool is tailored for individual use for enhancing personalized care. And while reviving George Canghuilem's important historical message about individuality in medicine, it seeks to combine impersonal knowledge and personal knowledge to improve medical science and practice.

Clinical Implications: The demand from physicians to be more attentive in assessing the suffering of their patients was introduced by Eric Cassell (1982) as one of the fundamental aims of medicine. This concern, taken into account by health professionals, is difficult to fulfill in the absence of appropriate measures. In addition, the focus on end-of-life suffering has excluded, in the vast majority of cases, patients with chronic pathologies and in particular those with chronic pain whose suffering is part of everyday life. This chapter therefore presents the conceptual shift necessary for the development and validation of a simple and effective clinical tool to diagnose suffering, allowing a more precise assessment and individualized care for chronic pain patients, applicable to other chronic illnesses.

Keywords Human suffering · Pain · Chronic pain · Suffering assessment tool · Phenomenology · Experimental studies · Clinical studies · Philosophy and history of medicine · Eric Cassell · Georges Canghuilem · Arthur Kleinman

---

S. Bustan (✉)
University of Paris Diderot, Institut Humanités, Sciences et Sociétés (IHSS), Paris, France
http://www.suffering-pain.com

S. van Rysewyk (ed.), *Meanings of Pain*,
https://doi.org/10.1007/978-3-030-24154-4_3

1 Introduction: Phenomenology for the Practical Life of Evaluating Suffering in Pain Patients

Going through life, we cannot avoid asking why some suffer more than others or collapse in situations where others are spared. In response, we need to define a baseline and comparison criteria to assess different sufferings, which may lead to a controversy regarding our right to judgment over others' endured agony. Stating that the landscape of suffering takes various forms and is subjective by nature can challenge any pretense to answer these metaphysical questions. However, having investigated "who is more likely to suffer under pain" using experimental means, we found that psychological traits revealed a certain disposition to suffer but nevertheless, that contextual circumstances may modulate or alter these predicted tendencies [1]. Pain does not condemn us; but in order to get better we need to be diagnosed so that subsequent suffering is adequately addressed. Thus, in spite of the frequent confounding of pain and suffering in nearly every language or discipline, it is important to educate people to differentiate them, and to instruct professionals to evaluate them separately.

Pain and Suffering steer the living of chronic patients practically every day. When facing them in healthcare systems, it quickly becomes clear that the ultimate goal *to diagnose human suffering using measurable means for clinical purposes* is valuable to improve the well-being of patients, providing a more thorough understanding and adequate management of persistent suffering in the context of major diseases. This approach does not intend to unnecessarily expose hidden wounds that must remain protectively sealed, but to recognize harm in a comparative way, and in large populations of patients, with the ultimate goal of improving care. The challenge is to find the right methodology for conducting such evaluations. My ultimate goal is to inscribe the "Suffering Diagnosis Tool" in the day to day assessment in hospitals and psychiatric clinics. This would demonstrate the reasons a combined modality of philosophy and medicine may finally provide a standardized form of measurement of human suffering while remaining attentive to the individual (for better personalized care) and to inter-individual variations across culture, language, and gender (for world-wide institutionalized use).

In this introductory chapter to the "Suffering Diagnosis Tool" project, and a second in this book series [2], I explain why an integrative approach that employs phenomenology to diagnose human suffering in science and medicine can be the necessary conceptual and methodological breakthrough. This approach has been implemented in experimental pain research with healthy subjects [3–5] and also with chronic patients at the hospital [6]. I applied it to the on-going validation of a pain-related suffering assessment tool that seems to offer an innovative evaluation of human suffering related to pain and may be potentially applicable to other chronic illnesses such as cancer, diabetes, neurodegenerative conditions or psychiatric disorders [7]. This theoretical discussion will make use of a demonstrated observation and testing based on the study case of pain-related suffering. However, it is important to underline that its implications go beyond any specific disease or mental condition, taking "pain" to represent an "illness" and generally speak of an "ailment-related suffering."

Perhaps what is most important to note when seeking to apply phenomenology to the scientific and clinical assessments of suffering is: first the knowledge gap it fills; second, the benefits of such uncommon methodological choice; third and most critically, the required paradigmatic shift it entails.

Regarding this change in paradigm, the use of phenomenology presumes a fundamental modification in the basic conceptual approach that distinguishes itself from the reigning theories and practices of our time. We no longer look for the core properties of suffering that could be measured individually and that combined form a general picture of its extent and nature, as commonly done for studied constructs (for example, personality, trauma, empathy) and in accordance with traditions adapted from psychology. For example, Price, Wade and colleagues [8–11] describe their experimental protocols to study "pain-related suffering" as an umbrella category encompassing key negative emotions, often occurring in persistent pain conditions. The term "suffering" is never mentioned to the participants during the evaluation process and is collected indirectly through the scores of five selected negative emotions supposedly underlying this construct: fear, frustration, anger, anxiety, depression. The main concern with this classical methodological approach is that the choice of a particular list may not properly represent pain-related suffering and may leave out major emotional and cognitive variables (catastrophism, mental coping, hopelessness, helplessness, for example) that are equally associated with the experience of suffering [1, 12].

Both in philosophy and in science, the mainstream views present each time a specific type of problem as discussed further, while altogether they seem to follow the same trend of defining suffering in a rather reductive manner and always through an intermediate variable. In examining these views closely, it seems that if we want to give even an abstract explanation of the construct we are evaluating, we could look for the *inner content* of suffering:

As a *first option*, we can do so by revealing the essence of suffering and evaluating each manifestation as depicting this invariable factor. However, this option does not escape entanglement since, according to my philosophical theory, suffering is a double-natured phenomenon naming two sorts of experiential elements, a feeling and a condition, with a possible shift from one to another and a continuous need to first identify which of these two essences best represents the suffering state of the patient [6, 13, 14]. The main difficulty consists in the immediate identification of the suffering nature-type of a person we have yet to diagnose.

In a *second* alternate *option*, we may set aside the reference to the essence or pure form of the phenomenon. Instead, we decompose this complex concept by analyzing its vital constituents in order to scrutinize its manifestations with the greatest attention. So far, this second and more common option, as already mentioned, provides an inventory of primary negative emotions or traits presumably composing suffering. It overlooks, however, the fact that we end up *adjusting* this rich human phenomenon to a restricted scope of attributes that may not describe an authentic picture of the complex experience of one's suffering. Along with this "short list problem" there is also the "intermediate problem." Suffering was in fact included in many theoretical models of pain [8–10, 15–20], but they all seem to classify suffering under the name of something else that is supposed to be broad enough to represent it. Its assessment in experimental and clinical studies of pain remains therefore limited not only due to lack of viable methods, but also owing to a perpetuated conceptual difficulty that does not really tackle the crux of the matter. It is therefore unsurprising that no golden rule or consensual definition of suffering is available to date, and more specifically of pain-related suffering [21, 22].

By principle, I find *compromising* any form of *attempt to define suffering through a content*, either via an invariable (essence) or key constituents (emotions). In the process of assessing the suffering of an ill person, it seems impossible to determine the one thing, or philosophically speaking *a thing-in-itself*, that could serve as the main substance of suffering. The lack of one universal essence that an observer can easily identify from diverse behavioural manifestations plays an important role in the puzzles this phenomenon raises. As a consequence, studying such a double headed phenomenon may seem problematic for people from both natural sciences and the humanities who seek to target the single referential element of suffering or even a core list of representative contents. Instead, my phenomenological interpretation considers that its dual nature calls for creative forays into this merged philosophical-psychobiological investigation, when aiming at the most diversified display of such a complex percept. Putting aside the single substance or key substances approach will not only prevent the reductive trap of focusing on physical suffering for merely discovering physical suffering, but it will also provide a point of departure for something deeper and finer, allowing the emergence of a maximal range of markers of this *construct*.

The answer to this dilemma requires a change of perspective by studying the way suffering functions and by focusing on the *modes of its experience*. If we want to tackle the issue and propose a measure to diagnose human suffering related to chronic pain or to other long-term maladies, the benefit of the phenomenological approach consists in totally changing the way we go about studying it, and in two ways:

First, we keep the idea of decomposing the concept of suffering mentioned above but *analyze the dynamic structure of this composite instead of its components* through a sequence of evaluations of the way the "lived experience of suffering" operates at different levels. Our task is to provide an objective methodology capturing the subjective state of suffering. It is thus a method that is valid for everyone while accounting for the suffering happening to oneself and the way it evolves (required for a long-term patient follow-up). For that we need to focus on the "how" instead of on the "what" and we are not to ask, "What does your suffering consists of or what makes you suffer?" We are only to ask, "How do you experience suffering on various levels?" In short, I propose to deconstruct suffering into a handful of fundamental principles, and that the constellation of basic components is replaced by that of basic experiential modes (dimensions). This approach would clarify the ultimate concept of suffering as well as each individual suffering, while avoiding the trap of an arbitrary list of distinctive contents.

Second, the embedded phenomenological method allows to *approach suffering directly*, as opposed to the classical indirect form of inquiry that investigates a concept without ever mentioning the term; this is one of the most consequential aspects of this project. The scientific and clinical practices are naturally preoccupied by the optimal and least intrusive form of inquiry to access the experience of suffering [2]. Advocates of the indirect approach may be reluctant to ask a patient directly if they suffer as such questioning may be psychologically brutal and intensify one's agony by bringing it to the surface. In addition, indirect questioning consists in exploring particular patterns of which the patient himself is *unaware*. This is the rationale behind screening tests, like depression for example, in people who seem to be significantly "down" but cannot really tell if they are depressed. Symptoms of this distress are detected through indirect evidence when asking them about losing interest or pleasure in doing things, feeling bad about themselves, feeling tired or with little energy, and so on. However, the first thing I noticed after having developed and tested both formats via newly constructed indirect and direct pain-related suffering questionnaires, is that the direct approach engenders a process of insight into coherence and self-mirroring. Patients, especially with chronic conditions like pain, know that they are suffering, but often not its extent nor which aspects are most prevailing. When directly evaluated, the responses of patients were very positive not only because they considered this form of questioning less technical than the frequent series of adjectives and descriptive statements, but also because this mirroring process allowed them to actively deconstruct their own experience and realize which suffering dimensions are most dominant and why. The direct investigation is therefore not more difficult than a traditional indirect assessment form using implicit statements, adjectives, or associated emotions to obtain a proper detection. In fact, I suggest combining both forms. But a direct approach does allow to openly investigate each dimension and magnitude of suffering by explicitly inquiring with the patient in both a straightforward and systematic manner about the intensity and nature of their suffering at all fundamental levels. In this respect, the first to encourage a straightforward discussion with patients about their suffering was the American physician Eric Cassell, formulating recommendations without a standardized methodology and guiding caregivers who are unable to free themselves sufficiently how to flexibly deal with these concerns [2, 23].

In order to present a more structured approach that may reveal what most physicians want to know but often do not dare to ask, *we must change perspectives and operate a displacement through the phenomenological method*, adopting a new orientation that exposes suffering in a more direct and more well-rounded way without giving up on traditions in psychology or neurobiology of emotions. This chapter is therefore largely concerned with explaining the legitimacy of such a conceptual shift by gradually demonstrating: first, the three major definitions of pain-related suffering and their respective biases; second, how this broad term is parceled into manageable parts whereby phenomenology enacts a reversal in the mode of exposition as it focuses on the way the phenomenon of suffering unfolds, instead of focusing on its distinctive features; thirdly, my fan model serves as the visual support; and last, following Georges Canguilhem's reminder of the role of individuality in medicine, I will show how this integrated phenomenological-clinical approach provides us with an objective method to assess Suffering and Pain-related Suffering that is tailored for individual use and investigating personal suffering.

2 Defining Suffering/Pain-Related Suffering: Biases and Confusions

Present-day science, medicine, or even philosophy, do not provide a blueprint of "an absolute conception of suffering." At the same time, as part of the overall effort to treat the patient as a "whole," [24, 25], there were growing demands in recent years for physicians to become more attentive in the evaluation of their patients' suffering, introduced by Eric Cassell [26] as one of the fundamental goals of medicine. While these concerns have not fallen on deaf ears, they were difficult to address lacking appropriate simple-to-use methods for healthcare professionals. In addition, most of the focus was put on terminally ill patients in palliative care [27–30], neglecting the majority of cases of chronic patients whose suffering manifests as part of daily living rather than at the end of life. I therefore distinguish between the two: *terminal suffering* in end-of-life and threatening conditions [31, 32] requires coping with fatal decline and the finitude of death. *On-going or prolonged suffering* requires coping with fluctuating states of a permanent illness that may be present for decades, presenting a threat not to life but to the quality of life and well-being [6]. The urge for patient relief is identical for both suffering types, but it means appeasement for terminal suffering, while it implies the attenuation of a continuously torturous nonlethal state that is part of daily living for on-going suffering involved in chronicity like that of chronic pain.

Reviewing the pain literature on the concept of suffering, the most frequently cited definition for the past four decades was provided by Eric Cassell, referring to suffering as a severe **distress** generated by the imminent **threat** to the integrity of a person in pain [26]. Chapman and Garvin [33] described suffering as "a complex negative affective and **cognitive** state" provoked by the sense of **helplessness** when coping with the threatening pain. The first to explicitly include suffering as a separate aspect was Loeser's schematic four-dimensional pain model [15], depicted in nested circles (nociception, pain, suffering, and pain behaviors), stressing its cognitive character as "an emergent property of the human brain" that "is dependent upon consciousness" [34–36]. As mentioned earlier, Price, Wade and colleagues specifically speak of "pain-related suffering" and refer to it as an agglomeration of five distinctive negative emotions. Further along this line, Price and Barrell provided an in-depth account of pain-related suffering fundamental meanings, conveying an enduring **burden**, a permanent **harm** or the interruption of a person's ability to live [37]. In going from viewing suffering as an enhanced negative emotional and cognitive response of the individual, and thus as primarily a private-subjective experience, to include social factors which call to maintain the biopsychosocial model of chronic pain [34], the observance of suffering in clinical pain was attributed to a sense of great **isolation**, alarming **lack of understanding** and **dependence** on others [38, 39]. This extended view echoes to conceptions from social fields (ethnography, anthropology, sociology, history) and more particularly to the domain of social suffering [40], asserting that suffering is a **social** phenomenon shaped by social forces before being a personal experience inherent to the individual. Thus, pain science continues to define suffering without being able to label it in a way that allows for a precise determination of its influence on pain, inviting for a standardized view in order to assure a clinically useful model. Moreover, these accounts of pain-related suffering are problematic not only because of their multiple confounding definitions, but basically because the phenomenon of suffering is being continuously reduced to something else and not viewed on its own term.

To summarize, there are three major difficulties regarding the concept of suffering in pain research to date:

1. **Suffering** is viewed as a **distress caused by threat**, following the most referenced definition of Cassell, thus disregarding the very simple fact that one's suffering can be totally dissociated from a sense of menace (especially when the chronic condition is not life threatening) and cannot be limited to these aspects alone.
2. Or, in clinical practice, **suffering is admitted as a simple extension**—more intense, more severe—**of the unpleasantness** of the pain, **named** "a secondary emotional state" or "**secondary unpleasantness**" [41] with long-term implications of pain associated with threat, loss, and imminent damage to oneself. Historically, pain has been viewed consistently as composing two dimensions: a sensory (intensity) one and an emotional (unpleasantness) one. However, we may observe elevated levels and long-lasting effects involving higher affective and cognitive processes proportional to the extent of damage rather transcribed as suffering. Pain-related suffering expresses an increased emotional and cognitive burden that the unpleasantness of the painful sensation cannot adequately reflect, revealing its "unbearable," "unsustainable" character [17, 42], often associated with "misery," "despair," [43] "disability" [44, 45] or even mal-être (ill-being). This calls for a radical change in the way we understand pain and clinically evaluate it. In this respect, our experimental results confirm that pain-related suffering is a distinct yet an integral component of pain, and thus a third aspect to be assessed along with the pain intensity and unpleasantness as illustrated below in my first 2014 Fan Model. Therefore, we may consider suffering as a complex phenomenon that exceeds that of pain but when associated with this condition, it should to be included as part of the pain definition under the designation of "pain-related suffering."
3. Last, while **suffering** was indeed recognized by some scientists and clinicians as an independent complementary constituent of pain it was still restricted, as explained earlier, to the evaluation of an agglomeration of **five dominant negative emotions** (fear, anger, frustration, anxiety and depression) [9–11] and **labeled "prolonged emotions related to pain"** [8, 37, 46, 47]. In the context of this perspective, we then notice the arbitrary choice of variables chosen to

assess pain-related suffering, questioning all efforts to validate a key components group to qualify it. For example, in the largest database search on suffering in various patient groups and particularly in acute and chronic pain populations, specific negative emotions including the ones identified above have been contested, arguing that "depression/hopelessness was not considered to be necessarily equivalent to suffering although depression could conceivably be considered an instantiation of suffering. As such, the search did not include those headings" [12]. Further along the line, this selective approach for depicting suffering appears to be even more arbitrary, when studies in advanced cancer and palliative care patients reporting pain rather choose to evaluate it through "emotional distress, anxiety, loss of control, and perceived coping and spiritual distress items such as loss of dignity and hopelessness" [48]. These contradictory choices of the core variables underlying pain-related suffering may elicit significant errors and reflect randomness in its diagnostic approach.

To avoid systematic gaps, we may either agree on the right emotional-cognitive key variables from the various lists depicting this construct, though I believe that to start with, such an approach would not capture the phenomenon of our study in its most encompassing form. Or, we may opt for a totally different solution when shifting the focus onto the construct as a whole, either through the mid-term solution of evaluation of suffering levels [20][1] or through the assessments of pure suffering dimensions using phenomenology, as shown below in my second 2014 Fan Model [6].

Finally, it may be of further help to clarify another major bias which challenges the endeavor of measuring suffering in patients: the confusion of suffering with pain in science and everyday speech. Research often attributes the overlap between pain and suffering to "ordinary language: we speak of being in pain due to the loss of a friend, but we declare that we suffer from a toothache" [49] and possibly to a genuine conceptual inability to distinguish between them in cases where suffering is identified with "painfulness" of any sort. Insistence on their differentiation often results in a dualistic yet judicially viable [50] and *very common categorization attributing pain to the physical realm and suffering to the mental realm*. Publications following Fordyce's historical appeal to refrain from the frequent confounding of suffering and pain [18] led to a supposed agreement in pain science about their treatment as two correlated but distinct phenomena [29, 33, 51, 52], admittedly avoiding the fall back into the Cartesian mind/body separation. In practice, this decision barely applies.

In examining the decoupling of suffering from pain while determining their 'inner relation', it seems that the lack of consent maintains their confusion over a definition of pain-related suffering, in addition to their frequent exchangeability both in

[1]This mixed solution still falls between the cracks since the suffering levels are evaluated through a direct measurement of the degree of suffering and an indirect measurement of its correlation to various experiential dimensions (physical, psychological, spiritual), each including a basic group of components.

literature and clinical work [53]. In Fishbain and colleagues' review mentioned above, suffering synonyms included the terms: "mental pain, psychological pain, mental defeat" [12]. These reflect the frequent swap in various scientific domains such as neuroscience and mindfulness (labeling suffering "social pain") [54, 55], psychology (labeling suffering "psychological pain") [56], psychiatry and suicidology (labeling suffering "mental pain") [57–59]. In hospice clinical practice, Saunders coined the famous term "total pain" encompassing physical, mental, social and spiritual components of pain in order to address suffering [60–63]. These various accounts of suffering may contradict what doctors, scientists, philosophers or psychologists say to one another when employing the words "pain" and "suffering." We may well admit that in order to make sure professionals share a clear view of patients' suffering, better communication and understanding among the various fields, it is necessary to avoid the persistent interchange of pain and suffering.

3 The Important Parallelism Between Pain and Suffering

I construct a theory that extrapolates what patients who suffer refer to, whether expressly stated or vaguely implied. To propose a methodology that is valid for empirical and clinical enquiry, we need a consensual guiding definition of suffering and, more particularly, of pain-related suffering. But instead of thinking in terms of a Kantian type regulative idea that only echoes in every manifestation and serves as an ideal conceptual reference, we rather seek a "standard measure for suffering," a *structural* framework used as the explanatory basis for singular suffering, which can incessantly change and adapt to the disease evolution. How can this be accomplished? And how can we retrieve the given suffering from its changing modes of appearance and continuous transformation *without* presupposing the 'pure object' of the phenomenon we want to explain? This may be achieved in accepting that by focusing on the lived modalities and the medium through which suffering is experienced, we rely on a *stable invariant structure* shared by all the manifestations of suffering and *not on a conceptual invariant or essence.*

In this regard, I do not believe that there is an overarching metaphysical idea encompassing all our interrogations on human suffering. And paradoxically, not grasping suffering through an investigation of its essence may seem contradictory to the announced use of phenomenology as a qualitative research method in this field of healthcare, since phenomenology is based on the study of essences [64, p vii, 65]. As Husserl, the father of this philosophical methodology, states:

> Pure phenomenology as science so long as it is pure [...] can only be essence investigation and not at all an investigation of being-there. [...] The particular can in its immanence be posited only as this—this disappearing perception, recollection, etc.—and if need be, can be brought under the strict essential concepts resulting from essential analysis [66, p. 183].

And yet, positing a suffering essence as identifier in a preliminary definition to which the patient's reported suffering must correspond (the *eidos* that presents itself throughout all the potential forms) is not suitable. This is due to the phenomenon's double nature, pre-reflective character [67], as well as to important variations in the meaning of "suffering" in the various languages [68]. In my view, the phenomenon of suffering has two essences and is unusual in the landscape of theories on emotions due to its lack of a sole *quid*. Hence, it is difficult to decide during evaluation what term best represents the patient's suffering: whether a feeling (episodic), a condition (existential, chronic) or the transitioning from one to the other. A person may be experiencing surging suffering that erupts and overwhelms him for a period of time due to an emotional crisis, being fired or fighting cancer. But when the suffering evoked is no longer considered to be a hurting episode but rather an existential condition that becomes permanent (duller, yet not less tormenting), the infliction reflects a transient feeling turning into a way of being.

There is an *important parallelism* between *acute pain associated with suffering as a feeling, and chronic pain associated with suffering as a condition.* The difference between the two suffering essences is a function of both time (episodic versus long lasting), and of endurance (interrupted versus continuous) and a load becoming heavier and heavier because it is continuously present at the back of the mind with no apparent respite. Suffering as a condition undergoes fluctuations in accordance with its cause or the person's general state: intense yearning for the deceased child, the worsening of the chronic disease, or a generally increased vulnerability that make the unpleasant turn unbearable, for example. The aim in working with patients is to lead them towards a reconciliation with the oppressive pain so as to learn to live with it for attenuating their suffering. Prior diagnosis of the disease associated suffering is therefore crucial. But when trying to understand the gravity of what overwhelms a person, this duality of principle challenges the suffering evaluation procedure.

In effect, acute post-surgical pain is acceptable during the recovery period. When transitioning into chronic pain, as it happens to millions of people every year [69], the resulting temporary suffering mutates into a permanent agony affecting the person's entire well-being; its nature is thus totally different. Responsiveness to the illness may be a function of a prior health condition as well as differences in inter-individual reactions to the actually lived experience, making the factors list responsible for the degree of suffering both unlimited and arbitrary. This may happen when the bittersweet news of overcoming a life-threatening illness, as for the breast cancer survivors I evaluated, announce that the pain that developed during treatment is there to stay. Cure from cancer grants a natural sense of victory and relief from the suffering caused by reported feelings of change in body image, fear, diminished attractiveness and intellectual function [70]. But after winning over cancer to then develop chronic pain, some patients suffer severely from helplessness and defeat over an invincible affliction that cannot be healed; as reported to me by a patient: "If you ask me, I'd take the cancer back right now if I could only get rid of this crap of pain." Others may consider instead that being alive is a gift worthy of the burden imposed by chronicity and are thus not shattered by the pain. The challenge may also come from the radical adjustment required when moving from curable cancer to chronic pain, having to readjust their coping mechanism from fighting to resilience to reduce suffering. Therefore, the changeable dynamic when transitioning from a feeling of suffering to a condition of suffering does not discredit the attempt to understand or even systematically measure human suffering. It does however challenge the method to decipher and effectively evaluate it when attempting to propose a viable theoretical approach followed by the development of a reliable tool to assess suffering both in different individuals, and across various maladies. For these reasons, *my use of the phenomenological attitude avoids measuring every suffering account against a pre-defined concept or essence, as this may lead to a wrong interpretation of patients' authentic experience.*

4 The Phenomenological Shift: Assessing the Modalities of Suffering

More generally, our vision of suffering is often determined by elements from our personal, family, socio-cultural, or religious biography. Yet as we accumulate experiences throughout life, our own idea of suffering continuously develops, and its actual experience often supersedes any a priori conception. There is no termination in the forming of the percept and our already established idea of suffering may be totally overturned following a hurtful experience exceeding our threshold, and the failing of our coping mechanism. Being already familiar with suffering or with our own estimated strength does not protect us from being thrown deeper into the pit of torment. And for the investigators or clinicians, knowing the severity of one's condition does not guarantee a more precise estimation of their agony. In this evaluation, room must therefore be made for the unknown and unexpected, accepting a full possession by a sense of excess ("to suffer, it is to suffer too much," says Paul Ricœur) [49, p. 68] that may also consume our capacity to forbear and endure. Room must also be made to embody human suffering without creating a scientific all-inclusive inventory of its manifestations. The latter is often hard to accept, especially when seeking a framework of thought giving us all the options beforehand, so as to follow up the regained and declined states of patients with the hope of providing a long-term prediction test to identify and thus relieve their suffering. Admitting its *contingent and context-sensitive aspects*, the challenge consists in the assessment of a subjective experience (personal suffering) on the grand scale, using an objective measurement suited for large populations, and including an open horizon to account for possible, unforeseeable, fluctuations. This is an advantage that an open method based on phenomenology can provide.

Obviously, transferring theory to practice raises controversies regarding the right account of experiential philosophy. The vital point here, I re-emphasize, concerns the investigation of suffering on the basis of a predefined notion, versus letting the percept arise out of the patient's life-world as it unfolds experientially, without presupposing what it is meant to explain, what it is supposed to be, and basically how the patient is expected to suffer. A systematic diagnosis of suffering in medical settings with large cohorts requires the concept of "suffering" to be valid and verifiable beyond dispute, assuring an unfalsifiable methodology for the empirical procedure applied to evaluate it. *An objective reference to a subjective notion like suffering* seems necessary in order to be able to assess the negative impact of an illness on a patient's life as portrayed by suffering. The field of experiential suffering as a whole has no definite boundaries nor points of view. Therefore, the adopted approach must remain *context sensitive* so as to fit the individual experience, language, culture and the perception of suffering under changing circumstances. Within the right philosophical framework of thought, we can establish a universal and objective approach based on *a few fundamental principles*, leaving out the inexhaustible open-endedness of the totality of the conceptual schemes of suffering. How then to proceed to achieve this?

The phenomenological break-down of this notion is based on the principle of *dimensionality*. This principle explicitly interrogates the patient about their ongoing suffering through a request to assess its previously defined key dimensions: physical suffering, emotional suffering, mental suffering, social suffering, existential suffering and spiritual suffering. As already stated, the leading philosophical thread consists in a change of focus; rather than asking, "what is suffering?", we explore "how is suffering experienced?" [2, 13]. This approach accounts for suffering not through the distinctive *features* or *contents* representing its experience, but rather via its principal *lived dimensions* underlining the required *phenomenological shift* I adopted throughout the testing of the suffering diagnostic tool for chronic pain and related-cancer patients. The operationalized definitions and validated methodology of the primary tool for assessing pain-related suffering will appear in a separate paper.

My original Fan Model (Fig. 3.1) [6] exposes the three pain measurement components.

The idea of a fan, with its natural movement of opening from bottom (pain axis) to up (suffering axis), illustrates the correlation between pain and suffering, whereby more pain leads to more suffering. This basic model is used to support our claim that pain should be assessed through the measure of its intensity and unpleasantness, as well as of its related suffering. The yellow and black crossing lines inside the intensity/unpleasantness/suffering crescents show that the experimentally tested correlations between these three pain components, following our laboratory results, are not necessarily proportionally related among themselves. This means that an increase in the pain intensity definitely leads to higher unpleasantness and suffering, but not necessarily at a constant ratio. In patients, we can notice this non-linear relation when increasing pain intensity inducing in some instances terrible suffering, and not in others. These different dependencies result thus from many other environmental and psycho-physiological factors and influences.

In the more complete second fan model presented here (Fig. 3.2) [6], we see that suffering, as pain, is a complex percept. Therefore, instead of speaking of qualities, types or attributes of pain, we extend upon Melzack and Casey's [71] historical and established account that a typical pain experience has distinct physical (sensory discriminative), emotional (affective motivational) and mental (cognitive evaluative)

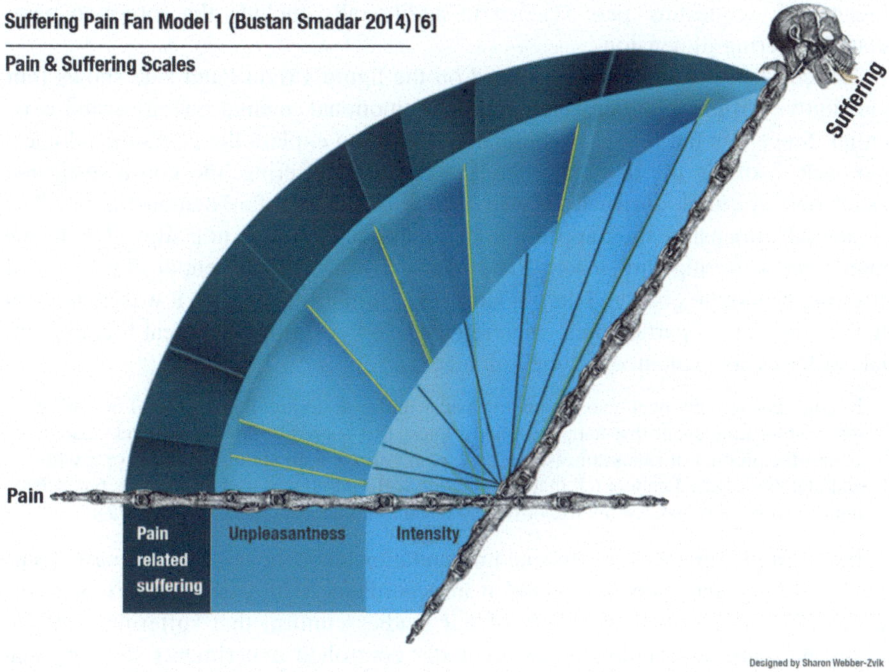

Fig. 3.1 Smadar Bustan Fan Model 1, 2014

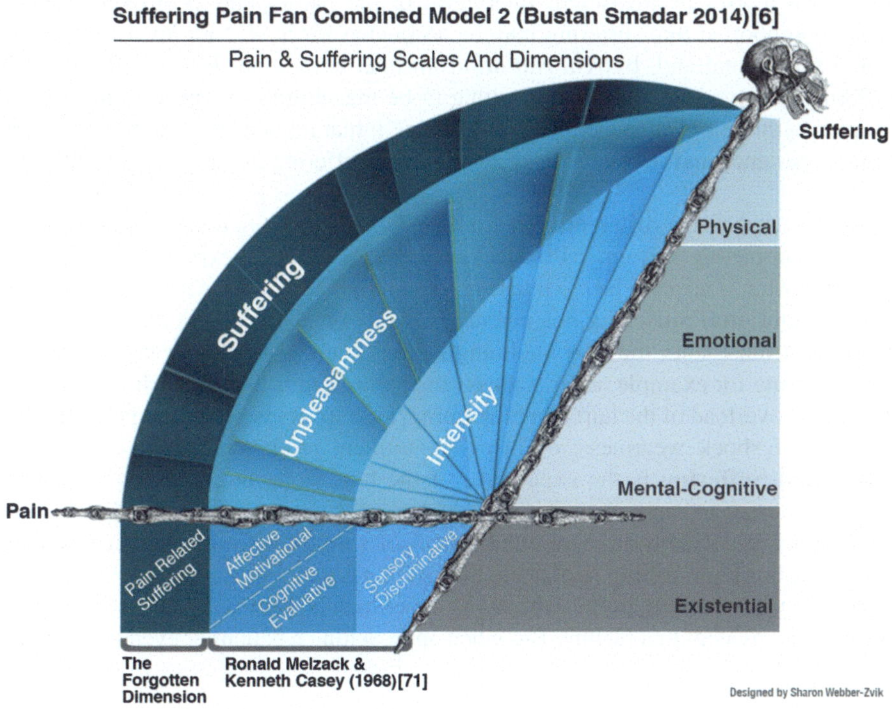

Fig. 3.2 Smadar Bustan Fan Model 2, 2014

dimensions, requesting pain science to additionally include the forgotten pain-related suffering dimension.

The construct of suffering presented on the figure's right-hand side shows four constitutive suffering dimensions: physical, emotional, mental-cognitive and existential. Since this fan model was initially adapted to explain the phenomenological approach inspiring the laboratory investigation on suffering, the non-investigated social and spiritual dimensions were put aside in the fan demonstration. The existential dimension thus appears under the x axis and is not aligned with the other three suffering dimensions. The reason is that the existential dimension of suffering cannot be observed during laboratory pain protocols, but it was mentioned by the interviewed participants when referring to the absence of a real life, existential, suffering as explained by one volunteer:

> It didn't damage my health so I cannot consider this as suffering. It was painful but still it was a short time event that happened, and I know that sooner or later it will end. Also, it wasn't the product of illness or disease, so I wasn't afraid at all about my health or my life and that this is why I cannot tell that it was really suffering. It was painful for sure, but as it didn't involve any risk for my life or for my health, I could stand it voluntarily [72].

Interestingly, this was an unusual interpretation since the majority of participants rated suffering and then interpreted it in accordance with casual suffering occurrences [73]. The statistical results of our studies confirm that suffering emerged during the pain stimulation protocols under controlled experiments showing that suffering is inherent to the pain experience, to be evaluated alongside pain intensity and unpleasantness. More generally, we have shown that human pain-related suffering can indeed be measured. Interviews of the participants post experimental sessions supported that suffering may be accounted for in all four levels: physical, emotional, mental and, by *reductio ad absurdum*, existential. But in order for the suffering observed in a laboratory setting to be meaningful, we need to be able to adopt a methodological principle for measuring it that remains truthful to its wholesome experience and to the "real," raw reality of suffering in day to day healthcare systems.

In this respect, the different model dimensions play distinctive but complementary roles depicting suffering: **Physical suffering** is a bodily experience of feeling hurt from what is taken to be too intense and no longer tolerable as in cases of pain or injury. **Emotional Suffering** is a negative affective reaction of the hurt to 'the too much' that must stop, the overwhelming feeling that can no longer be endured or faced causing for example anger, sadness or shame. **Cognitive Suffering** expresses the mental overload of the hurt that is alarming, alerting or else saturating the mind as in cases of shock weariness, or incessant torment reducing cognitive function. **Existential Suffering** is the experience of the hurt as an ongoing condition that has become second nature as in cases of loss, chronicity or trauma. These four dimensions of suffering involve either forms of surging suffering that erupts suddenly (feeling), or lasting suffering that dominates a person's entire existence (condition). Obviously, in the laboratory experimental pain only arouses episodic suffering equivalent to a feeling. But when approaching a patient or even a volunteer in an experimental pain study, the open circulation among the various dimensions of what is simply called "suffering" should be taken into consideration as it can better explain the meanings of hurtful experiences and identify the dominant dimension (s) at any given time.

According to this phenomenological approach, the value accorded to truth is based on the most singular way of living one's suffering. This is reflected in the distribution of the results of measurements across the various dimensions. No concept or essence of suffering is previously imposed here, and every participant or patient brings his or her own idea of suffering (possibly an evolving one) to the assessment. It is thus important to evaluate each dimension separately, as well as their combined relations so as to identify both the outstanding aspect and the combinations causing one's suffering.

Unlike the common belief, I found that patients are profoundly aware of their suffering (albeit in cases of alexithymia) but avoid formulating complaints about this complex issue with clinicians, who concentrate on the ailment alone. Inevitably, as the suffering has to be somehow expressed, it is disguised and confused with the disease symptoms. Chronic pain patients, for example, could express their suffering through reports of maximal pain levels (9–10/10), an unlikely high score for people enduring pain on a daily basis over long periods of time, possibly decades. Consequently, the standard apparatus used to account for suffering and its various aspects should be redesigned for more accurate forecasts, and to prevent drug over-prescriptions in the absence of a clear distinction between the illness (chronic pain, cancer) and the suffering (its negative impact). To provide an accurate and complete portrayal of personal suffering, phenomenology may thus allow us to methodologically incorporate suffering as a concept into a standard examination procedure of patients, with the measurement of its severity in the various dimensions of my model.

While foretelling is vital for clinical prognosis, the accuracy of prediction for suffering relies on the collection of knowledge through these phenomenological guiding principles, aiming to avoid all disparity between the expected organization and the actual organization of suffering within a person. Tested individually, the division into dimensions of the phenomenon of suffering exposes its internal experiential structure, and the different singular parts that compose it. In order for a person's suffering to be fully understood, it is sometimes necessary to follow up with a verbal exchange to insure scoring is compatible with the patient's report. For example, a patient whose condition has significantly worsened through successive treatment failures, could score high on Physical Suffering (being totally disabled), on Emotional Suffering (feeling anxious and distressed since hopeless) as well as on Mental Suffering (experiencing mental exhaustion), but low on Existential Suffering, despite clear signs of despair. Upon inquiring with the patient, we may discover that a religious faith may provide them with higher guidance and confer resilience, buffering for example end-of-life thoughts. However, this does not imply that their suffering is not alarming or extreme. The tool itself, inspired by the integrative approach exposed here, seeks to provide a standardized format based on clear definitions, and a combination of methods that can identify such possible misinterpretations.

As an example, the behavioral map framed by my Fan Model finds a very good reference in Arthur Kleinman's book entitled *Suffering, healing and the Human Condition*, in the case of Howard Harris, "The Vulnerable Police lieutenant" [74, p. 60–74]. This man, in his late fifties, for the past 20 years has been suffering from lower back pain after lifting a heavy piece of equipment during his job on a construction site. He has been unsuccessfully treated with every medication available and consulting any possible variety of pain experts. Kleinman describes the police lieutenant's posture and behavior as a walking testimony to his constant pain:

> In one hand he carries everywhere with him a white cushion shaped for his lower spine. His other hand touches the back of each piece of solid furniture as if he wanted to be sure which could be relied on to support him in the event that his back were to give way and he were suddenly to fall [74, p. 61].

This account emphasizes the anticipatory nature of suffering associated with a long-lasting threat experience, which specifically characterizes the existential dimension and is by no means equivalent to the anticipation of a perceived stimuli by a subject during a pain experiment, sitting blocked in a chair, waiting in the laboratory for the plastic bullet inside the transparent tube of the Impact Stimulator device to heat again the middle phalanx of his finger. In both cases the suffering state is a direct outcome of the possible nociception of the expected pain. However, while we can argue that a comparison between the manifestations of all other three suffering dimensions (physical-emotional-mental) can bring about some equivalence between the eruptive feeling of suffering during a pain experiment (feeling) and the suffering state of the chronically sick person undergoing another cycle of intense pain attack (feeling), the existential dimension of his suffering (condition) bears an added metaphysical value. The menacing aspect of Howard's vulnerable existentiality requires looking at the total sum of interactions to better understand what predetermines his reality in a way that risks amplifying other suffering occurrences:

> "Indeed", Kleinmann writes, "Howie doesn't respond so much as anticipate. He waits for the pain. He seeks out its earliest sensation. He attempts to 'catch it early', 'keep it from developing', 'prevent it from getting worst'" [74, p. 63].

Further reading of this description reveals that even though the pain pertains all day, every day, its intensity is limited and could seem manageable (3 to 4 in a scale of 10). But Howard's wait for the irruption of severe episodes lasting several days shape his daily experience of pain and consequently his entire personal, professional, marital, and parental life [74, p. 63]. Not only that the patient feels "vulnerable" in fear of losing his job and being considered a "malingerer," his disability makes him feel "spineless" and "self-conscious" due to his doubting surrounding—his colleagues, his family and even the family physician—causing isolation and minimizing unnecessary interactions in order to avoid any physical burden or social stress. An assessment of suffering in a patient like Howard, who is relatively low on pain intensity scores but who clearly suffers, would allow us to deconstruct his experience by evaluating the different dimensions of his experience so that we could point to the dominant aspect directing his anxiety and deciding what is the priming factor or factors: the incapacitation, the loosening of his social ties or rather the fear of what he cannot control.

Having met 265 patients in the course of the validation of the suffering tool, I realized that the contribution of phenomenological analysis is crucial to our understanding of the lifeworld of the suffering patient. This is not only because it remains attentive to its continuous unfolding within an open horizon of experiences, thus allowing to examine the patient's improvement or worsening in due course following treatment. But it also portrays one's modes of suffering without being bound by a canonical definition, allowing for what we least expect to be finally revealed.

5 An Integrative Method that Objectively Captures Subjective Suffering

In exploring *how the phenomenon of suffering is given*, within each of its principal dimensions in a particular person and on a particular occasion, we discover the unique interplay between our mind, body, and the world that causes our suffering. In this respect, the integrated phenomenological investigation provides a plausible account of the individual that is transposable to generalized claims about the derivation, nature, significance, and implications of the conditions associated with great suffering. The *epistemological structure explicitly adopting a six-dimensional concept of suffering* is amenable to objective assessment applicable to large patient cohorts, and regarding which we do not draw a line between what is normative and what is not. We concentrate instead on what it means for the sick individual to be suffering and how. In calling to understand the subjective experience, we may compare individual sufferers by detecting a regularity in the interactions. This may explain the variations in a patient's distressed reaction to a repeated illness, why these individuals differ in their responses to an identical illness, or whether similar traumatic events play different roles in various social groups.

To this view, understanding morbidity requires personalized knowledge, especially in an era dominated by molecular biology and genetics as the basis for explaining a pathology. Attention to *singularity* through the search for molecular mechanisms whereby each patient will be better diagnosed and treated, owing to his personalized genes' mapping, remains insufficient without the systematic diagnosis of suffering, both for getting "the big picture" and the details of the patients' life and health. This pursuit is obviously intended to improve personal care and act for humanity in medicine. But the more immediate purpose of using an explanatory phenomenological tool to diagnose individual suffering is truth-oriented to enhance professional efficiency, seeking to combine impersonal knowledge with personal knowledge, so as to improve both medical science and practice. It consists in identifying the manner a disease behaves in a person and at all levels, including the internal subjective reality of their suffering, so that this singular observation is objectified as general understanding that can reliably serve the ill. This integrated approach takes into account the daily medical practice of seeing patient after patient, coordinating the measure of the singular account against the background of numerous consultations, and subsequently unifying the numerous separate realities into a collective account.

Finally, this last point emphasizing singularity brings new values on the management of pain through the evaluation of individual suffering based on sound historical perspectives. From Georges Canguilhem's remarks on the question of singularity, we learn that this idea already played a major role in medicine's history, and its reminder here could help us avoid repeating the same mistakes all over again. In his book "The Normal and the Pathological," the twentieth century French philosopher and historian of medicine highlights the significance of *individuality* when describing the illness as a global experience that happens first and foremost to the single person [75, p. 224]. In crediting Leriche's work on the "disease" of pain, Canguilhem considers that the anatomical character and the "physiological specificity" from which derives the pathology, cannot be seen without the living being undergoing it:

> It seems quite important to us that a doctor recognizes in pain a phenomenon of total reaction which makes sense, which is a sensation only at the level of concrete human individuality [75, p. 97].

What constitutes any pathological state is the way it is experienced by someone, consequently yielding different reactions to the same physiological dysfunction or sensations [75, p. 94]. His definitions of the normal, the pathological, of states of health, and of disease, or even his views on clinical practice, are rooted in the response to the singularity of the patient [76, 77]. Canguilhem's work provides an observation and understanding of the norms of the ill's life. It also provides an important historical lesson about science and medicine, and on what progress in medical standards may be achieved with the development of worldviews. Looking back at the history of the epistemological revolution in medicine from the 19th to the turn of the twentieth century, we notice the subsequent change of its focal point from disease to health, and the shift from pure medicine of cure to preventive medicine. At this pivotal period, the practice of medicine became essentially scientific and populational, introducing measures of public health for large populations. This was the case, for example, for the reforms carried by the social hygiene movement (hygienism), or the free mandatory vaccinations (the French Act passed in 1902). In his sweeping philosophical analysis, Canguilhem states that the consequence was "bracketing of the single patient, singular object, elective, of the attention and intervention of the clinician" who became less attentive to the subjective experience of the sick person [78]. In response, the philosophic analysis of Canguilhem regretfully turns into astonishment about the undeniable though clearly ignored precedence in medicine:

> There has always been a moment when, all things considered, the practitioner's attention has been drawn to certain symptoms, even solely objective ones, by men who were complaining of not being normal—that is, of not being the same as they had been in the past—or of

> suffering. If, today, the physician's knowledge of disease can anticipate the sick man's experience of it, it is because at one time this experience gave rise to, summoned up, that knowledge. Hence medicine always exists *de jure*, if not *de facto*, because there are men who feel sick, not because there are doctors to tell men of their illnesses. The historical evolution of the relations between the physician and the sick man in clinical consultation changes nothing in the normal, permanent relationship of the sick man and disease [78, p. 93].

This reminder of the priority of individuality in human medicine and the price for not acknowledging it, is striking when it comes to suffering. This is so for two main reasons: first, because the frequent confounding between the severity of the disease and suffering does not allow us to see where the former ends, and where the particular disposition of suffering starts, as a subjective negative reaction with regard to the disease. Second, medicine does not exist without the sickness which in return does not exist without the sick. But the patient, especially in chronicity, is not really sick if he does not suffer, otherwise he simply carries a disease. Recognizing the "singular" way of being in order to provide better care does not only consist in seeing the sickness through the sick, but also through their most singular way of experiencing it.

This chapter therefore explained the impossibility to identify a generic character of suffering identical to all people, and to tie it down to one depicting and conceptual invariant to provide a simple definition applicable to all forms and states. Instead, the big shift conducted by phenomenology proposes an operationalized solution, that could better benefit medicine as it exposes the *singular fashion of suffering*, while acknowledging the singularity of the sick patient without betraying the need for large scale populational perspectives.

Acknowledgments I wish to thank Javier Escartin for his kind help in revising this paper and to Simon van Rysewyk for his editing.

References

1. Bustan S, Gonzalez-Roldan A, Schommer C, Kamping S, Löffler M, Brunner M, et al. Psychological constructs and contextual factors associated with the assessment of pain and related-suffering. PLoS One. 2018;13(7):e0199814.
2. Bustan S. A scientific and philosophical analysis of meanings in studies of pain and suffering. In: van Rysewyk S, editor. Human meanings of pain. New York: Springer; 2016. p. 107–28.
3. Bustan S, Gonzalez-Roldan AM, Kamping S, Brunner M, Löffler M, Flor H, et al. Suffering as an independent component of the experience of pain. Eur J Pain. 2015;19(7):1035–48.
4. Brunner M, Löffler M, Kamping S, Bustan S, Gonzalez-Roldan AM, Anton F, et al. Assessing suffering in experimental pain models. Z Psychol. 2017;225(1):45–53.
5. Löffler M, Brunner M, Kamping S, Kleinböhl D, Bustan S, Anton F, et al. The impact of controllability on pain and suffering. Pain Reports. 2018;3(6):1–10.
6. Bustan S. Peut-on mesurer la souffrance liée à la douleur? [Can we measure pain-related suffering?]. Douleur et Analgésie. 2017;30(4):230–8.
7. Bustan et al. manuscript in preparation.
8. Price D. Psychological and neural mechanisms of pain. New York: Raven Press; 1988. 241 p.
9. Bush FM, Whitehill JM, Martelli M. Pain assessment in temporomandibular disorders. Cranio. 1989;7:137–43.

10. Wade JB, Dougherty LM, Archer CR, Price DD. Assessing the stages of pain processing: a multivariate analytical approach. Pain. 1996;68:157–67.
11. Wade JB, Hart RP. Attention and the stages of pain processing. Pain Med. 2002;3:30–8.
12. Fishbain DA, Lewis JE, Gao J. The pain suffering association, a review. Pain Med. 2015;16:1057–72.
13. Bustan S. Suffering. Mafte'akh. 2011;3:151–178. Available from: http://din-online.info/pdf/mft3.pdf.
14. Bustan S. Suffering – a premise for the social and political thought. Zeitschrift für kritische Sozialtheorie und Philosophie. 2015;3(2):379–417.
15. Loeser JD. Perspectives on pain. In: Clinical pharmacology and therapeutics. Turner P, Padgham C, Hedges A, editors. Macmillan: Palgrave; 1980. p. 313–316.
16. Clark WC, Carroll JD, Yang JC, Janal MN. Multidimensional scaling reveals two dimensions of thermal pain. J Exp Psychol Hum Percept Perform. 1986;12:103–7.
17. Clark WC, Ferrer-Brechner T, Janal MN, Carroll JD, Yang JC. The dimensions of pain: a multidimensional scaling comparison of cancer patients and healthy volunteers. Pain. 1989;37:25–32.
18. Fordyce WE. Pain and suffering: a reappraisal. Am Psychol. 1988;43(4):276.
19. Wade JB, Dougherty LM, Hart RP, Rafii A, Price DD. A canonical correlation analysis of the influence of neuroticism and extraversion on chronic pain, suffering, and pain behavior. Pain. 1992;51(1):67–73.
20. Krikorian A, Limonero JT, Román JP, Vargas JJ, Palacio C. Predictors of suffering in advanced cancer. Am J Hosp Palliat Med. 2014;31(5):534–42.
21. Büchi S, Sensky T, Sharpe L, Timberlake N. Graphic representation of illness: a novel method of measuring patient's perceptions of the impact of illness. Psychother Psychosom. 1998;67:222–5.
22. Bueno-Gómez N. Conceptualizing suffering and pain. Philos Ethics Human Med. 2017;12(1):7.
23. Cassell EJ. The nature of suffering and the goals of medicine. New York: Oxford University Press; 1991. 336 p.
24. Jansen LA, Sulmasy DP. Proportionality, terminal suffering and the restorative goals of medicine. Theor Med Bioeth. 2002;23(4–5):321–37.
25. Botbol-Baum M. Peut-on répondre à la souffrance d'un point de vue médical? [Can we reply to suffering from a medical point of view?]. In: Traité de bioéthique. Toulouse: ERES; 2010. p. 469–79.
26. Cassell EJ. The nature of suffering and the goals of medicine. N Engl J Med. 1982;306 (11):639–45.
27. Ruijs CD, Onwuteaka-Philipsen BD, van der Wal G, Kerkhof AJ. Unbearability of suffering at the end of life: the development of a new measuring device, the SOS-V. BMC Palliat Care. 2009;8(1):16.
28. Krikorian A, Limonero JT. An integrated view of suffering in palliative care. J Palliat Care. 2012;28(1):41.
29. Krikorian A, Limonero JT, Maté J. Suffering and distress at the end-of-life. Psychooncology. 2012;21(8):799–808.
30. Krikorian A, Limonero JT, Corey MT. Suffering assessment: a review of available instruments for use in palliative care. J Palliat Med. 2013;16(2):130–42.
31. Quill TE, Byock IR. Responding to intractable terminal suffering: the role of terminal sedation and voluntary refusal of food and fluids. Ann Inter Med. 2000;132(5):408–14.
32. Cassell EJ, Rich BA. Intractable end-of-life suffering and the ethics of palliative sedation. Pain Med. 2010;11(3):435–8.
33. Chapman CR, Garvin J. Suffering and its relationship to pain. J Palliat Care. 1993;9(2):5–13.
34. Loeser JD. Pain and suffering. Clin J Pain. 2000;16(2):S2–6.
35. Loeser JD, Melzack R. Pain: an overview. Lancet. 1999;353(9164):1607–9.
36. Marchand S. The phenomenon of pain. Seattle: IASP Press; 2012. p. 14.

37. Price DD, Barrell J. Inner experience and neuroscience – merging both perspectives. Boston, MA: MIT Press; 2012. p. 159–81.
38. Vélez MC, Palacio C, Moreno AI, Krikorian A. Psychological and family-related facts of suffering in patients with chronic diseases. Tech Reg Anesth Pain Manag. 2013;17(1):7–10.
39. Charmaz K. Loss of self: a fundamental form of suffering in the chronically ill. Sociol Health Illn. 1983;5(2):168–95.
40. Kleinman A, Das V, Lock M, Lock MM, editors. Social suffering. Berkeley: University of California Press; 1997. 434 p.
41. Fields HL. Pain: an unpleasant topic. Pain. 1999;82:S61–9.
42. Clark WC, Janal MN, Carroll JD. Multidimensional pain requires multidimensional scaling. In: Loeser JD, Chapman CR, editors. Issues in pain measurement. New York: Raven Press; 1989. p. 285–326.
43. Melzack R, Wall PD. Challenge of pain. New York: Basic Books; 1983. 368 p.
44. Scott W, Trost Z, Bernier E, Sullivan MJ. Anger differentially mediates the relationship between perceived injustice and chronic pain outcomes. Pain. 2013;154(9):1691–8.
45. McCracken LM. Anger, injustice, and the continuing search for psychological mechanisms of pain, suffering, and disability. Pain. 2013;154(9):1495–6.
46. Price D. Psychological and neural mechanisms of the affective dimension of pain. Science. 2000;288:1769–72.
47. Price D. Central neural mechanisms that interrelate sensory and affective dimensions of pain. Mol Interv. 2002;2:392–403.
48. Krikorian A, Limonero JT, Vargas JJ, Palacio C. Assessing suffering in advanced cancer patients using pictorial representation of illness and self-measure (PRISM), preliminary validation of the Spanish version in a Latin American population. Support Care Cancer. 2013;21 (12):3327–36.
49. Ricoeur P. La souffrance n'est pas la douleur [Suffering is not pain]. In: Von Kaenel JM, editor. Souffrances corps et âme, épreuves partagées [Suffering body and soul, shared hardships]. Autrement. 1994;2:58–69.
50. Wissler RL, Evans DL, Hart AJ, Morry MM, Saks MJ. Explaining pain and suffering awards: the role of injury characteristics and fault attributions. Law Hum Behav. 1997;21(2):181.
51. Chapman CR, Gavrin J. Suffering: the contributions of persistent pain. Lancet. 1999;353 (9171):2233–7.
52. Gómez Sancho M, Grau JA. Dolor y sufrimiento al final de la vida pain and suffering at the end of life [Pain and suffering at the end of life]. Madrid: Aran; 2006.
53. Fordyce WE. Pain and suffering: what is the unit? Qual Life Res. 1994;3:S51–6.
54. Preacher KJ, Hayes AF. SPSS and SAS procedures for estimating indirect effects in simple mediation models. Behav Res Methods. 2004;36(4):717–31.
55. Masten CL, Morelli SA, Eisenberger NI. An fMRI investigation of empathy for 'social pain' and subsequent prosocial behavior. NeuroImage. 2011;55(1):381–8.
56. Jollant F, Olié E. La douleur psychologique [Psychological Pain]. Douleur et Analgésie. 2018;31(3):122–8.
57. Orbach I, Mikulincer M, Sirota P, Gilboa-Schechtman E. Mental pain: a multidimensional operationalization and definition. Suicide Life Threat Behav. 2003;33(3):219–30.
58. Orbach I, Mikulincer M, Gilboa-Schechtman E, Sirota P. Mental pain and its relationship to suicidality and life meaning. Suicide Life Threat Behav. 2003;33(3):231–41.
59. Orbach I. Mental pain and suicide. Isr J Psychiatry Relat Sci. 2003;40(3):191.
60. Saunders C. The symptomatic treatment of incurable malignant disease. Prescr J. 1964;4 (4):68–73.
61. Saunders C, Baines M. Living with dying: the management of terminal disease. Oxford: Oxford University Press; 1983. 80 p.
62. Baines MJ. Cancer pain. Postgrad Med J. 1984;60(710):852.
63. Clark D. Total pain, disciplinary power and the body in the work of Cicely Saunders, 1958–1967. Soc Sci Med. 1999;49(6):727–36.

64. Merleau-Ponty M. Phenomenology of perception. Smith C, translator. London: Routledge; 1995/1945. 544 p.
65. Dahlberg K. The essence of essences – the search for meaning structures in phenomenological analysis of lifeworld phenomena. Int J Qual Stud Health Well Being. 2009;1(1):11–9.
66. Mccormick P, Elliston F. Husserl Shorter Works. Notre Dame, Indiana: A University of Notre Dame Press; 1981. 440 p.
67. Bustan S. On the limits of knowing suffering and pain. In: Fernandes I, editor. Creative Dialogues: narrative and medicine. Newcastle upon Tune: Cambridge Scholars Publishing; 2015. p. 259–73.
68. Bustan S. Voicing pain and suffering through linguistic agents: Nuancing Elaine Scarry's view on the inability to express pain. Subjectivity. 2016;9(4):363–80.
69. Correll D. Chronic postoperative pain: recent findings in understanding and management. F1000Res. 2017;6:1054.
70. Moryl N, Carver A, Foley K. Suffering in the patient, family, and physician. In: Kufe DW, Pollock RE, Weichselbaum RR, editors. Holland-frei cancer medicine cloth. Hamilton: BC Decker; 2003. 2008 p.
71. Melzack R, Casey KL. Sensory, motivational, and central control determinants of pain: a new conceptual model. In: Kenshalo D, editor. The skin senses. Springfield, IL: Thomas; 1968. p. 423–43.
72. Participant n. 47, PASCON working package 1, Luxembourg; 2013.
73. Bustan S, Gonzalez-Roldan A, Odero A, Baumann M, Kamping S, Flor H, et al. The generation of suffering-related items: a process integrating consensus experts and Sorensen's similarity analysis. 2019, under review.
74. Kleinman A. The illness narratives: suffering, healing, and the human condition. New York: Basic Books; 1988. 304 p.
75. Canguilhem G. The normal and the pathological. Fawcett CR, Cohen RS, translators. New York: Zone Books; 1991. 336 p.
76. Lefève C. La philosophie du soin [The philosophy of care]. La Matière et l'esprit. 2006;4:25–34.
77. Lefève C. De la philosophie de la médecine de Georges Canguilhem à la philosophie du soin médical [From the philosophy of medicine of Georges Canguilhem to the philosophy of medical care]. Rev Metaphys Morale. 2014;2:197–221.
78. Canguilhem G. Le statut épistémologique de la médecine [The epistemological Status of Medicine]. History Philos Life Sci. 1988;10(Suppl):22–3.

Chapter 4
"Pain Takes Over Everything": The Experience of Pain and Strategies for Management

Marie Crowe, Deb Gillon, Cate McCall, and Jennifer Jordan

Abstract This chapter explores the personal experience of pain from its biological underpinnings to strategies people identified for managing this experience. The somatic experience of chronic pain describes the biological processes involved in pain and how this can become a chronic experience with psychological and social implications. The personal experience of pain is explored through a systematic review of research of qualitative experiences. We found that the experience of pain was similar despite its etiological underpinnings—whatever the biological cause there were similarities in the personal experience. Participants in the studies identified five themes that described these personal experiences: (1) body as obstacle; (2) disrupted sense of self; (3) invisible but real; (4) unpredictability; and (5) keeping going. This section of the chapter is followed by the findings of a systematic review of how older people learn to manage their pain experiences: "adjusting to the inevitable," "doing it my way without medication" and "the importance of support in managing the struggle." The chapter concludes by discussing some of the strategies that can be used to manage the self in pain: support for self-management, medication, exercise and psychological interventions (mindfulness and cognitive behavioural therapy).

Clinical Implications: Many people manage their chronic pain by actively balancing the competing forces of hope and despair. There are similarities in this experience of chronic pain across a range of conditions which has implications for the development of pain management strategies and interventions that address the pain experience and not just the biological condition. Learning to manage the self in pain involves acceptance of pain as on-going and a part of who they are; keeping connected with others; keeping occupied through meaningful activities; getting meaningful support; and developing new meaning in life. Clinical interventions need to a focus on the person's sense of self, strategies for maintaining hope, strategies that provide relief for the distress associated with pain; and providing

M. Crowe (✉) · D. Gillon · C. McCall · J. Jordan
Centre for Postgraduate Nursing Studies, Department of Psychological Medicine, University of Otago, Christchurch, New Zealand
e-mail: marie.crowe@otago.ac.nz

S. van Rysewyk (ed.), *Meanings of Pain*,
https://doi.org/10.1007/978-3-030-24154-4_4

people with a sense of control over their experiences both with the pain itself and in their encounters with the medical profession and the use of medication.

Keywords Pain · Chronic pain · Qualitative · Elderly · Transdiagnostic · Patient experience · Long term conditions · Coping

1 The Somatic Experience of Chronic Pain

Pain is an experience that is difficult to quantify but can be conceived of as an output of the brain, a complex construct associated with nociceptive data purposed with interpreting evidence in order to identify dangers to the body and requisite protective behaviours. Pain can be described in terms of the physiology of nociception and the conscious cognitive and emotional perception of the experience. Noxious stimuli, chemical, thermal or mechanical, of sufficient intensity will cause tissue disruption that provokes inflammatory and immune processes with the release of inflammatory mediators. The pain system also involves a descending pathway which is involved in the regulation and modulation of pain impulses, functioning in a manner that inhibits or controls the ascending pathway. Pain that persists beyond the expected healing time may indicate the transition to chronic pain where pain is felt but no longer accurately signals damage as the pain system becomes increasingly sensitive, imprecise and uninformative [1].

The aetiology of chronic pain is a complex, multifactorial process that may have been initially provoked by an inflammatory, infectious, ischemic, idiopathic, iatrogenic, or autoimmune event. The meaning of pain as it relates to complex cognitive and affective states such as hopelessness, fear, anxiety, distress, and depression significantly influence the degree of suffering experienced by individuals and influences the secondary concerns such as functional disability, fatigue, sleep disruption, lowered immune function, maladaptive stress responses, social displacement, and work disruption [2, 3]. Irrespective of the aetiology or classification, the chronicity of pain transgresses the biological usefulness of nociceptive feedback and perception as the chronic pain complaint becomes a pathological entity in its own right, often with a disproportionate relationship to the provoking event with disturbing consequences. In this way pain is cognitively modulated by a shift in perceived meaning from a protective function to an overprotective state that may be associated with activity and pain related fear and catastrophizing.

The mechanistic causes of associated affective, cognitive, and behavioural components are linked to long-term neuroplastic and neurotransmitter adaptations. The combined effects of which are described in terms of a biopsychosocial condition in which the complex interplay between nociception and pain undergoes changes that entwine significant suffering and pathological adaptations of the pain system architecture. Sensitisation of the physiological mechanisms of nociceptive transmission and neuroplastic changes in higher cerebral functions are linked to hypersensitivity to pain and the triggering of pain from stimuli that do not normally provoke pain [4].

2 The Personal Experience of Chronic Pain

Because pain is difficult to quantify, we were interested in exploring the personal experience of chronic pain generally rather than as specific to a chronic condition. While physiological conditions may trigger the pain experience, the nature of that experience is influenced by a range of personal and social factors. It is a highly symbolic experience. The treatment of the underlying condition may require bio-medical intervention, but the pain experience may respond better to a biopsychosocial approach which views pain and disability as a complex and dynamic interaction among physiological, psychological and social factors [5].

A focus on the pain experience across conditions provides the opportunity to learn more about pain from a patient-centred perspective to add to the rather limited bio-medical one. It is important to understand the person's experience of pain because "when the narratives of chronic pain patients are unheard or dismissed as unimportant, key pieces of evidence that may shape diagnosis and treatment are lost" [6, p. 166].

> Nobody likes pain, believe me... With this, how long does it go? When is it going to stop? I can't handle this no more ... Pain is dominant, it, it took over everything. Your brain becomes so taken and your mind, you think pain, you see pain, you feel pain, you live pain. You're just lost. It's constantly your focus in life [7].

Two of the main impacts of living with chronic pain are a poorer quality of life and a greater risk of developing depression [8]. The clustering of pain and depression occurs at a prevalence between 30% and 60% [9, 10]. This co-occurrence is possibly related to direct effects of cytokines on the neuronal environment [11]. There is also converging lines of evidence that four key neurotransmitters are involved in both pain and depression (serotonin, norepinephrine, substance P, and cortico-releasing factor) [12]. When pain is severe, impairs function and/or is refractory to treatment, it is associated with more depressive symptoms and worse depression outcomes [13]. The general consensus is that persistent pain is more likely to lead to depression than vice versa and that patients with more severe, frequent, and enduring pain are at risk for more severe depression [12].

Pain challenges the Cartesian mind/body duality at the centre of the medical model [14]. This framework positions the person experiencing pain as having either a disordered body or a disturbed mind. In the absence of objective evidence of a disordered body, the medical model defaults to a disturbed mind. Jackson [15] suggests that people who experience chronic pain are positioned within this model as transgressing divisions between mind and body which confounds the codes of morality surrounding sickness and health. Their status within the medical model becomes ambiguous and can result in medical skepticism, stigmatization and often inappropriate treatment.

Given the prevalence, associated disability, cost to individuals and their families and restricted access to specialist services we were interested to identify commonalities in the pain experience. We undertook a qualitative meta-synthesis of research into people's pain experiences to identify whether there were similarities in the pain experience, irrespective of its cause [16]. This synthesis produced five themes that were common to the pain experience across conditions: (1) body as obstacle; (2) disrupted sense of self; (3) invisible but real; (4) unpredictability; and (5) keeping going.

2.1 *Body as Obstacle*

This theme captured the sense of persuading an unwilling body and the intrusiveness created by the experience of chronic pain. Participants in the studies described the obstacles they encountered and how they regarded their bodies, or specific parts of their bodies, as posing an obstacle to their ability to function in everyday life. For many the body was regarded as a barrier that had to be overcome. Prior to the experience of pain, many people had taken their bodies for granted and were not consciously aware of the pain-free body. The experience of pain had taken away the taken-for-grantedness of having a functional body capable of providing comfort. Pain often led to inflated attention to somatic responses and a sense of constant surveillance. This hyper-awareness of the body often led to an objectification of the body and a sense that it was in some way separate from the self. Participants tended to talk about or describe their body as if it was not a part of their subjectivity, but rather something external or other than their sense of self.

> It's all the same, all over me I have to make myself do most things and drag this bloody thing (his body) around with me. I don't see it much different, except its broken and no-one can fix it [17].

2.2 *Disrupted Sense of Self*

This theme refers to both an experience of the present self as different from the past self, and the experience of a self as separate from the body. It captured the experience of redefining what was normal now and how they perceived themselves in the present as compared to the past. The theme captured a sense of the loss of a normal life and a disappointment with the body as the source of this problem. There was also a sense of disconnection from others because the participants said they felt that their pain set them apart. Campbell and Cramb [18] found that their participants reflected on how as a result of the chronic pain they were very different from their former selves and that when participants had to curtail employment and social activities they felt as though the pain and not themselves was in control of their life. Most people experience their life with pain as less than their life before.

Williams [19] found that the chronic pain experience involved the unpleasant and relentless presence of a body that was "not me" and the persistent disruption of many automatic and mundane activities that could not be avoided. He suggested that this contradiction between the painful body and the preferred self could represent an important obstacle to therapeutic rehabilitation if not acknowledged or resolved. The negative impact of pain on a person's sense of self is also supported by Eccleston et al. [20] who suggested that central to the experience of prolonged suffering is a fundamental challenge and threat to identity.

> I mean normally, a normal person isn't aware of their legs because they just say right walk, you know their brain tells them to walk and they walk, whereas when you're in pain you're aware of them all the time [18].

2.3 *Invisible But Real*

This theme captures the frustration of being disabled by an experience not visible to others. It encompasses both an attitude of stoicism and a sense of not being believed; and described how many participants kept their experience of pain to themselves rather than risk boring others or being disbelieved. This theme of disbelief was particularly relevant to participants' encounters with health professionals. Many felt they were not believed because pain could not be seen or objectively measured. The process of medical diagnosis is strongly reliant on making that which is invisible to the naked eye visible via blood tests and x-rays. In the absence of visible evidence of pain, the moral character of the patient became a central issue [21].

Rhodes et al. [22] drew attention to the complex inter-relationship between the biomedical paradigm's need to diagnose visible pathology and the private experience of pain. Because precise diagnosis of the cause is beyond medicine's current capabilities many people with chronic back pain feel marginalised and report feeling as if others perceive their reports as untrustworthy or fallacious. They suggested that there is a strong cultural reliance on seeing into the body as central to confirm and normalize patients' symptoms. They suggest that patients whose findings are negative or inconclusive must face disjunction between their experience of the reality of the body that has become an ongoing negative and constraining influence in their lives and the absence of medical evidence for this.

Pain however does not always have an easy medical explanation, but it is very real to the participants. Monsivais and Engbretson [6] have suggested that tension related to the veracity of pain arises when the practice of medicine, which is designed to localize the pathology to a discrete part of the body that can be made visible, does not have techniques to identify pain in a material form. Pain is invisible and thus difficult to prove to medical practitioners reliant on empirical measurement.

> Pain cannot be imagined and just because there is no medical diagnosis does not mean that the pain is not physical [20].

2.4 Unpredictability

This theme described the experience of staying ready and keeping alert for the pain to recur. It was associated with a sense of lack of control of one's body and the need to surrender to one's fate. Many participants could not manage their pain or were not able to predict when it would occur. The participants who experienced unpredictable pain described needing to be vigilant, staying prepared and always calculating the risk of engaging in certain activities. As a result of this unpredictability, the pain seemed illogical, incomprehensible and overwhelming. Many participants identified how they could not predict what actions would increase the pain and believed that the pain diminished or increased independently of what they did. Participants described how the unpredictability of pain changed their behaviour and the way that they thought about themselves, as they needed to be prepared for pain to occur at any time. The impact of unpredictability has been identified by Corbett et al. [23] who found that uncertainty surrounding the diagnosis and/or effective treatment made it difficult for people to construct a frame of reference as to the nature and course of the condition.

> Even when I wasn't doing something I'd end up in agony and it would be very frustrating because it would be like I don't understand, I haven't done anything. I should be able to, not make it go away, but manage it a bit better but there are times when it doesn't work like that [21].

2.5 Fighting to Keep Going

This theme captured the strategies participants used in order to keep going despite the struggle with pain. The process was described as a balancing act between two contradictory forces in which many participants were engaged: hope/resignation, dependency/withdrawal, past/present self, suicide/fight, normal/different. Maintaining the balance involved an ongoing negotiation and acceptance of pain. Participants described coping by comparing themselves to others who were worse off, making connections with others and assigning a personal meaning to their experiences. The process required acceptance of pain as part of their lives, getting support from others, adopting new roles, keeping occupied, learning preventative strategies and knowing how to use medications. The strategies they described for keeping going included finding support, learning to endure, assigning meaning to the experience and keeping occupied. A key to this experience was the participants' active engagement in the constant and dynamic balancing act associated with maintaining a sense of hope that it was worth fighting to keep going.

> I'm very conscious I don't want to be one of these people that live off their injury for the rest of their lives because I couldn't imagine anything more damaging to the self-esteem and mentally debilitating than that so I have that sort of fear in my back of my mind that I'm trying to get past, it's a good motivator shall we say [21].

This review found that there are similarities in the experience of chronic pain across a range of conditions that have implications for the development of transdiagnostic pain management strategies and interventions.

3 Learning to Live with Chronic Pain

Although these findings suggested that many people manage their chronic pain by actively balancing the competing forces of hope and despair, we wanted to identify some of the specific strategies used in this process. In particular, we were concerned to identify how older people manage pain because as people grow older they are more at risk of chronic and life-limiting pain from a variety of causes [24]. One of the characteristics of pain in older people is that there is a very poor correlation with evidence of tissue pathology [25].

For older people, the experience of pain has a considerable impact on autonomy and independence in undertaking activities of daily living [26]; quality of life [27]; disability [28, 29]; activity restriction [30]; increased morbidity [31]; and mortality [32]. Pain imposes considerable burden on older people, their families and the healthcare system. Bernfort et al. [33] have found an association between health service use and the severity of chronic pain in older people and identified that the more severe pain experienced, the more expensive and extensive resource use.

We conducted a qualitative meta-synthesis into how older people cope with non-malignant chronic pain [34] and identified three themes across those studies: "adjusting to the inevitable," "doing it my way without medication" and "the importance of support in managing the struggle."

3.1 Adjusting to the Inevitable

This theme described a coping strategy that involved adjusting one's lifestyle to the inevitability of living with an aging body. Participants talked about accepting that their bodies were frailer and more fragile, which restricted their participation in some activities. They adjusted their lifestyle accordingly and adapted to living with pain by controlling their movements and safe-guarding pain-free functions. For some, this involved pacing their activities and establishing daily rhythms to make the most of the activities of which they were capable. While the pain was experienced as frustrating, for some participants others accepted its inevitability as part of the aging process.

> You've had your three score years and ten and you've got that far, and anything after that is a bonus. I'm on bonuses: you can't change it, I can't take part in a 100 yards sprint, no chance. So you resign yourself to the fact, you know, you just take it easy and you'll be alright (Larry, 80) [30].

3.2 Doing It My Way Without Medication

This theme revealed a reluctance among older people to take medication for pain. They explained this as related to either a disappointment with previous results of taking medication, or that they considered taking medication as a bad thing because of the side effects, long-term consequences, potential for addiction. They described a general reluctance to take multiple medications for multiple health problems. Many saw medical interventions including medication, as a last resort and one that may result in the forfeiture of their independence. Participants generally preferred to cope with their pain using self-administered remedies. A strong desire to remain independent and in control of what happened to them was the impetus behind their avoidance of medical remedies. Some felt responsible for the pain and therefore responsible in managing it. Other participants were often stoic and resolute about managing without assistance, particularly from health services. Common strategies for coping with the pain involved keeping busy, becoming involved in diversional activities and maintaining social activities.

> Doctors just grunt and write out a prescription without telling you anything. They should hand us a leaflet with the information if they aren't going to talk to us [35].

3.3 The Importance of Support in Managing the Struggle

Participants described coping with pain as a daily struggle that was personal to each individual and sometimes a lonely experience. Developing confidence in, and perseverance with self-management of the pain were regarded by some as important in order to cope with the pain. Support was vital to enable participants to manage their daily struggle. While most did not want medical interventions, some did want support from health services. This support was identified as the provision of information on which to base their decisions about how to best manage the pain. Some wanted to be active participants in any healthcare decisions, whereas others identified peers as important sources of information.

> When I'm awake at night I panic when I have cramps in the legs and I feel depressed and distressed. I would like to be able to talk to someone on the phone to comfort me, someone to listen to my overload [7].

The findings suggest that older people with chronic pain used mostly emotion-focused coping strategies and although they perceived the problem as mostly unmodifiable they were active in utilizing strategies that brought some relief. Most had accepted the inevitability of pain and while they did not seek out medical solutions to the problem, they were reasonably active in seeking out alternative remedies.

From the descriptions of pain experiences across conditions and the descriptions of older people's coping strategies, we identified that people in pain need support, information and strategies for self-management. Some of the successful strategies that people had used were learning to endure, perseverance, finding meaning in a life lived in pain and keeping occupied. Specifically, this involved keeping busy, use of diversional activities, pacing oneself, home remedies, maintenance of social relationships and activities, and maintaining daily routines. What patients expect when they enter the healthcare system is affirmation that their pain is real, that they will be given a diagnostic label and that they will be provided with education or interventions related to current best practices [35]. People with chronic pain conditions have identified that they need validation, they need to rediscover their personhood, they need to gain control and have faith in those providing their care [36]. Clarke and Iphofen [37] also identified the importance of being listened to and heard by healthcare professionals.

4 Managing the Self in Pain

The participants described coping with pain as a process of self-management that involved an on-going balancing act between acknowledging the loss of their old life and functioning and developing a new life and a sense of self that integrated pain. This process of balancing is on-going; and equilibrium may not always be maintained. There was a constant struggle between wanting to keep going and wanting to give up; between hope and despair. Regardless of the condition triggering the pain experience, the participants in the studies in our reviews described how pain impacted on them at the fundamental level of the self. Their sense of self was altered by the experience with a sense of a rupture between who they were and who they are now.

Learning to manage the self in pain involves:

- An on-going negotiation of learning to accept pain as a part of their life.
- Keeping connected with others.
- Keeping occupied and establishing meaningful routines.
- Getting meaningful support at both a personal and healthcare level.
- Adopting new roles to promote new meanings in life.
- Learning how to use medication.

The findings from our reviews suggest that interventions for transdiagnostic pain need to incorporate:

1. A focus on the person's sense of self, strategies for maintaining hope
2. Strategies that provide relief for the distress associated with pain; and
3. Providing people with a sense of control over their experiences both with the pain itself and in their encounters with the medical profession and the use of medication.

4.1 Long Term Support for Self-Management

Because chronic pain is unlikely to be cured and is long-term, the support provided for people to self-manage their experience needs to also be long-term. Many people receive on-off short-term support from the health system, but that support needs to be similar to other long-term conditions i.e., long-term. Turk and Okifuji [38] have suggested that people with chronic pain need continuous care available to them with on-going maintenance support. This support needs to be individualised, taking account of the unique nature of the pain, the person's living situation, the resources available to them, their coping strategies and strengths, the long-term needs, and the degree of informal or peer support available to them.

The support would take different forms at different stages of the experience ranging from education about strategies and medications that may be helpful, and how and when to use them, to providing an opportunity to talk about the experience without fear of judgment. It needs to be accessible and well-informed about current evidence for pain management. Its delivery must reflect the needs of people experiencing pain rather than the needs of the organisation. A range of delivery modes should be made available: face-to-face, telephone, video-link, email and using new technologies as they become available.

Many traditional self-management programmes take a medical model that implies people need to be educated rather than person-centred approach with an emphasis on the person's experience and their expertise [39]. Holm [40] has suggested that a first-person perspective and the patient's understanding of their own lived body is central to successful self-management. It was found by Ross et al. [41] that people need to find out for themselves what strategies worked, and that professional advice was only sought-after trial and error and consultation with family and friends, because participants in that study were somewhat skeptical about the prescribing practices of clinicians.

4.2 The Role of Medications

A multi-dimensional approach to pain assessment has been recommended with self-report continuing to be the most valid and reliable method of assessing pain in cognitively intact older people and those with mild to moderate dementia [42]. Taking a person-centred approach, taking time and talking with older people using open ended questions and hearing stories about their pain experiences has been found to be crucial to successful communication and strategies, which can be incorporated into teaching older people how to effectively use an appropriate assessment pain tool (Numeric rating scale or verbal descriptor scale with cognitively intact older people and those with mild to moderate dementia). The selected tool should have sufficiently large print, be used with good lighting and then used consistently by all staff to assess effectiveness of any pain-relieving intervention such as medication.

In practice the complexities of the evolution and maintenance of chronic pain and the uncertainties around appropriate management lead to a lack of satisfaction in both patients and clinicians. The overarching principles of pharmacotherapy interventions in the management of chronic pain conditions include, an individualized and multidisciplinary approach and a multimodal strategy in relation to pharmacological and non-pharmacological interventions. Depending on individual patient and condition characteristics, the treatment approach may include drugs from the following classes: antidepressant, anticonvulsant, opioid and non-opioid analgesia, non-steroidal anti-inflammatory drugs, corticosteroid, antispasmodic, local anesthetic and specific disease modifying drugs. A multimodal approach is most often deployed in order to maximise therapeutic effects while minimizing adverse effects and events, with harm reduction a significant consideration underpinning the optimisation of non-opioid therapy in preference to opioid based treatment [43]. There is wide agreement that chronic pain requires a multimodal management strategy utilizing a coordinated collaborative approach drawing on a variety of clinical disciplines. It is acknowledged that the gains associated with pharmacological interventions in combination with non-pharmacological strategies may be effective in the order of a 30% reduction in pain. A multimodal approach to pharmacotherapy involves the combined use of analgesic drugs and co-analgesic drugs, although useful guideline documents support practice, there continues to be a significant lack of consensus about what constitutes effective care.

4.3 Developing Knowledge About Medication Use

Older people are not a heterogenous group and may also have hearing, eyesight, language, concentration, cognitive or dexterity issues, which need to be considered in any health education discussion about how to use pain medications and where to seek further help and support and pharmacists can often be a useful resource [44]. Individualised information about medication packaging options and dose administration aids, simplification of medication regimens and simple reminder cues tailored to usual routines have also been shown to assist older people in using their medication [45]. Much of what we know about the use of pain relief medication in older people is drawn from studies conducted with younger and middle-aged people, as few studies recruit older participants and in particular the oldest-old, those >85 years. Whilst a few older people obtain good relief that is better than a 30–50% reduction in pain intensity, from the use of pain-relieving medication, the vast majority do not, and for many the side effects of the treatment can be more harmful due to co-existing multi-morbidities and interactions with existing treatments. Furthermore, there is developing evidence of the ineffectiveness of some traditionally used medications such as paracetamol for osteoarthritis, a common painful condition experienced by large numbers of older people [46]. Trialing the effectiveness of any pain-relieving medication, measuring the effect and stopping if no or little response is important for the older person to understand as well as health professionals. Where possible, having open discussions with older people so they are aware that response to analgesic medication can be quite variable, with some responding well; i.e., within 1 to 2 weeks and others gaining no benefit at all is important to their understanding of the need for ongoing review and also avoiding unnecessary polypharmacy.

Providing such information can help reassure the older person who may be anxious or worried about issues of addiction. Health education in medication use for older people can further include discussion of: how the body changes its ability to handle drugs with age; increased sensitivity to some drugs such as analgesics; why only one new medication should be initiated at a time and the need to allow sufficient time between dose escalation; close monitoring requirements to detect any side effects; how several different classes of medications and therapies can work synergistically to provide better pain relief at lower doses than a single drug alone often with fewer side effects; how existing disease and their treatments may impact choices available to clinicians to treat pain in older people; how the type of pain influences the timing of pain relieving medication and the importance of reporting changes and response to treatments [24]. Using a person-centred approach to managing medication through gaining an understanding of the older persons knowledge of their chronic condition and its meaning; creating options to give the older person an opportunity to be involved in shared decision making and selecting a regimen which fits their needs are suggested useful strategies for helping optimize medication management and adherence with the older person [47].

4.4 Exercise

Exercise was found to be the preferred non-pharmacological option among older people with chronic pain [48]. The choice of exercise needs to promote strengthening, endurance, flexibility and balance [24]. The evidence produced from studies examining the impact of exercise on chronic pain is indicative rather than conclusive, but allows for some positive associations to be drawn [49]. The effects of exercise interventions on chronic pain may be reviewed in terms of specific effects, such as, reduction in pain intensity, frequency, duration and harm, and general effects such as quality of life, psychological effects, self-efficiency and self-management.

The general beneficial effects of exercise in the management of chronic pain may be associated with behavioural activation and distraction. Indeed, the collateral effects may afford important benefit to the individual in that exercise may result in improved posture and stability with reduced risk of fall, weight loss, increased muscular strength and flexibility with mitigation of deconditioning and improvement of disease specific symptom profiles [50]. Importantly, exercise interventions for chronic painful conditions has not been associated with significant harm, including increased pain [49].

Exercise is associated with significant health benefits including reduced depression and anxiety, improved cognitive function, improved sleep and in the older person improved functional health [50, 51]. The positive impact of exercise on physical and psychological wellbeing may be recruited towards the individuals' effort to maintain a positive equilibrium throughout the experience and management of chronic pain. Importantly, exercise may promote positive changes within important secondary pathologies associated with chronic pain, fear of movement, pain catastrophizing, anxiety and nervous system sensitisation, all of which significantly drive pain and disability [52]. In the absence of evidence based directions regarding exercise modality and dose; that is, exercise type, intensity, frequency and length, researchers advise exercise that is individualised and meaningful to the patient and is applied within a biopsychosocial approach. There is sufficient evidence that exercise is associated with desirable health outcomes and some albeit cautious confirmation of the benefits of exercise in pain reduction and improved physical function across a variety of specific conditions.

4.5 *Psychological Interventions*

Traditionally, psychological interventions for management of chronic pain have been from the cognitive and behavioural models and it is interesting to note that a Cochrane Library meta-analysis of psychological therapies for the management of chronic pain (excluding headache) in adults [53] only included cognitive and behavioural therapies. That meta-analysis found that these therapies had small to moderate effect sizes when studies were conducted without a control and small effect sizes with a control.

4.6 *Mindfulness*

A systematic review and meta-analysis of mindfulness interventions for pain [54] found small effects for pain, depression and quality of life. Mindfulness interventions may fit better with the issues identified in our reviews because of its compassionate stance towards the self and because it was originally designed to specifically help people experiencing pain. Mindfulness can be described as bringing one's complete attention to the experiences occurring in the present moment, in a non-judgmental and accepting way [55]. An effect of mindfulness is that it may modulate subjective experience of pain and disability, thus helping people to cope better [56]. It is a form of meditation practice, an attentional stance and a shift in attitude and perspective that allows the person to see her or his long-term condition in a way that is less anxiety-provoking [57]. There are many ways that mindfulness may promote self-management of long-term conditions associated with pain: decreased perception of pain severity; reduced stress, anxiety and depression; diminished need for medication; enhanced ability to reflect on choices; improved adherence to treatment; increased motivation for lifestyle change; enriched interpersonal relationships and social connectedness; and alterations to biological pathways affecting health such as the autonomic nervous system, neuroendocrine function and the immune system [56].

A review of the use of mindfulness-based stress reduction [58] identified that there was emerging evidence to suggest that it was an effective intervention for improving physical health status (functional health and well-being) in chronic pain. These improvements may relate to what Ludwig and Kabat-Zinn [56] hypothesized as a modulation of the subjective experience of symptoms which enables better coping. Support for the use of mindfulness interventions in clinical populations for people with co-occurring physical and psychological symptoms has been identified by Hofmann et al. [59] who proposed that the effects of mindfulness therapies may not be diagnosis-specific but may address transdiagnostic processes occurring in multiple disorders. It may improve the experience of physical symptoms such as pain by improving the emotional and evaluative dimensions that underlie general aspects of well-being. There is convincing evidence that mindfulness interventions can improve depressive and anxiety symptoms. (often associated with chronic pain) [59–61].

4.7 Cognitive Behavioural Therapy for Pain

Cognitive Behavioural Therapy (CBT) has been applied to pain management for over five decades (e.g., [62]) and is recognised as a gold standard evidence-based treatment [63]; Canadian Agency for Drugs and Technologies in Health, 2016). CBT utilises a biopsychosocial model and tends to be at the core of multimodal pain management programmes [64].

The CBT formulation holds that the way people appraise their circumstances of the onset and ongoing occurrence of pain has a fundamental impact on their experience of pain, and subsequent feelings and behaviours influencing their coping and ability to live with chronic pain. People may associate pain with damage and so fear of further damage may lead them to reduce activity, resulting in physical deconditioning, and reduced engagement in normal life activities. They may have been desperately seeking a medical cure but instead have become demoralised, depressed and hopeless with constrained life functioning and poor quality of life.

CBT typically comprises components such as: psychoeducation about the physiology and characteristics of pain; presenting the CBT model (that appraisals affect emotional reactions and behaviours); and strategies to promote active self-management to enable clients to live the best life possible in the presence of a chronic pain condition. Individuals are taught to identify and monitor their own thoughts, feelings and physical sensation including pain, and behaviours to increase awareness of contextual factors that modulate pain experience and the consequent impacts. Key skills include cognitive restructuring strategies to enable individuals to modify their appraisals of their pain (e.g., to reduce pain catastrophizing which may be exacerbating their distress), and behavioural strategies such as problem-solving and behavioural activation to improve self-efficacy, address avoidance and build new and meaningful activities and a sense of control over their lives [65]. It may include strategies such as breathing exercises or relaxation, pacing and managing flare-ups, and sleep hygiene. Contemporary CBT models may also utilise mindfulness approaches to thoughts as well as to sensations [66].

CBT may be delivered via the internet or in face-to face individual sessions but is commonly delivered in groups. Group approaches are particularly helpful in providing support and reducing the sense of isolation. CBT programmes for pain are as effective in older persons as in other adult groups [67]. Effect sizes are moderate and around a third experience significant benefit [64]. Meta analyses of CBT for chronic pain indicate that it is most effective for reducing psychological distress (anxiety, depression), reducing sense of disability and improving quality of life [53]. Although some studies also report change in the pain experience [67], evidence for change in this domain is mixed (see [64]). Qualitative studies of CBT pain programmes support these findings, with many individuals reporting a fundamental and persisting shift in attitudes and a shift to an active self-management stance, which reduces the sense of powerlessness. In one long-term follow-up, participants reported changing the way they live their lives with improved quality of life.

5 Conclusion

Chronic pain is often experienced as an assault on the self. It takes over everything in the biological, psychological and social aspects of the person. Many people manage their chronic pain by actively balancing the competing forces of hope and despair. There are similarities in this experience of chronic pain across a range of conditions which has implications for the development of pain management strategies and interventions that address the pain experience and not just the biological condition. Learning to manage the self in pain involves acceptance of pain as on-going and a part of who they are; keeping connected with others; keeping occupied through meaningful activities; getting meaningful support; and developing new meaning in life. Clinical interventions need to a focus on the person's sense of self, strategies for maintaining hope, strategies that provide relief for the distress associated with pain; and providing people with a sense of control over their experiences both with the pain itself and in their encounters with the medical profession and the use of medication. These findings from qualitative studies may not be generalizable to the wider population and reflect an historical context to managing pain. Further research is required to examine the clinical effectiveness of transdiagnostic pain interventions.

References

1. Moseley GL, Butler DS. Fifteen years of explaining pain: the past, present, and future. J Pain. 2015;16(9):807–13.
2. Baliki MN, Apkarian AV. Nociception, pain, negative moods, and behavior selection. Neuron. 2015;87(3):474–91.
3. Ballantyne JC, Sullivan MD. Intensity of chronic pain—the wrong metric? N Engl J Med. 2015;373(22):2098–9.
4. Latremoliere A, Woolf CJ. Central sensitization: a generator of pain hypersensitivity by central neural plasticity. J Pain. 2009;10(9):895–926.
5. Scascighini L, Sprott H. Chronic nonmalignant pain: a challenge for patients and clinicians. Nat Clin Pract Rheumatol. 2008;4(2):74–81.
6. Monsivais DB, Engebretson JC. Cultural cues: review of qualitative evidence of patient-centered care in patients with nonmalignant chronic pain. Rehabil Nurs. 2011;36(4):166–71.
7. Maly MR, Krupa T. Personal experience of living with knee osteoarthritis among older adults. Disabil Rehabil. 2007;29:1423–33.
8. Currie S, Wang J. Chronic back pain and major depression in the general Canadian population. Pain. 2004;107(1–2):54–60.
9. Arnow B, Blasey C, Lee J, Fireman B, Hunkeler E, Dea R, et al. Relationships among depression, chronic pain, chronic disabling pain and medical conditions. Psychiatr Serv. 2009;60:344–50.
10. Blair MJ, Robinson RL, Katon WJ, Kroenke K. Depression and pain comorbidity: a literature review. Arch Intern Med. 2003;163:2433–45.
11. Walker AK, Kavelaars A, Heijnen CJ, Dantzer R. Neuroinflammation and comorbidity of pain and depression. Pharmacol Rev. 2013;66:80–101.
12. Campbell LC, Clauw DJ, Keefe FJ. Persistent pain and depression: a biopsychosocial perspective. Biol Psychiatry. 2003;54(3):399–409.
13. Gambassi G. Pain and depression: the egg and the chicken story revisited. Arch Gerontol Geriatr. 2009;49:103–12.
14. Cohen M, Quintner J, Buchanan D, Nielsen M, Guy L. Stigmatization of patients with chronic pain: the extinction of empathy. Pain Med. 2011;12(11):1637–43.
15. Jackson JE. Stigma, liminality, and chronic pain—mind-body borderlands. Am Ethnol. 2005;32(3):332–53.
16. Crowe M, Whitehead L, Seaton P, Jordan J, McCall C, Maskill V, et al. Qualitative meta-synthesis: the experience of chronic pain across conditions. J Adv Nurs. 2017;73(5):1004–16.
17. Osborn M, Smith JA. Living with a body separate from the self. The experience of the body in chronic benign low back pain: an interpretative phenomenological analysis. Scand J Caring Sci. 2006;20(2):216–22.
18. Campbell C, Cramb G. 'Nobody likes a back bore'—exploring lay perspectives of chronic pain: revealing the hidden voices of nonservice users. Scand J Caring Sci. 2008;22(3):383–90.
19. Williams S. Chronic illess as biographical disruption or biographical disruption as chronic illness? Sociol Health Illn. 2000;22:241–57.
20. Eccleston C, Williams C, Rogers W. Patients' and professionals understandings of the causes of chronic pain: blame, responsibility and identity protection. Soc Sci Med. 1997;45:699–709.
21. Crowe M, Whitehead L, Gagan MJ, Baxter GD, Pankhurst A, Valledor V. Listening to the body and talking to myself—the impact of chronic lower back pain: a qualitative study. Int J Nurs Stud. 2010;47(5):586–92.
22. Rhodes L, McPhillips-Tangum C, Marham C, Klenk R. The power of the visible: the meaning of diagnostic tests in chronic back pain. Soc Sci Med. 1999;48:1189–203.
23. Corbett M, Foster N, Ong B. Living with low back pain. Soc Sci Med. 2007;65(8):1584–94.
24. British Geriatrics Society. Guidance on the management of pain in older people. Age Ageing. 2013;42:i1–i57.

25. Keefe FJ, Porter L, Somers T, Shelby R, Wren AV. Psychosocial interventions for managing pain in older adults: outcomes and clinical implications. Br J Anaesth. 2013;111(1):89–94.
26. Jakobsson U, Hallberg IR, Westergren A. Overall and health related quality of life among the oldest old in pain. Qual Life Res. 2004;13(1):125–36.
27. Wu M, Brazier JE, Kearns B, Relton C, Smith C, Cooper CL. Examining the impact of 11 long-standing health conditions on health-related quality of life using the EQ-5D in a general population sample. Eur J Health Econ. 2015;16(2):141–51.
28. Eggermont LH, Leveille SG, Shi L, Kiely DK, Shmerling RH, Jones RN, et al. Pain characteristics associated with the onset of disability in older adults: the maintenance of balance, independent living, intellect, and zest in the elderly Boston study. J Am Geriatr Soc. 2014;62 (6):1007–16.
29. Silva AG, Queiros A, Cerqueira M, Rocha NP. Pain intensity is associated with both performance-based disability and self-reported disability in a sample of older adults attending primary health care centers. Disabil Health J. 2014;7(4):457–65.
30. Mackichan F, Adamson J, Gooberman-Hill R. 'Living within your limits': activity restriction in older people experiencing chronic pain. Age Ageing. 2013;42(6):702–8.
31. van Hecke O, Torrance N, Smith BH. Chronic pain epidemiology and its clinical relevance. Br J Anaesth. 2013;111(1):13–8.
32. Torrance N, Elliott AM, Lee AJ, Smith BH. Severe chronic pain is associated with increased 10 year mortality. A cohort record linkage study. Eur J Pain. 2010;14(4):380–6.
33. Bernfort L, Gerdle B, Rahmqvist M, Husberg M, Levin LA. Severity of chronic pain in an elderly population in Sweden-impact on costs and quality of life. Pain. 2015;156(3):521–7.
34. Crowe M, Gillon D, Jordan J, McCall C. Older peoples' strategies for coping with chronic non-malignant pain: a qualitative meta-synthesis. Int J Nurs Stud. 2017;68:40–50.
35. Lansbury G. Chronic pain management: a qualitative study of elderly people's preferred coping strategies and barriers to management. Disabil Rehabil. 2000;22(1–2):2–14.
36. Howarth M, Warne T, Haigh C. Pain from the inside: understanding the theoretical underpinning of person-centered care delivered by pain teams. Pain Manag Nurs. 2014;15(1):340–8.
37. Clarke KA, Iphofen R. Believing the patient with chronic pain: a review of the literature. Br J Nurs. 2005;14(9):490–3.
38. Turk DC, Okifuji A. Pain terms and taxonomies. In: Loeser D, Butler SH, Chapman JJ, Turk DC, editors. Bonica's management of pain. 3rd ed. London: Lippincott Williams & Wilkins; 2001. p. 18–25.
39. Crowe M, Whitehead L, Gagan M-J, Baxter D, Panckhurst A. Self-management and chronic low back pain: a qualitative study. J Adv Nurs. 2010;66(7):1478–86.
40. Holm S. Justifying patient self-management – evidence based medicine or the primacy of the first person perspective. Med Health Care Phil. 2005;8:159–64.
41. Ross M, Carswell A, Hing M, Hollingworth G, Dalziel W. Seniors' decision making about pain management. J Adv Nurs. 2001;35:442–51.
42. Schofield P. The assessment of pain in older people: UK national guidelines. Age Ageing. 2018;47(1):i1–i22.
43. Busse JW, Craigie S, Juurlink DN, Buckley DN, Wang L, Couban RJ, et al. Guideline for opioid therapy and chronic noncancer pain. Can Med Assoc J. 2017;189(18):E659–E66.
44. George J, Elliott RA, Stewart DC. A systematic review of interventions to improve medication taking in elderly patients prescribed multiple medications. Drugs Aging. 2008;25(4):307–24.
45. Topinkova E, Baeyens JP, Michel JP, Lang PO. Evidence-based strategies for the optimization of pharmacotherapy in older people. Drugs Aging. 2012;29(6):477–94.
46. Gaskell H, Derry S, Moore RA. Treating chronic non-cancer pain in older people—more questions than answers? Maturitas. 2014;79(1):34–40.
47. Marcum ZA, Hanlon JT, Murray MD. Improving medication adherence and health outcomes in older adults: an evidence-based review of randomized controlled trials. Drugs Aging. 2017;34 (3):191–201.

48. Park J, Lavin R, Couturier B. Choice of nonpharmacological pain therapies by ethnically diverse older adults. Pain Manag. 2014;4:389–406.
49. Geneen LJ, Moore RA, Clarke C, Martin D, Colvin LA, Smith BH. Physical activity and exercise for chronic pain in adults: an overview of Cochrane reviews. Cochrane Database Syst Rev. 2017;4:CD011279.
50. Kredlow MA, Capozzoli MC, Hearon BA, Calkins AW, Otto MW. The effects of physical activity on sleep: a meta-analytic review. J Behav Med. 2015;38(3):427–49.
51. Martinez-Gomez D, Guallar-Castillon P, Garcia-Esquinas E, Bandinelli S, Rodriguez-Artalejo F. Physical activity and the effect of multimorbidity on all-cause mortality in older adults. Mayo Clin Proc. 2017;92(3):376–82.
52. Booth J, Moseley GL, Schiltenwolf M, Cashin A, Davies M, Hubscher M. Exercise for chronic musculoskeletal pain: a biopsychosocial approach. Musculoskel Care. 2017;15(4):413–21.
53. Williams AC, Eccleston C, Morley S. Psychological therapies for the management of chronic pain (excluding headache) in adults. Cochrane Database Syst Rev. 2012;11:CD007407.
54. Hilton L, Hempel S, Ewing BA, Apaydin E, Xenakis L, Newberry S, et al. Mindfulness meditation for chronic pain: systematic review and meta-analysis. Ann Behav Med. 2017;51 (2):199–213.
55. Kabat-Zinn J. Full catastrophe living: using the wisdom of the body and mind to face stress, pain and illness. New York: Dell; 1990. 722 p.
56. Ludwig DS, Kabat-Zinn J. Mindfulness in medicine. JAMA. 2008;300:1350–2.
57. Kabat-Zinn J. Mindfulness-based interventions in context: past, present and future. Clin Psychol Sci Pract. 2003;10:144–56.
58. Crowe M, Jordan J, Burrell B, Jones V, Gillon D, Harris S. Mindfulness-based stress reduction for long-term physical conditions: a systematic review. Aust NZ J Psychiat. 2016;50(1):21–32.
59. Hofmann SG, Sawyer AT, Witt AA, Oh D. The effect of mindfulness-based therapy on anxiety and depression: a meta-analytic review. J Consult Clin Psychol. 2010;78(2):169–83.
60. Bohlmeijer E, Prenger R, Taal E, Cuijpers P. The effects of mindfulness-based stress reduction therapy on mental health of adults with a chronic medical disease: a meta-analysis. J Psychosom Res. 2010;68(6):539–44.
61. Khoury B, Lecomte T, Fortin G, Masse M, Therien P, Bouchard V, et al. Mindfulness based therapy: a comprehensive meta-analysis. Clin Psychol Rev. 2013;33:763–71.
62. Fordyce WE. Behavioral-science and chronic pain. Postgrad Med J. 1984;60(710):865–8.
63. Ehde DM, Dillworth TM, Turner JA. Cognitive-behavioral therapy for individuals with chronic pain: efficacy, innovations, and directions for research. Am Psychol. 2014;69(2):153–66.
64. Wilson IR. Management of chronic pain through pain management programmes. Br Med Bull. 2017;124(1):55–64.
65. Turk DC, Wilson HD, Cahana A. Treatment of chronic non-cancer pain. Lancet. 2011;377 (9784):2226–35.
66. Hughes LS, Clark J, Colclough JA, Dale E, McMillan D. Acceptance and commitment therapy (ACT) for chronic pain: a systematic review and meta-analyses. Clin J Pain. 2017;33 (6):552–68.
67. Lunde M, Nordhus IH. Combining acceptance and commitment therapy and cognitive behavioral therapy for the treatment of chronic pain in older adults. Clin Case Stud. 2009;10:125–43.

Chapter 5
Changing Pain: Making Sense of Rehabilitation in Persistent Spine Pain

James E. Eubanks, Michael E. Farrell, Brandon S. Barndt, Chandler L. Bolles, Maria Vanushkina, and James W. Atchison

Abstract When acute pain persists beyond the expected healing time following an injury, important neurological changes occur that allow pain to transition from adaptive to maladaptive. Spine pain has become an important global problem, with significant increases in prevalence, disability, and subsequent healthcare costs over the past several decades. Low back pain is now the number one cause of disability in the world. Because of the magnitude of the effect of low back pain, and especially chronic low back pain, it has become imperative that we embrace the best available evidence and clinical sensibilities as we work with patients to find appropriate solutions. Intrinsic to the successful care of persons with spine pain is the acknowledgment that the experience of pain is a biopsychosocial one. There is no universal *experience* of pain and thus our solutions must accommodate variation in the meanings of pain. Experiential (qualitative, subjective) knowledge of spinal pain can be integrated with our understanding of spinal pain neurobiology (quantitative, objective) in rehabilitation contexts to improve health outcomes. Ultimately, the rehabilitation of persons with spine pain exists at the intersection of the objective and subjective goals of care.

J. E. Eubanks (✉)
Department of Physical Medicine and Rehabilitation, University of Pittsburgh Medical Center (UPMC), Pittsburgh, PA, USA
e-mail: eubanksje@upmc.edu

M. E. Farrell
Department of Rehabilitation Medicine, Georgetown University/Medstar National Rehabilitation Hospital, Washington, DC, USA

B. S. Barndt
Department of Physical Medicine and Rehabilitation, Temple University Hospital/MossRehab, Philadelphia, PA, USA

C. L. Bolles
Logan University, College of Health Sciences, Chesterfield, MO, USA

M. Vanushkina
Spine Center, New England Baptist Hospital, Boston, MA, USA

J. W. Atchison
Department of Physical Medicine and Rehabilitation, Mayo Clinic, Jacksonville, FL, USA

S. van Rysewyk (ed.), *Meanings of Pain*,
https://doi.org/10.1007/978-3-030-24154-4_5

Clinical Implications: Understanding the full biopsychosocial scope of spinal pain allows clinicians to strengthen their therapeutic alliance with patients, reinforce self-efficacy, identify patients at risk for poor outcomes and intervene early, stratify care appropriate to the individual's needs, reduce or prevent pain chronification, reduce direct and indirect costs to patients and society, and improve overall quality of life.

Keywords Spine · Back pain · Exercise-based therapy · Chronic pain · Biopsychosocial · Pain rehabilitation

1 Introduction

> Nature is not mute. It eternally repeats the same notes which reach us from afar, muffled, with neither harmony nor melody. But we cannot do without melody. . .It is up to us to strike the chords, to write the score, to bring forth the symphony, to give the sounds a form that, without us, they do not have.
> —François Jacob, The Statue Within

François Jacob represents the consummate scientist, someone unshakably dedicated to the pursuit and interpretation of foundational knowledge that can bring people together as we seek to better understand the world in which we coexist. His work illustrates the rigorous and disciplined pursuit of science and a life imbued with meaning. Jacob was the originator of the idea that DNA transcription and subsequent protein translation governs the actions of the enzymes within our bodies. We need enzymes for our very survival. Protein translation, when responding to the needs of its surroundings, takes basic codes and converts them into something functional for the organism. However, enzymes can become dysfunctional or serve no purpose at all. In extreme cases, they cause diseases. Pain, too, is like this.

Pain is an interpretation of the myriad inputs and outputs that arise within the nervous system. Intuitively, we think of pain as something that happens *to us*. While noxious stimuli can indeed arise from the body interacting in a harmful way with the world—as in the case of tissue injury—the brain itself can generate, amplify, and alter the neurological inputs and outputs that give rise to the perception that we ultimately call "pain." Tissue damage is sufficient, but not necessary, for the perception of pain, which is also dependent on memory and prior experiences [1]. In the end, pain, like protein translation, can serve a useful, functional purpose, but sometimes pain becomes dysfunctional, or serves no purpose at all.

The late Craig D. Brigham, MD (1954–2013), was an orthopedic spine surgeon in Charlotte, North Carolina, USA, who dedicated his career to spine care and patient education. He worked tirelessly to translate the basic knowledge of twenty-first century medicine and healthcare into something useful for patients with spine pain. He was his region's spinal deformity expert but would take spine cases from acute low back pain to persistent spine pain. Importantly, Dr. Brigham understood the diverse and multifaceted nature of pain. He spent as much time talking patients out of unnecessary procedures as he did performing the necessary ones. First and foremost, he was a patient advocate who operated through a biopsychosocial framework. As defined by George Engel, MD, the biopsychosocial approach systematically addresses biological, psychological, and social factors and their interactions in understanding health, illness, and health care delivery [2].

Dr. Brigham would offer four key concepts to the patients who would seek out his advice:

1. An honest assessment of the clinical problem, including psychosocial confounders. He refused to overlook these just because they were more of a challenge, and he understood that each person must be considered within their biopsychosocial matrix.
2. Specificity of diagnosis. When this is not possible, as in the case of non-specific low back pain, we must say so and inform the patient that full disclosure is better than guessing or trying to fit them into a particular box without the requisite evidence.
3. An understandable, "plain language" explanation of the clinical problem(s) and solution(s).
4. Realistic reassurance and, most of all, the *hope* that patients can participate in the improvement of their own lives within the limits of their condition. Conveying a message of self-efficacy is critical to success.

The following are patient responses that Dr. Brigham catalogued after asking the common opening question: "Tell me about your back pain."

Responses:

- "I was doing all the wrong things for my back, you know, like lifting heavy things that I probably shouldn't have while working in the yard."
- "I took a yoga class to try to stop my years of back spasms. After the first session put me in the worst spasm I've ever had, I saw clinician who told me that yoga can put tremendous pressure on your discs. I'll never do yoga again."
- "I have 'degenerative disc disease.' For years doctors told me that when my back pain got bad enough, it would be time for surgery."
- "I know why my back hurts. I've got bulging discs and advanced arthritis."
- "I was told that I was lifting all wrong."
- "It must be from moving furniture last year during our move. I can't think of anything else I did to rupture a disc in my back."

The above statements have something in common: they all reflect typical *misunderstandings* about spine problems. These misunderstandings and the ease with which we propagate them stems from the narrative fallacy, the idea that we cannot just look at a series of facts but feel compelled to weave a story out of them. These myths contribute to the fear that drives inappropriate or unsupported treatments and keeps our care of patients with spine pain from entering the twenty-first century scientific world. Though many patients learn and accept these myths, evidence suggests that they also affect the beliefs and subsequent advice given by clinicians [2].

Four Back Pain Myths:

1. Severe pain reflects serious injury
2. Lifting is bad for your back
3. An MRI explains back pain
4. A single treatment can cure back pain

The discussion that follows will address these myths and more. Pain is an interpretation of our moment-to-moment psychoneurological status. Pain can result from injury, but it can also result from physiological changes that are not associated with injury at all. This is difficult for us to reconcile due to the naturally oppressive potential of pain. When we perceive pain, our evolutionary wiring is such that our attention converges on it. This is the very point of pain when it serves a functional purpose: to capture our attention and motivate us to escape a threat, or to avoid using an injured body part while it heals. While pain is unpleasant, it is vital for our very survival. Without pain, such as in congenital insensitivity to pain and anhidrosis (CIPA), also known as hereditary sensory and autonomic neuropathy type IV, we have no early warning system and affected persons often die in childhood.

The spine is like any other part of the body in that it is constructed to move, twist, bend, and lift. Like the myth that pain equates to injury, the popular idea that lifting is bad for the back is an intuition that does not pan out. Data supports that exercising the back is an effective management strategy for back pain [3]. Finally, the notion that structure consistently predicts pain has been debunked through careful research. In fact, the majority of persons who have an MRI will show disc bulges and osteoarthritic changes, yet they are asymptomatic [4]. Surprisingly, while degenerative changes in the spine correlate well to age, the peak of back pain occurs in middle age for the majority of people [5]. In order to understand what we might do to address the inconsistencies between our beliefs and behaviors as they relate to spine pain, let us take a look at what we know about pain and the experience of the patient with spine pain, especially when it is persistent.

2 Defining Pain

> An unpleasant sensory and emotional experience associated with actual or potential tissue damage, or described in terms of such damage.
> —International Association for the Study of Pain (IASP) [6]

When attempting to define pain, it is important to note that there is a distinction between pain and nociception. Pain, by convention, requires an interpretation of our real or perceived sensations. Although these distinctions may seem nuanced, they have led to ongoing debate over the ideal definition of pain within the literature. In a paper published by Cohen, Quinter, and van Rysewyk, a new definition of pain was suggested to supplant the definition used by the IASP: "Pain is a mutually recognizable somatic experience that reflects a person's apprehension of threat to their bodily or existential integrity." Gone are the words "unpleasant" and "tissue damage," which the authors feel are unable to reflect the full scope of the meaning of pain. They argue that the IASP definition "privileges the stance of the observers over that of the experiencer," and "implies equivocation about its veracity as a form of distress when there is no obvious nociception." The authors argue that their new definition is compatible with the IASP definition but is "more philosophically sound, biologically relevant, clinically applicable, and meaningful for people experiencing pain and for health care professionals who engage with them" [7].

A rebuttal from Rolf-Detlef Treede, chair of neurobiology at the Center for Biomedicine and Medical Technology Mannheim, points out that the IASP definition is sufficiently complete insofar as it "does not require the presence of a noxious stimulus for pain to exist; the third part of the definition clearly states that anything that hurts is pain. The verbal report by the patient has to be trusted, but the management of pain will differ according to the underlying causes and mechanisms" [8]. Philosopher Murat Aydede at the University of British Columbia proposes the following simplification of the IASP definition of pain as: "An unpleasant sensory and emotional experience that results from actual or impending tissue damage, or is correctly describable in terms of such damage." Aydede claims that "the definition is an invitation to vigorously develop research programs designed to help the health-care professionals to bridge the epistemic gap between the pain itself and the means with which they can detect or assess its presence or absence" [9].

It is challenging to provide a single, all-encompassing definition of pain. The fact is that pain is adaptive but can become maladaptive. The neurobiological facet of pain is often the focus of biomedical interventions, yet the patient's perception of pain is an interdependent phenomenon. One's psychosocial context influences the characterization of pain and the meaning that is assigned to it. We will examine this psychosocial context further in the next section.

3 Pain and Psychosocial Context

Pain perception has a plethora of modifying factors, for which two main pathways have been described. The first is the bottom-up pathway, where peripheral nerves detect noxious stimuli and signal via the dorsal root ganglion (DRG), leading to afferent signaling that travels up the spinal cord to higher-order cortical and pre-cortical structures in the brain. These regions play a key role in the interpretation of incoming afferent signaling in order to synthesize the experience of pain. This nociceptive pathway can be viewed as an early-warning system or a set of physiological protections intended to inform us about environmental dangers [10].

Adaptive and protective pain is an extension of this afferent pathway involving peripheral sensitization to discourage overuse of injured structures [10]. Inflammation plays a key role in this process. Sensory nerves can respond to injury under normal circumstances by generating an action potential when directly stimulated, but they also can become peripherally sensitized (i.e., non-proportionally to a stimulus or even in the absence of an external stimulus). In acute injury, peripheral sensitization can be viewed as contributing to a functional pain as it serves as protection for the injured structure. Take for example a newly sprained ankle. The pain experienced following an acute ankle sprain serves the purpose of limiting or preventing weight bearing. When this sensitization persists beyond the point that the ankle is healed, it serves as the basis for neuropathic and dysfunctional pain, which we can collectively call pathological pain as it is not protective or functional [10]. This propagation of nociception from normally non-noxious stimuli, known as allodynia, is thought to occur due to pathological changes in sodium channel function and neurotransmitter trafficking, as well as through demyelination and structural alterations to the nerve during regeneration following an injury [11].

The second pathway is the top-down pathway, where higher-order centers, such as the prefrontal cortex (PFC), thalamus, and limbic structures project onto and modify the signaling of lower-order centers, leading to a modification of the afferent signaling coming in from peripheral nerves [11]. This is also known as descending modulation. Cognitive-emotional input involves elements of motivation and learning, further informs perceptions of pain and is involved in the transition from acute to chronic pain [12, 13].

The cognitive-emotional dimensions of pain enable learning to take place as new neuronal behaviors take hold in the CNS. Just as any experience affords opportunities for learning new behaviors, the experience of pain can as well. In fact, Marc Lewis has suggested that brain changes in addiction likely correspond to a learning process rather than a pathological one [14]. Similarly, learning appears to be integral to the chronification of pain. It is known that changes in the corticolimbic region (i.e., mesolimbic dopamine system, amygdala, and medial prefrontal cortex) lead to cellular and behavioral maladaptations [12]. This chronification process is so powerful that it affects motivation and reward, and it is an obstacle for patients with persistent spine pain who seek resolution. Adding to the complexity, behavioral changes are critical to the successful management of chronic pain as well as its development. This may explain the magnitude of the obstacles many patients with chronic pain face.

To further examine the dynamic relationship between the bottom-up and top-down pain pathways, functional magnetic resonance imaging (fMRI) is currently being used to map key neural pathways that transmit nociception and that are intrinsic to the various brain regions involved in the perception of pain. Hashmi et al. have used fMRI to describe neural correlates of spine pain, showing that acute/subacute back pain "is limited to regions involved in acute pain, whereas in the chronic back pain group, activity is confined to emotion-related circuitry" [15]. Importantly, they also observed that in those with back pain who experienced resolution of their pain, brain activity decreased in the areas of the brain related to emotions. However, in the persistent back pain group they also saw decreased activity in acute pain circuits and increased activity in emotion-related circuitry [15].

Activity of specific brain regions in response to pain appears to vary with context when studied via imaging. Pain is modulated in the CNS through a variety of neurobiological mechanisms—including endogenous opioids, endocannabinoids, and other neurotransmitters and neuromodulators—and is directly shaped and interpreted through the lens of patient expectations [16]. This is known as the subject-expectancy effect, and it is a large component of placebo and nocebo effects. As David Spiegel, MD, puts it, "placebo and other non-pharmacological analgesic treatments affect brain pathways that modulate overall pain perception more than reduce nociceptive input" [17, p. 1309]. Depending on the circumstances of the expectations that a patient has, different portions of the PFC, anterior cingulate cortex (ACC), and insula are activated. This variance in neurological activation in the brain plays an important role in matching experience with expectations. An example would be the promise of pain relief implicit in receiving a "pain" medication, which oftentimes enhances the analgesic effect of the medication beyond the intrinsic pharmacological effect. Due to the deeply personal, subjective nature of pain, it is highly susceptible to placebo and nocebo effects [18]. Notably, baseline patient expectations play such an important role in the experience of spine pain that they appear to directly influence treatment outcomes in low back pain [19].

The process through which pain is influenced by the context of the experiencer explains the connection between social expectations, or norms, and pain. Studies have demonstrated that pain is, in a sense, transmissible when the social group with which one identifies strongly demonstrates pain behaviors. In fact, the partners of those with musculoskeletal pain have a higher risk of developing musculoskeletal pain themselves [20]. This phenomenon is thought to originate in higher level associative cortices as our brains strive to align our experiences with our expectations. Thus, the modeling of society and those around us play a crucial role in the net experience of pain [16]. These challenges raise a valid question: what should the clinician do? (Fig. 5.1)

It is not surprising that we do not find agreement in the literature on one sole definition of pain. When we consider how the experience of pain can be both functional and dysfunctional, top-down and bottom-up, adaptive and pathological, or often a mix with multiple patterns occurring simultaneously, it is no wonder why attempts to find an all-encompassing definition have led to disagreement. Similarly,

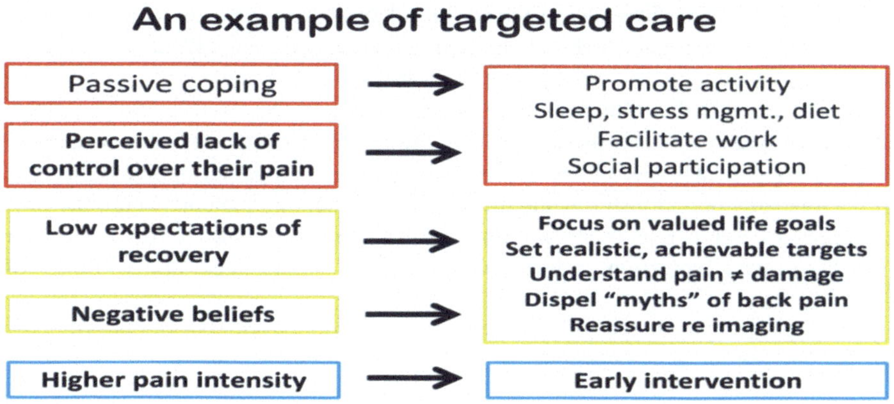

Fig. 5.1 What should the clinician do? (Image by Derek Griffin, adapted from Chen et al. [21])

the prescription of generic treatment regimens for back pain is often as challenging as agreeing on one all-encompassing definition.

4 Chronic Spine Pain Management as Patient-Centered Active Care

To better accommodate the inherent variability of each patient experience, a diverse set of treatment options are offered to manage chronic spine pain. Haldeman and Degenais have described it as a "supermarket approach." As they explain, "Patients with chronic low back pain are finding it increasingly difficult to make sense of the growing list of treatment approaches promoted as solutions to this widespread problem. Their confusion is compounded by the financial and emotional cost of previous failed attempts" [22, p. 1]. One of the persistent problems is the focus on advice and treatments that are unproven or that lack sufficient evidence to support their use in clinical practice.

In 2001, the World Health Organization (WHO) International Classification of Functioning, Disability and Health (ICF) encouraged clinicians to embrace a patient-centered, active approach to rehabilitation focused on the contexts of an individual's life where they find meaning [23]. This includes identifying an individual's functional needs and goals within the home, workplace, and the community. Such an approach to pain care is both patient-centered and active in that the patient is ultimately in control of their pain management strategies.

Using focus groups, Dima et al. outlined four dimensions that form the foundation of patients' beliefs about back pain therapies: credibility, effectiveness, concerns and individual fit [24]. For patients, understanding the cause of their pain within the context of a trusting relationship was a primary concern. Understanding is viewed as a "prerequisite for meaningful engagement with treatment decision making." Echoing this, Chou et al. performed a systematic review of patient perceptions of their needs of health care clinicians in the setting of back pain management, and found that those with back pain seek patient-centered care, active involvement with their treatment plan, and a good relationship with the treating clinician [25].

It is imperative that clinicians help patients navigate the overwhelming "supermarket" of spine care. This includes encouraging an active approach to management and helping patients develop a better understanding of their pain. Implicit in this type of doctor-patient relationship is the difficult task of helping patients weigh the pros and cons of alternative treatment options. This is best achieved by a team or interdisciplinary approach, which places patient preferences above those of each clinician.

5 The Promise of Interdisciplinary, Multimodal Pain Management

The arsenal of treatment options in pain medicine can be classified as active or activating treatments and passive treatments. Active treatments are immediately accessible by the patient when the need arises to manage their pain and rely on the patient participating in these techniques on their own time. This approach offers the potential to influence both biophysical factors such as strength and motor control, as well as psychosocial factors such as self-efficacy and fear-avoidance with the goal of promoting long-term functional recovery. Passive treatments, such as medications and procedures, can provide relatively rapid-onset pain relief but require the patient to act in the role of recipient. The passive role has the potential to reinforce feelings of powerlessness and displace the responsibility of pain management onto the clinician. Instead, the goal should be to shift chronic low back pain patients into a more empowering, active model of care and use passive modalities to facilitate participation when needed.

Several major organizations have taken steps to promote an algorithmic approach to the care of low back pain patients that addresses all aspects of the biopsychosocial model. In 2016, the National Institute for Health and Care Excellence (NICE), a UK-based organization, published clinical guidelines intended for use by the National Health Service (NHS) for management of low back pain [26]. The recommended initial assessment is designed to stratify the likelihood of recovery and identify barriers to good functional outcomes like fear-avoidance, workplace dissatisfaction, active litigation, and psychological disorders. Clinical tools such as Keele's STarT Back Tool may be used to assist clinicians in establishing an initial evidence-based care pathway. Once the likelihood of recovery is stratified, patients with high levels of motivation and few barriers may be educated about their pain and reassured of the promising recovery rates from back pain experienced by the majority of people. Group exercise is recommended on an as needed basis in these low risk groups. In the absence of red flags–such as acute traumatic injuries, cancer, infection, inflammatory disease and cauda equina syndrome–a referral to a specialist is often not needed.

The American College of Physicians (ACP) issued their guidelines in the *Annals of Internal Medicine*, titled "Noninvasive Treatments for Acute, Subacute, and Chronic Low Back Pain: A Clinical Practice Guideline from the American College of Physicians" [27]. For chronic low back pain, the updated ACP guidelines focus on the reliance on interdisciplinary active rehabilitation, which includes exercise, stress management (via techniques for cognitive activation such as mindfulness, tai chi, yoga, and progressive relaxation), cognitive behavioral therapy (CBT) or manual therapy. These guidelines encourage clinicians to assess and address fear and activity avoidance, low mood and social withdrawal, the belief that back pain is potentially severely disabling, and the belief that passive treatment will help. Consequently, a successful patient-centric approach to chronic low back pain empowers patients with appropriate education to correct erroneous beliefs, focuses on building confidence, and provides a pain management selfcare tool-box of exercises and behavioral techniques that can be employed independently from the clinic setting.

6 Patient Education and the Therapeutic Alliance

Disability associated with chronic low back pain is highest in working class individuals in low-income and middle-income countries [28]. Owing to the peak incidence in middle age, it is a large economic burden due to lost wages. Evidence-based guidelines for the management of back pain emphasize a multimodal approach which includes addressing behavioral change. This is in part due to the chronification of pain being attributed to differences in central pain-processing systems as well as operant conditioning [29]. Pain neurophysiology education (PNE) is theorized to decrease sensitization and disrupt negative behaviors such as kinesiophobia and fear avoidance, which are both associated with an increased risk for the development of chronic pain. PNE consists of, but is not limited to, delineating between tissue damage and pain, the promotion of physical activity, and the avoidance of negative thoughts and stress through positive coping behaviors [30]. Pardo et al. found that patients who underwent PNE and therapeutic exercise (TE) demonstrated improved pain intensity, disability, psychosocial factors, and physical performance when compared to patients who received TE alone [31]. These results were maintained at 3-month follow up. Although pressure pain thresholds increased in both the combination group (PNE + TE) and the single intervention group (TE), the increase was greater in the combination group. The time and cost-effectiveness of this intervention is noteworthy as this study used only two group sessions of PNE compared to other studies using one-on-one education.

A characteristic of the human experience is our unique ability to communicate *how* pain impacts our daily lives. It is not enough for us to alert others to our pain, we also want them to *understand* through personal experience or analogy "what it is like to have *my* pain." Pain has an important role in human activity and in the organization of human reactions. Although pain is a personal experience, much more is at stake than the sensory registration of a private occurrence. Pain is associated with the strong need to *tell* someone else that the pain exists. The experience of pain is articulated with other experiences to which concepts can be applied. For example: "What has caused the pain?", "How long did it last?", "Is it sharp or dull, burning, boring, aching, stinging, throbbing or nagging?", and "Where do you feel the pain?" Pain is always articulated in a system of concepts by which we make conscious, intelligent, and informed responses to what is happening. This facilitates sharing of the experience with others and seeking help in understanding and managing the pain. Karos et al. argue that "pain is a fundamentally social and threatening human experience because it challenges several basic needs": the need for autonomy, the need to belong, and the need for justice/fairness [32, p. 1690]. The drive to reclaim these needs is an important theme for clinicians who treat chronic pain to know as it provides a basis for mutual understanding between patient and clinician.

Edward Bordin introduced the concept of a working alliance between a change agent and a change seeker. This union was described as “the collaborative and affective bond between therapist and client” [33]. Bordin identified the concept of a therapeutic alliance as being composed of three qualities: an agreement on goals, an assignment of task, and the development of bonds. Of particular salience is the position by Horowitz in which he claimed, “the therapeutic alliance is not only a prerequisite for therapeutic work, but often may be the main vehicle of change” [33, p. 255].

In 2018, Miciak reasoned the four central tenants of an effective alliance are presence, receptivity, a genuine demeanour, and a commitment to the patient [34]. Jorge Fuentes upheld the significance of a healthy working alliance between patients and clinicians when he assessed interferential current therapy on patients with low back pain with and without a strong alliance [35]. Not only did the results reveal a statistically significant difference in pressure pain thresholds in patients designated to the alliance group, but patient compliance to treatment was also strengthened compared to controls. Though interventions such as interferential current are discouraged in the management of acute or chronic low back pain [29], this work supports the notion that *contextual* factors may play a potent role in the perception of pain [35].

Clinician trust has also been linked to higher levels of self-efficacy [36]. According to Bandura et al. self-efficacy (SE) is defined as “personal confidence to carry out an activity with the aim of successfully achieving a desired outcome” [37]. Higher levels of SE are associated with a better prognosis in chronic musculoskeletal pain [38]. Additionally, SE facilitates treatment adherence and could possibly mitigate the transition from acute to chronic pain [38, 39].

Underpinning a strong alliance between a patient and clinician is the need to address any preconceived understanding of the patient’s condition [40]. A notion that Dr. Craig Brigham stressed when giving advice and direction to his spine care patients was using understandable language. Poor coping strategies, negative affect, and fear avoidance behaviors have all been independently associated with poor outcomes after a bout of low back pain [29, 41]. Equally as important as addressing misguided beliefs is understanding the role the clinician’s words play in the multifactorial aspects of recovery. Information conveyed to patients may inadvertently amplify symptoms and heighten distress [40, 42, 43]. Darlow et al. showed that off-hand statements made by clinicians can influence recovery expectations (an independent predictor of poor outcomes) [40]. Likewise, this work showed that clinicians can contribute to avoidance behaviors and disability by providing pathoanatomic explanations for back pain. “When clinicians focus on informing patients about what they should not do, this can give the impression that avoidance is important for recovery” [40]. A positive valence associated with particular avoidance behaviors (such as linking movement with pain) perpetuates operant conditioning, thus maintaining the undesirable behavior. Excessive reliance on avoidance behaviors increases pain-related fear [29].

Clinicians should elicit and promote life goals, which help provide the energy for patient action. However, the view that “controlling” pain is necessary in order to achieve other goals may be detrimental [44]. Healthcare professionals should appreciate that oversimplifying a person's unique experience of pain risks reinforcing an inflexible model of health behavior change. Importantly, the job of the clinician is instead to nurture change by identifying and promoting the amalgamation of factors unique to a person that strengthen their motivation to act. Several components of a successful rehabilitation process include being believed by health professionals, receiving support and adequate explanations, and acquiring strategies to influence the pain, disability and psychological distress [45]. Mark D. Sullivan argues that restoring the capacity for personally meaningful action in patients with chronic pain is what can change the individual from a patient back into a *person* [46]. Acceptance and Commitment Therapy (ACT) is an example of a behavioral approach that emphasizes realistic acceptance of function regardless of pain. ACT is now a recommended management approach showing efficacy in chronic pain.

Poor health literacy, defined as "the ability to seek, understand, and utilize health information" is associated with suboptimal medical condition knowledge as well as limited self-management skills [41]. Perhaps the greatest asset to a patient is not one particular intervention but a heightened understanding of their condition. An awareness of personal goals as a main focus of patient health should be endorsed by health professionals. Arguably, a patient becomes healthier when he or she is better able to pursue and achieve his or her vital personal goals [46].

Chronic low back pain remains an enigma for spine care professionals. Embodying this problem is a deficit in the ability to classify patients according to diagnostic, prognostic, or treatment-based parameters [47, 48]. Chronic low back pain subsumes a heterogeneous population of people with presentations that span a wide range of potential causes and influences. Historically, classifying patients based on a dualistic model of pathophysiology has been proposed and thus guided treatment. The body gets dissected into constituent parts where biological and psychological components become separate entities. Unfortunately, the confluence of coexisting conditions makes putative diagnoses such as "facet syndrome" ill-informed. Fairbank et al. [47] sought to unravel the paradox of chronic low back pain classification in 2011. He found that while current classification systems exist, few demonstrate validity and those that do show little to no difference in outcomes compared with controls [47]. Even more recently, Riley et al. found that "Studies with moderate to high methodological quality that have a low risk of bias do not support [movement-based classification] as being superior to general exercise or [guideline-based care] in the treatment of nonradicular [chronic] LBP" [49].

"The holy grail of classification systems is one that is comprehensive, easily applicable, reliable, directs treatment, and is predictive of outcomes of the directed treatment" [47, p. 41]. Annie O'Conner and Melissa C. Kolski highlight the potential utility of classification systems in their text, *A World of Hurt* [50]. They emphasize that musculoskeletal pain is perpetuated and sustained by copious factors ranging from socioeconomic status to genetic predisposition. Perhaps one of the more salient points introduced in their first chapter is the high variability in care, which classification systems ultimately attempt to reduce. There are many attempts at classifying pain by mechanism, biophysiological origin or amplification method, but so far only a few classification terms have been shown to have true discriminative validity in practice and none have yet been shown to be predictive of outcomes despite growing efforts. Promising results from a study published in *The Clinical Journal of Pain* showed preliminary discriminative validity of nociceptive, peripheral neuropathic, and central pain [51]. However, further validation is imperative before use in clinical practice is recommended.

Interestingly, medical specialty plays a role in the facilitation of diagnostic testing [52]. Cherkin and colleagues authored "Who You See Is What You Get," in which they noted that for similar patient presentations, neurosurgeons were more likely to order spinal images whereas physiatrists were three times more likely to order electrodiagnostic testing [53]. This can have negative downstream effects for patients and society. Of note, early investigation into putative "causes" of low back pain not only incur a significant cost to the patient but increased utilization of spine imaging in the United States. This correlates with a 2- to 3-fold increase in surgical rates [53].

7 Exercise in the Management of Spine Pain

Exercise is among the most effective strategies to reduce health-related disability and associated health care costs [54]. It has additionally been shown effective in the management of acute, subacute and chronic spine pain [55]. Cohen and Rainville explain: "By eliminating impairments in back function, altering fears and beliefs about pain, and reducing disability, patients with chronic low back pain can achieve meaningful improvements in their quality of life" [55]. James Rainville, MD, chief of Physical Medicine and Rehabilitation at New England Baptist Hospital (NEBH) and assistant professor at Harvard Medical School, leads a multimodal, interdisciplinary, exercise-based "Back Bootcamp" at NEBH. He and Lisa Childs, PT, founded the NEBH Spine Center in 1997 and have been offering their back-pain rehabilitation program since that time. Dr. Rainville explains to patients that back pain is especially challenging because symptomatic improvement is often not linked to a specific injury. Instead, he frames back pain as a neurological healing process rather than a structural one. This is evidenced most directly by the fact that back pain is seen in those with *and* without significant structural changes on diagnostic imaging.

Further evidence of the disconnect between structure and pain is seen with normal aging. The spine naturally degenerates as we age, and thus an entirely structural approach would suggest that pain increase with age. But this is not what we see. Instead, most back pain is reported in middle age. Adding to this, important work by Michele Crites Battie, Ph.D., has demonstrated that degenerative changes in the spine are determined primarily by genetics [56]. In 2017, Daniel Belavy et al. presented the first data in humans that exercise can be beneficial for intervertebral discs [57].

Dr. Rainville emphasizes that the nervous system must adapt to the pain, again explaining why patients with significant structural changes in their spine, including severe spinal deformities, may experience no pain at all. Their nervous systems have positively adapted. Consequently, an intensive pain rehabilitation program helps those whose nervous systems that have not adapted by a process of gradual recalibration through graded exposure.

Early movement-based therapy has been associated with a significant reduction in the usage of opioids and larger improvements in functional outcomes and pain ratings [58–66]. Multimodal functional restoration programs that depend on exercise-based therapies significantly improve physical and psychological variables and increase the number of those with chronic spine pain who return to work [67]. Interestingly, minimizing one's perception of their own functional disability improves treatment outcomes, suggesting that individual perceptions and experiences are necessary factors in obtaining a positive outcome [68]. Furthermore, considering individual factors such as one's circumstances in life (i.e., sociodemographics, workplace conditions, and individual motivation) is important in predicting treatment success [68]. This complements psychological variables known to portend poorer outcomes, such as anxiety, depression, and fear. These variables are associated with illness behaviors and poor self-efficacy in chronic pain [69, 70].

One area of success that speaks to the potential of exercise as therapy is its success in modifying the experience of those with fibromyalgia. A recent Cochrane review evaluated the literature on aerobic exercise in fibromyalgia. When comparing aerobic exercise to standard care, moderate-quality evidence was found to suggest that aerobic exercise improved health-related quality of life (HRQL) and that it was beneficial for pain and function over the long-term. Notably, no significant differences between types of aerobic exercises were found. When comparing aerobic exercise to non-exercise-based therapies, exercise showed statistically significant improvement in HRQL and pain [71] (Fig. 5.2). As Nijs et al. [73] explain:

> Applying contemporary pain neuroscience to improve care for people with CLBP includes identifying relevant pain mechanisms to steer intervention, addressing sleep problems and optimising exercise and activity interventions. This approach includes cognitively preparing patients for exercise therapy using (therapeutic) pain neuroscience education, followed by cognition-targeted functional exercise therapy [73, p. 108].

8 The Science of Exercise in Spine Pain Management

The neurobiological underpinnings of exercise-induced analgesia have been well studied recently [74]. In the presence of tissue injury, neurological changes in the CNS and peripheral nervous system (PNS) occur and may amplify pain processing [10]. Preliminary human studies and animal studies have demonstrated that physical exercise can alter nociception both directly, through nervous system pathways, and indirectly, through alterations to the immune system.

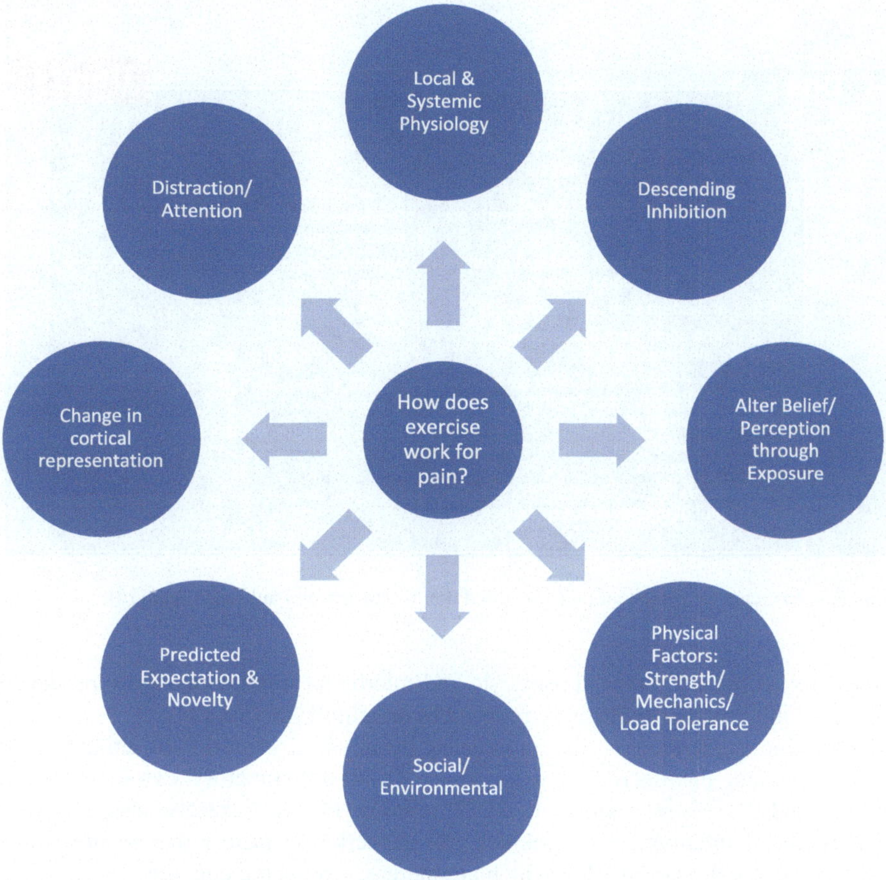

Fig. 5.2 How does exercise work for pain? (Image by Ben Cormack [72])

The immune system is modulated by physical exercise, particularly the balance between inflammatory and anti-inflammatory phenotypes has been demonstrated to be strongly affected by exercise [74]. Sedentary conditions are associated with a preponderance of inflammatory cytokines, whereas physical activity is associated with the opposite. This effect is seen in both the periphery and in the CNS. Physically active animals with nerve injuries demonstrate a higher proportion of M2 (anti-inflammatory cytokine releasing) to M1 (inflammatory cytokine releasing) type macrophages in their muscles, whereas the opposite is seen in injured animals that are sedentary.

In the CNS, neuroglial cells are largely responsible for modulating both inflammatory and anti-inflammatory cytokines. Exercise seems to push the body towards a more anti-inflammatory state, as reduced glial cell activation, reduced inflammatory cytokines, and increased anti-inflammatory cytokines were observed in the DRG in animals with nerve injuries that engaged in physical activity, in comparison to

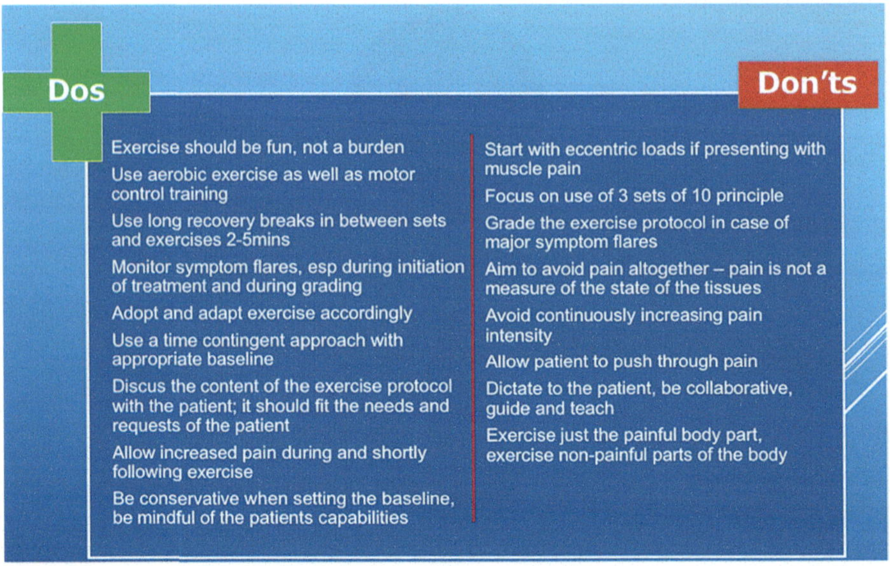

Fig. 5.3 Exercise for persistent pain: do's and don'ts. (Image by Paul Lagerman [76])

sedentary animals. Enhanced astrocyte and microglial immunoreactivity produced by nerve injury was significantly reduced by treadmill running [74].

Cytokine expression is also modulated by physical exercise. In animals with nerve damage, the decreased expression of anti-inflammatory cytokines (IL-4, IL-1ra, and IL-5) is reversed by treadmill running. Physical exercise also decreases expression of inflammatory cytokines and transcription factors that regulate IL-1 beta, NF kappaB, and NLRP3, which are increased by nerve damage [74].

The sedentary condition has been associated with multiple changes in the CNS that lead to increased sensitivity to pain. Exercise programs have been long associated with decreased pain, and recent work by Alzahrani et al. has shown an inverse association between the volume of physical activity in humans and the odds of developing chronic spine pain [75]. A meta-analysis of highly active athletes compared to normally active adults demonstrated reduced pain sensitivity overall in the athletes, and a 12-week exercise program demonstrated increased pain thresholds and decreased temporal summation in people with osteoarthritis [74] (Fig. 5.3).

In addition to exercise programs, psychological interventions, such as CBT are associated with decreased pain and better outcomes for patients with chronic pain, thus multimodal pain programs that include exercise commonly integrate CBT into their rehabilitative approach. A review article that assessed the role of CBT in pain management in non-specific low back pain concluded that "CBT is a beneficial treatment for chronic nonspecific low back pain" and that integrating CBT with other approaches may "represent the future direction of management of chronic back pain" [77].

Psychological well-being has only more recently risen to such prominence in the literature in the consideration of its role in pain management. As an example, one summary of Cochrane reviews to determine the effectiveness of physical activity and exercise interventions for adults with chronic pain found that "only 5 of the 21 reviews had psychological well-being as a variable" [78]. However, things are clearly changing in the literature. A randomized controlled trial (RCT) from 2013 found that exercise and CBT together, as part of a multi-modal intervention, was superior than "exercise alone in improving quality of life and reducing disability and fear avoidance beliefs" [79]. Some authors have posited that exercise "produces positive benefits that subsequently promote cognitive restructuring, increases self-efficacy by encouraging self-management, attenuates rumination through increases attentional demands of exercise, and decreases pain through activation of descending inhibitory systems" [74].

When it comes to exercise, no specific type has shown superiority over another. As Rainville et al. state, "The evidence-based medicine paradigm would suggest that, in the absence of strong evidence for the superiority of any [specific] exercise strategies, the decision for which specific type of exercise should be a matter of patient preference and clinician experience" [3]. Instead, reinforcing exercise participation by finding ways to promote the corresponding health behavior change may be more important than finding the "right" exercise [80]. Heather Kroll, MD, suggests that addressing the following can improve exercise participation (Fig. 5.4):

- Identify the baseline level of function at which the patient can be successful.
- Determine a rate of progression that is safe and achievable.
- Stick to a realistic schedule each day.
- Redirect the patient to achievable levels of exercise participation if they fall off course.
- Remind patients that activity levels are not pain-contingent, and that a major goal is to remove pain as the guiding control over their activity level. Instead, control should come from a rational process of understanding and reasonable activity in their daily lives [81].

9 Medical Treatments and Interventions: When Is It the Right Time?

> We must all die. But that I can save him from days of torture, that is what I feel is my great and ever new privilege.
>
> —Albert Schweitzer.

When considering the individual's physical and emotional experience of pain, along with social and economic costs of pain, it is not surprising why physicians have made continual efforts to understand and control it [83]. Historically, it has been the perceived burden to find and treat pain *at its cause* that has ultimately led to

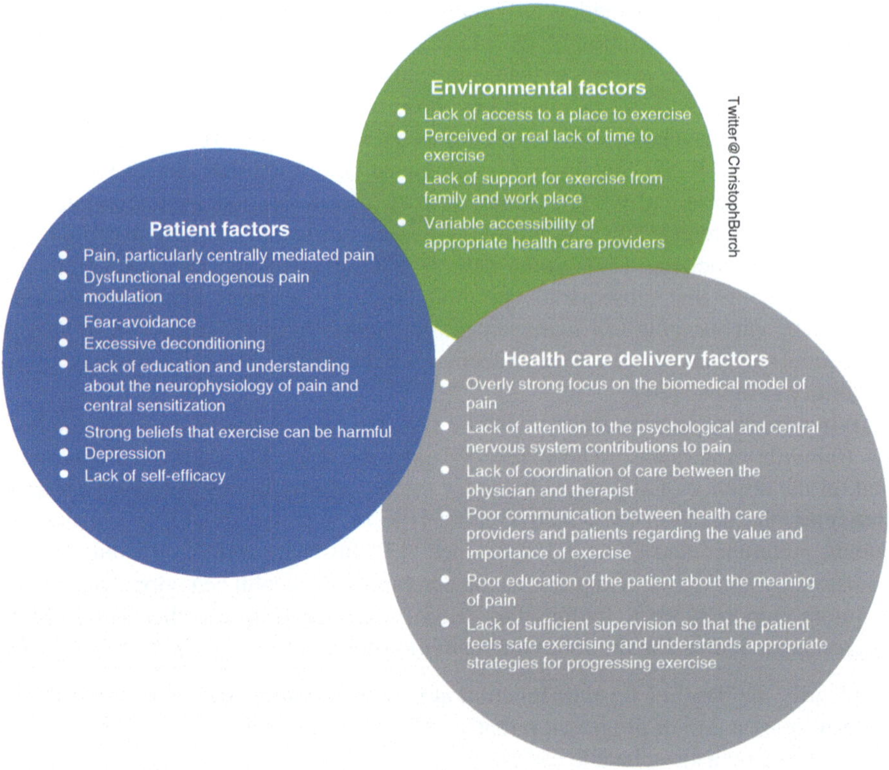

Fig. 5.4 Exercise therapy for chronic pain. Barriers to exercise participation. Image by Christoph Burch [82]

the medical subspecialty of interventional pain management. Since inception to date, it remains uncertain how interventional spine procedures fit within our ever-evolving understanding of the complexities of back pain. More specifically, what is the clinical utility of interventional spine procedures within an interdisciplinary model using a biopsychosocial approach for the management of back pain? To better answer these questions, let us first consider the history and evolution of regional anesthesia.

The origins of interventional pain management can be traced back to the first applications of regional anesthesia in the late 1800s. It is not surprising that physicians of the day were diligently focused on finding anatomic structures which could be the potential causes of pain. The predominant theory of pain origin during this period was the previously mentioned "bottom-up" theory centered around Descartes' idea that nociception should be viewed as a warning signal conveyed by specific nerve fibers from the periphery to the brain [83]. Using this simplistic cause and effect model, physicians suspected that nociceptors could be isolated and their signal altered to provide relief from the experience of pain. The first clinical application of this theory has been credited to Dr. Karl Koller, a colleague of Sigmund Freud. As a practicing ophthalmologist, Dr. Koller demonstrated how topical application of cocaine could interrupt nociception and cause local anesthesia. Not long after this discovery, the first clinical operation under regional anesthesia was performed in 1814 using cocaine superficially on the eye [84]. Word of his success spread rapidly throughout the medical community. In turn, others began to conceptualize new applications for chemical anesthetics which could be aimed at targeting pain inducing structures throughout the body.

Given the frequency of back pain, interventional pain procedures would eventually move on to include the spine. Improved understanding of clinical anatomy increased the application of injectable anesthetic compounds to include intraspinal injections. As early as 1899, descriptions of therapeutic nerve blocks emerged touting successful treatment of sarcoma-related pain in the leg using spinally administered cocaine [85]. As it became more widely understood that spinal nerves correspond to areas of pain as felt throughout the body, the key pieces of information were in place to lead to the discovery and implementation of the modern diagnostic nerve blocks and intraspinal injections. Physicians now understood that all sensory nerves converge within the spinal cord and the spine became a prime target for physicians looking to permanently alleviate and control pain throughout the body. However, the effectiveness of interventional spine procedures hinged on the concept that pain exists solely as a bottom-up phenomenon driven by tissue pathology. As spinal procedures evolved so did our understanding of the complex nature of spine related pain. This growing understanding called into question the clinical utility of interventional spine procedures as they were initially intended.

The notion that spine related pain can be effectively managed with treatments that solely focus on a "bottom-up" approach to pain control has not been supported by the literature. This observation, in conjunction with the temporary nature of interventional spine procedures has led to the recent criticisms of their place within modern treatment of back pain. Yet, from a clinical perspective, it can be argued that their limitations are also in part their strength. Due to their temporary nature, perhaps the most useful clinical outcome of interventional spine procedures is their ability to target, and more specifically diagnose, painful structures within the spine. Within the context of better patient care, helping to establish the cause of pain is helpful for building a trusting relationship and fosters meaningful patient-clinician engagement.

Often the history and clinical examination of musculoskeletal conditions can lead to very broad differentials with several possible structures contributing to what the patient experiences as pain [83]. From the perspective of a clinician, it is pivotal to pinpoint an anatomical structural diagnosis and identify one or more painful structures for the proper management of chronic pain [86]. This is especially true in cases, such as low back pain, where diagnostic evaluation using history, physical examination, and imaging may provide limited information. Patients with non-specific back pain are often fearful, confused, or frustrated by the regular occurrence of pain where a specific pain generator cannot be identified. This uncertainty can lead to rumination, or fear avoidance behaviors, that interrupt rehabilitation efforts and foster negative beliefs about the nature of their back pain and chances for full recovery. Contemporary diagnostic interventional spine procedures can fit within the biopsychosocial model through their ability to provide invaluable clinical information. Through the accurate identification of a specific pain generator when present, patients can be educated and reassured about the prognosis of their condition, which is one of the cornerstones of an active care paradigm.

What is important to appreciate is that many of the modern procedures regularly utilized by interventional spine and pain physicians to treat back pain are often diagnostic by nature. This simply means that they are purposefully designed to isolate a specific anatomic structure and determine whether that structure is contributing to the patient's symptomatology. This includes targeting common sources of back pain with procedures such as medial branch blocks, epidural steroid injections, provocative discography, and sacroiliac joint blocks. Although this list is not all inclusive, it can help us to see how pain physicians may utilize these procedures within an interdisciplinary model of treating pain to rule out, or possibly rule in, causes of back pain.

One of the technical limitations of interventional spine procedures, as previously mentioned, is the temporary nature of pain relief provided due to the pharmacokinetics of the nerve blocking agents. Steroids have often been added to the injectate to extend the duration of pain control. However, since the introduction of neuraxial steroids, there has been controversy with regards to their pharmacologic mechanism of action, neural toxicity, and other side effects and complications [87]. As a result, many interventional spine practitioners hope that newer technologies may provide their patients with more sustained relief. This includes regenerative medicine procedures, neuroablative technology, as well as neuromodulation through a variety of means including spinal cord or dorsal root ganglion stimulation. In contrast with injections, radiofrequency ablation utilizes heat energy to create a semi-permanent lesion within sensory nerves to provide a more sustained and controlled pain relief. Neuromodulation, in the form of spinal cord stimulators, has also shown promise for long term modulation of some types of chronic pain. Better understanding of clinical anatomy, coupled with a desire to provide sustained relief of pain, has given rise to several promising new technologies that will undoubtedly influence the future utilization of interventional spine procedures.

Although many interventional spine procedures discussed are not curative, that should not negate their clinical utility within the context of an interdisciplinary model of pain management. There is tremendous clinical benefit in alleviating patient's uncertainty and concerns while simultaneously determining the source(s) of their pain. In addition to their diagnostic benefit, interventional procedures can provide targeted treatment to painful structures within the spine. In these situations, the temporary and focal relief of pain provided can promote greater efforts in ongoing rehabilitation. The ability to complete rehabilitation with less, or in some cases no pain, can reinforce healthy return to activity and help eliminate fear avoidance behaviors. In addition, interventional procedures can reduce the need for opioid pain medications when used judiciously [88, 89]. For patients hoping to decrease their dependence on oral pain medications, interventional spine procedures may offer hope of a future free of sedating medications and those with undesirable effects. Given their role in establishing a clearer diagnosis, coupled with an ability to mitigate pain and facilitate ongoing rehabilitation and conservative treatment, interventional spine procedures should be viewed as a useful treatment option within a modern interdisciplinary approach to pain. As newer technologies lead to more sustained relief, it may create an opportunity to expand the role interventional spine procedures play within the interdisciplinary model of spine care.

10 Conclusion

Taking our understanding of spine pain and applying it to patients remains the most important step in successful implementation of a spine pain rehabilitation program. We have identified several key aspects to a patient-centered, active rehabilitation approach to persistent spine pain:

1. An appropriate understanding of pain science that embraces an understanding of pain as a biopsychosocial phenomenon that cannot be reduced to structure alone;
2. A focus on the therapeutic alliance and the patient's functional needs and goals;
3. A rehabilitation strategy that employs exercise and cognitive-behavioral strategies as a means of decreasing or reversing maladaptive neurological changes; and
4. Adding the judicious use of medical treatments and procedures to rehabilitation strategies when there are clear indications to do so.

Though our understanding of spine pain and its management through rehabilitation should inspire optimism about the future of care for persons with spine pain, there are inherent limitations to our analysis. The subjective nature of pain and the outcome measures associated with it naturally limit our ability to define progress objectively. The evaluation of *pain* depends on the personal experience by the person with pain, not merely on the phenomenological interpretations of the clinical observers. We can, of course, evaluate improvement in function (i.e., the ability to perform self-care, ADLs, and family, occupational, recreational, and community roles). This focus on functional improvement is largely considered a satisfactory surrogate that suits the need for objective measurement. Nonetheless, bias remains a problem in the study of multimodal interventions like spine pain rehabilitation where placebo controls and the identification of specific elements of care that have the greatest effect are often not feasible. This is due to the interplay between numerous factors that are required to successfully address the experience of pain, which is itself multifactorial (biological, psychological, sociological).

Despite limitations and the need for greater reconciliation between biological and psychosocial factors through translational research, the future of spine pain care is bright. The opioid crisis gave us the opportunity to own our mistakes and learn that there is no single element that can palliate the experience of pain. It is an inherently and deeply interdependent phenomenon that requires respect. Such respect can be found through interdisciplinary, multimodal spine pain rehabilitation programs and spread through continued education for clinicians, patients, and communities alike.

Spine pain rehabilitation programs allow the cultivation of the skills needed to assist patients as they reframe their pain ("subjective") and learn new ways to prevent disability, restore function in their lives, and subsequently achieve their goals ("objective") by changing the activity of their nervous system.

References

1. Main CJ, Foster N, Buchbinder R. How important are back pain beliefs and expectations for satisfactory recovery from back pain? Best Pract Res Clin Rheumatol. 2010;24(2):205–17.
2. Borrell-Carrió F, Suchman AL, Epstein RM. The biopsychosocial model 25 years later: principles, practice, and scientific inquiry. Ann Fam Med. 2004;2:576–82.
3. Rainville J, Hartigan C. Exercise-based therapies for low back pain. In: Atlas SJ, section editor. UpToDate [Internet]. Waltham (MA): UpToDate Inc; 2019. [updated 14/02/2019; cited 05/03/2019]. Available from https://www.uptodate.com/contents/exercise-based-therapy-for-low-back-pain
4. Brinjikji W, Luetmer PH, Comstock B, Bresnahan BW, Chen LE, Deyo RA, et al. Systematic literature review of imaging features of spinal degeneration in asymptomatic populations. Am J Neuroradiol. 2015;36(4):811–6.
5. Videman T, Battie MC, Gibbons LE, Manninen H, Kapiro J. Associations between back pain history and lumbar MRI findings. Spine. 2003;28:582–8.
6. International Association for the Study of Pain (IASP). IASP Terminology [Internet]. 2017 [cited 25/02/2019]. Available from https://www.iasp-pain.org/Education/Content.aspx?ItemNumber=1698
7. Cohen M, Quinter J, van Rysewyk S. Reconsidering the IASP definition of pain. Pain Rep. 2018;3(2):e634.
8. Treede RD. The international association for the study of pain definition of pain. Pain Rep. 2018;3(2):e643.
9. Aydede M. Defending the IASP definition of pain. Monist. 2017;100(4):439–64.
10. Woolf CJ. What is this thing called pain? J Clin Investig. 2010;120(11):3742–4.
11. Dickenson A. Pain: the neurobiology of chronic pain states. Anaesth Intensive Care Med. 2016;17:436–9. https://doi.org/10.1016/j.mpaic.2016.06.010.
12. Taylor AMW. Corticolimbic circuitry in the modulation of chronic pain and substance abuse. Prog Neuropsychopharmacol Biol Psychiatry. 2018;87:263–8.
13. Lumley MA, Cohen JL, Borszcz GS, Cano A, Radcliffe AM, Porter LS. Pain and emotion: a biopsychosocial review of recent research. J Clin Psychol. 2011;67(9):942–68.
14. Lewis M. Brain change in addiction as learning, not disease. N Eng J Med. 2018;379(16):1551–60.
15. Hashmi JA, Baliki MN, Huang L, Baria AT, Torbey S, Hermann KM, et al. Shape shifting pain: chronification of back pain shifts brain representation from nociceptive to emotional circuits. Brain. 2013;136:2751–68.
16. Koban L, Jepma M, Geuter S, Wager TD. Review article: what's in a word? How instructions, suggestions, and social information change pain and emotion. Neurosci Biobehav Rev. 2017;81(Part A):29–42.
17. Spiegel D. Modulation of nociception in multiple brain systems-the strain in pain. JAMA Neurol. 2018;75(11):1309–10.
18. Price DD, Finniss DG, Benedetti F. A comprehensive review of the placebo effect: recent advances and current thought. Ann Rev Psychol. 2008;59(1):565–90.
19. Eklund A, De Carvalho D, Pagé I, Wong A, Johansson MS, Pohlman KA, et al. Expectations influence treatment outcomes in patients with low back pain. A secondary analysis of data from a randomized clinical trial. Eur J Pain. 2019;23(7):1378–89.

20. Campbell P, Shraim M, Jordan KP, Dunn KM. In sickness and in health: a cross-sectional analysis of concordance for musculoskeletal pain in 13,507 couples. Eur J Pain. 2016;20 (3):438–46.
21. Chen Y, Campbell P, Strauss VY, Foster NE, Jordan KP, Dunn KM. Trajectories and predictors of the long-term course of low back pain: cohort study with 5-year follow-up. Pain. 2018;159 (2):252–60.
22. Haldeman S, Dagenais S. A supermarket approach to the evidence-informed management of chronic low back pain. Spine J. 2008;8:1–7.
23. World Health Organization. International classification of functioning, disability & health (ICF). WHO: Geneva; 2001.
24. Dima A, Lewith GT, Little P. Identifying patients' beliefs about treatments for chronic low back pain in primary care: a focus group study. Br J Gen Pract. 2013;63(612):e490–8.
25. Chou L, Ranger TA, Peiris W, Cicuttini FM, Urquhart DM, Sullivan K. Patients' perceived needs of health care providers for low back pain management: a systematic scoping review. Sport J. 2018;18(4):691–711.
26. National Institute for Clinical Excellence. Low back pain and sciatica in over 16s: assessment and management. London: NICE; 2016.
27. Qaseem A, Wilt TJ, McLean RM, Forciea MA. Noninvasive treatments for acute, subacute and chronic low back pain: a clinical practice guideline from the American college of physicians. Ann Intern Med. 2017 Apr;166(7):514–30.
28. Hartvigsen J, Hancock MJ, Kongsted A, Louw Q, Ferreira ML, Genevay S, et al. What low back pain is and why we need to pay attention. Lancet. 2018;391(10137):2356–67.
29. Vlaeyen JWS, Maher CG, Wiech K, van Zundert J, Meloto CB, Diatchenko L. Low back pain. Nat Rev Dis Primers. 2018;14(1):52.
30. Pain neuroscience education: slides for supporting/illustrating your explanation. Pain in Motion [Internet]. 2016 [cited 13/04/2019]. Available from http://www.paininmotion.be/education/tools-for-clinical-practice
31. Bodes Pardo G, Lluch Girbés E, Roussel NA, Gallego Izquierdo T, Jiménez Penick V, Pecos Martín D. Pain neurophysiology education and therapeutic exercise for patients with chronic low back pain: a single-blind randomized controlled trial. Arch Phys Med Rehabil. 2018;99 (2):338–47.
32. Karos K, Williams AC, Meulders A, Vlaeyen JWS. Pain as a threat to the social self: a motivational account. Pain. 2018;159(9):1690–5.
33. Bordin ES. The generalizability of the psychoanalytic concept of the working alliance. Psychother Theory Res Pract. 1979;16(3):252–60.
34. Miciak M, Mayan M, Brown C, Joyce AS, Gross DP. The necessary conditions of engagement for the therapeutic relationship in physiotherapy: an interpretive description study. Arch Physiother. 2018;8(1):3.
35. Fuentes J, Armijo-Olivo S, Funabashi M, Miciak M, Dick B, Warren S, et al. Enhanced therapeutic alliance modulates pain intensity and muscle pain sensitivity in patients with chronic low back pain: an experimental controlled study. Phys Ther. 2014 Apr;94(4):477–89.
36. Lee YY, Lin JL. The effects of trust in physician on self-efficacy, adherence and diabetes outcomes. Soc Sci Med. 2009;68(6):1060–8.
37. Bandura A. Self-efficacy: toward a unifying theory of behavioral change. Psychol Rev. 1977;84 (2):191–215.
38. Martinez-Calderon J, Zamora-Campos C, Navarro-Ledesma S, Luque-Suarez A. The role of self-efficacy on the prognosis of chronic musculoskeletal pain: a systematic review. J Pain. 2018;19(1):10–34.
39. Berube M, Choinière M, Laflamme YG, Gélinas C. Acute to chronic pain transition in extremity trauma: a narrative review for future preventative interventions (part 2). Int J Orthop Trauma Nurs. 2017;24:59–67.
40. Darlow B, Dowell A, Baxter GD, Mathieson F, Perry M, Dean S. The enduring impact of what clinicians say to people with low back pain. Ann Fam Med. 2013;11(6):527–34.

41. Briggs AM, Jordan JE, Buchbinder R, Burnett AF, O'Sullivan PB, Chua JY, et al. Health literacy and beliefs among a community cohort with and without chronic low back pain. Pain. 2010;150(2):275–83.
42. Barsky AJ. The Iatrogenic potential of the physician's words. JAMA. 2017;318(24):2425–6.
43. Darlow B. Beliefs about back pain: the confluence of client, clinician and community. Int J Osteopath Med. 2016;20:53–61.
44. Crombez G, Lauwerier E, Goubert L, van Damme S. Goal pursuit in individuals with chronic pain: a personal project analysis. Front Psych. 2017;7:966.
45. Biguet G, Nilsson Wikmar L, Bullington J, Flink B, Löfgren M. Meanings of "acceptance" for patients with long-term pain when starting rehabilitation. Disabil Rehabil. 2016;38 (13):1257–67.
46. Sullivan M. The patient as agent of health and health care. Autonomy in patient-centered care for chronic conditions. Corby: Oxford University Press; 2017. 487 p.
47. Fairbank J, Gwilym SE, France JC, Daffner SD, Dettori J, Hermsmeyer J, et al. The role of classification of chronic low back pain. Spine (Phila Pa 1976). 2011;36(21 Suppl):S19–42.
48. Kamper SJ, Maher CG, Hancock MJ, Koes BW, Croft PR, Hay E. Treatment-based subgroups of low back pain: a guide to appraisal of research studies and a summary of current evidence. Best Pract Res Clin Rheumatol. 2010;24(2):181–91.
49. Riley SP, Swanson BT, Dyer E. Are movement-based classification systems more effective than therapeutic exercise or guideline based care in improving outcomes for patients with chronic low back pain? A systematic review. J Man Manip Ther. 2018;27(1):5–14.
50. O'Connor A, Kolski M. A world of hurt: a guide to classifying pain. St. Louis, MO: Thomas Land Publishers; 2015. 367 p.
51. Smart KM, Blake C, Staines A, Doody C. The discriminative validity of "nociceptive," "peripheral neuropathic," and "central sensitization" as mechanisms-based classifications of musculoskeletal pain. Clin J Pain. 2011;27(8):655–63.
52. Flynn TW, Smith B, Chou R. Appropriate use of diagnostic imaging in low back pain: a reminder that unnecessary imaging may do as much harm as good. J Orthop Sports Phys Ther. 2011;41(11):838–46.
53. Cherkin DC, Deyo RA, Wheeler K, Ciol MA. Physician variation in diagnostic testing for low back pain. Who you see is what you get. Arthritis Rheum. 1994;37(1):15–22.
54. Alford L. What men should know about the impact of physical activity on their health. Int J Clin Pract. 2010;64(13):1731.
55. Cohen I, Rainville J. Aggressive exercise as treatment for chronic low back pain. Sports Med. 2002;32(1):75–82.
56. Battie MC, Videman T, Kaprio J, Gibbons LE, Gill K, Manninen H, et al. The twin spine study: contributions to a changing view of disc degeneration. Spine J. 2009;9(1):47–59.
57. Belavy DL, Quittner MJ, Ridgers N, Ling Y, Connell D, Rantalainen T. Running exercise strengthens the intervertebral disc. Sci Rep. 2017;7:45975.
58. Horn ME, George SZ, Fritz JM. Influence of initial provider on health care utilization in patients seeking care for neck pain. Mayo Clin Proc Innov Qual Outcomes. 2017;1(3):226–33.
59. Horn ME, Brennan GP, George SZ, Harman JS, Bishop MD. A value proposition for early physical therapist management of neck pain: a retrospective cohort analysis. BMC Health Serv Res. 2016;16:253.
60. Stevans JM, Fitzgerald GK, Piva SR, Schneider M. Association of early outpatient rehabilitation with health service utilization in managing Medicare beneficiaries with nontraumatic knee pain: retrospective cohort study. Phys Ther. 2017;97(6):615–24.
61. Childs JD, Fritz JM, Wu SS, Flynn TW, Wainner RS, Robertson EK, et al. Implications of early and guideline adherent physical therapy for low back pain on utilization and costs. BMC Health Serv Res. 2015;15:150.
62. Liu X, Hanney WJ, Masaracchio M, Kolber MJ, Zhao M, Spaulding AC, et al. Immediate physical therapy initiation in patients with acute low back pain is associated with a reduction in downstream health care utilization and costs. Phys Ther. 2018;98(5):336–47.

63. Sun E, Moshfegh J, Rishel CA, Cook CE, Goode AP, George SZ. Association of early physical therapy with long-term opioid use among opioid-naive patients with musculoskeletal pain. JAMA Netw Open. 2018;1(8):e185909.
64. Huffman KL, Rush TE, Fan Y, Sweis GW, Vij B, Covington EC, Scheman J, Mathews M. Sustained improvements in pain, mood, function and opioid use post interdisciplinary pain rehabilitation in patients weaned from high and low dose chronic opioid therapy. Pain. 2017;158(7):1380–94.
65. Gilliam WP, Craner JR, Cunningham JL, Evans MM, Luedtke CA, Morrison EJ, Sperry JA, Loukianova LL. Longitudinal treatment outcomes for an interdisciplinary pain rehabilitation program: comparisons of subjective and objective outcomes on the basis of opioid use status. J Pain. 2018;19(6):678–89.
66. Gagnon CM, Scholten P, Atchison J. Multidimensional patient impression of change following interdisciplinary pain management. Pain Pract. 2018;18(8):997–1010.
67. Hildebrandt J, Pfingsten M, Franz C, Saur P, Seeger D. Multidisciplinary treatment program for chronic low back pain, part 1. Schmerz [Article in German]. 1996;10(4):190–203.
68. Pfingsten M, Hildebrandt J, Saur P, Franz C, Seeger D. Multidisciplinary treatment program on chronic low back pain, part 4. Prognosis of treatment outcome and final conclusions. Schmerz [Article in German]. 1997;11(1):30–41.
69. Ciccione DS, Just N. Pain expectancy and work disability in patients with acute and chronic pain: a test of the fear avoidance hypothesis. J Pain. 2001;2:181–94.
70. Martinez-Calderon J, Zamore-Campost C, Navarrow-Ledesma S, Luque-Suarez A. The role of self-efficacy on the prognosis of chronic musculoskeletal pain: a systematic review. J Pain. 2018;19(1):10–34.
71. Bidonde J, Busch AJ, Schachter CL, Overend TJ, Kim SY, Góes SM, et al. Aerobic exercise training for adults with fibromyalgia. Cochrane Database Syst Rev. 2017;6:CD012700.
72. Cormack B. How does exercise work for pain? Cor-Kinetic [Internet]. 2018 [cited 13/04/2019]. Available from https://cor-kinetic.com
73. Nijs J, Clark J, Malfliet A, Ickmans K, Voogt L, Don S, et al. In the spine or in the brain? Recent advances in pain neuroscience applied in the intervention for low back pain. Clin Exp Rheumatol. 2017;107(5):108–15.
74. Kathleen AS, Laura F-L, Marie HB. Exercise-induced pain and analgesia? Underlying mechanisms and clinical translation. Pain. 2018;1:S91.
75. Alzahrani H, Shirley D, Cheng SWM, Mackey M, Stamatakis E. Physical activity and chronic back conditions: a population-based pooled study of 60,134 adults. J Sport Health Sci. 2019;8 (4):386–93.
76. Lagerman P. Exercise for persist pain: do's and don'ts [Internet]. 2018 [cited 13/04/2019]. Available from https://thenakedphysio.com
77. Sveinsdottir V, Eriksen HR, Reme SE. Assessing the role of cognitive behavioral therapy in the management of chronic nonspecific back pain. J Pain Res. 2012;5:371–80.
78. Geneen LJ, Moore RA, Clarke C, Martin D, Colvin LA, Smith BH. Physical activity and exercise for chronic pain in adults: an overview of Cochrane reviews. Cochrane Database Syst Rev. 2017;4:CD011279.
79. Monticone M, Ferrante S, Rocca B, Baiardi P, Farra FD, Foti C. Effect of a long-lasting multidisciplinary program on disability and fear-avoidance behaviors in patients with chronic low back pain: results of a randomized controlled trial. Clin J Pain. 2013;29(11):929–38.
80. Bennell KL, Dobson F, Hinman RS. Exercise in osteoarthritis: moving from prescription to adherence. Best Pract Res Clin Rheumatol. 2014;28(1):93–117.
81. Kroll HR. Exercise therapy for chronic pain. Phys Med Rehabil Clin N Am. 2015;26 (2):263–81.
82. Burch C. Exercise therapy for chronic pain [Internet]. Available by request from: @christophburch on Twitter.
83. Manchikanti L, Boswell MV, Raj PP, Racz GB. Evolution of interventional pain management. Pain Physician. 2003;6(4):485–94.

84. Brown DL, Fink B. The history of neural blockade and pain management. In: Cousins MJ, Bridenbaugh PO, editors. Neural blockade in clinical anesthesia and management of pain. 3rd ed. Philadelphia: Lippincott-Raven; 1998. p. 3–34.
85. Tuffer T. Anesthesie medullaire chirurgicale par injection sousarachnoidienne lombiare de cocaine; technique et results. Sem Med. 1900;20:167.
86. Deyo RA, Rainville J, Kent DL. What can the history and physical examination tell us about low back pain? JAMA. 1992;268:760–5.
87. Manchikanti L. Role of neuraxial steroids in interventional pain management. Pain Physician. 2002;5(2):182–99.
88. Sharan AD, Riley J, Falowski S, Pope JE, Connolly AT, Karst E, et al. Association of opioid usage with spinal cord stimulation outcomes. Pain Med. 2018;19(4):699–707.
89. Gilmore CA, Kapural L, McGee MJ, Boggs JW. Percutaneous peripheral nerve stimulation (PNS) for the treatment of chronic low back pain provides sustained relief. Neuromodulation. 2018;18(1):94–103.

Chapter 6
"Let Me Be a Meaningful Part in the Outside World": A Caring Perspective on Long-Term Rheumatic Pain and Fear-Avoidance Beliefs in Relation to Body Awareness and Physical Activities

Helena Lööf

Abstract Pain is a key outcome measure in persons with rheumatoid arthritis (RA) or psoriatic arthritis (PsA). Improving physical and social functioning is an important treatment goal in patients with RA or PsA. High self-rated pain is linked to elevated fear-avoidance beliefs for engaging in physical activity. Some research has demonstrated that when people's attitudes and beliefs towards pain are negative, they often have elevated levels of body awareness. Many persons with moderate to severe self-rated rheumatic pain report a desire to be part of a wider social and active context, which is perceived to be a source of positive energy and meaning in life. Patients have suggested that physical activities can mediate pain reduction and a distraction from it. During periods of solitude or episodes of sedentary behaviour, inner bodily attention in some patients with RA or PsA is often directed towards the "problematic body" and self. The perceived "untrustworthy body" has to deal with many bodily memories and emotions. Contemplating the past (the former perceived "healthy body") may interrupt a person's ability to live life and trying to determine a future life with an unreliable body presents concerns for future consequences. This behavioural pattern can lead to maintained negative fear-avoidant behaviour and a more inactive life-style, resulting in new negative existential meanings of loss and helplessness. Despite suffering from rheumatic pain in this way, these patients point out their desire to participate in a wider social and active context without being subject to any form of discrimination, which is perceived as the main and new source of positive energy and meaning in life.

Clinical Implications: There is a common meaning of bodily intrusion and threat in persons with long-term rheumatic pain. Registered nurses can play a crucial role by helping their patients identify new understandings and cognitive meanings in

H. Lööf (✉)
Division of Caring Sciences, Mälardalens University (MDH), School of Health, Care and Social Welfare, Västerås, Sweden

Sophiahemmet University, Stockholm, Sweden
e-mail: Helena.loof@mdh.se

S. van Rysewyk (ed.), *Meanings of Pain*,
https://doi.org/10.1007/978-3-030-24154-4_6

relation to long-term rheumatic pain. Living a meaningful life in the active outside world is paramount for improving and maintaining general health and wellbeing in persons with RA and PsA. A sedentary lifestyle together with long periods of solitude can lead to additional health problems and negative meanings.

Keywords Body awareness · Fear-avoidance beliefs · Long-term rheumatic pain · Meanings of pain · Physical activity

1 Rheumatic Pain

Rheumatoid arthritis (RA) [1–3] and psoriatic arthritis (PsA) [4–6] are common inflammatory rheumatic diseases in the general population. RA and PsA are characterised by stiffness, pain, swelling and tenderness of the joints and surrounding ligaments and tendons. However, individuals with PsA are likely to suffer from psoriasis before joint disease has developed. Pain is an important outcome measure in people with RA and PsA. RA is characterised by symmetric and erosive arthritis that typically involves small and medium-sized joints [1, 3]. PsA may present as mild monoarthritis or severe polyarthritis (it may also involve the axial skeleton and the entheses). PsA is common among patients with psoriasis and may result in permanent joint damage and disability [4, 5]. Moreover, PsA and RA are disorders with other health consequences (beyond joint function and joint destruction). These rheumatic diseases can damage different organs, including skin, eyes and blood vessels. Persons with RA and PsA have an increased risk of developing other diseases (e.g., cardiovascular disease and type 2 diabetes) [7].

Early diagnosis and treatment can relieve pain and inflammation, which may achieve disease remission and prevent progressive joint involvement and damage [3, 6, 8]. Evidence shows that rheumatic pain can be constant or intermittent, as well as localised or spread across a wide area and radiate from a central point [9]. Long-term illness tends to increase the gap between what patients and clinicians want to manage (the treatment goal or goals) and what they actually accomplish [10]. Improving physical and social functioning in daily life is an important treatment goal for patients with RA and PsA [11].

2 Meanings of Pain

The International Association for the Study of Pain (IASP) [12] defines pain as an "unpleasant sensory and emotional experience that is associated with actual or potential tissue damage, or described in such terms." This definition of pain was the first to recognise that pain is a personal experience incorporating sensory and emotional dimensions and perceived personal meanings (i.e., "associated with actual or potential tissue damage, or described in such terms"). Consequently, the IASP's definition of pain represents a serious challenge to traditional physiological accounts of pain that favour tissue damage or pathology to the exclusion or near exclusion of experiential patient factors (such as the meaning of pain to the person experiencing pain). A large number of experimental and clinical studies have helped to further describe the experience of pain from the perspective of the patient [13–21]. These studies have revealed that the common experiential factors of immediate pain include: (1) sensory qualities similar to those that appear during tissue damage, or during stimulation that would result in tissue damage if it were maintained over time, (2) a feeling of unpleasantness and/or other negative emotions, and (3) a common meaning of intrusion or threat.

In immediate pain, the common meaning of intrusion might convey physical injury or harm, designating that the body of the person with pain is physically damaged or perceived to be damaged following nociceptive stimulation (or stimulation perceived to be nociceptive stimuli), or it could denote being personally consumed or overwhelmed by the pain [17]. If pain persists, further negative existential meanings might develop in patients in response to the pain itself or to the negative consequences of pain on her or her life. These additional, more cognitive meanings may include interruption in the ability to live a normal life, concern for future consequences or enduring a burden over an extended period [17]. External negative events, such as (in some cases) inappropriate interactions with physicians or other health care professionals might fuel these cognitive meanings [22]. Patients' continued deliberation on these events may increase the meaning of threat such that their significance now broadens from merely the pain *per se* to threaten the integrity of the patient as a person [23]. Such a development of events may lead to person-focused or existential meanings of loss, helplessness, injustice, loss of independence or being reduced to nothing because of the pain [17]. In recognition of this knowledge, pain has been redefined to be a mutually recognisable somatic experience reflecting a person's apprehension of threat to his or her bodily or existential integrity [24].

This chapter intends to review some of the common meanings of pain reported by patients with rheumatic pain. To briefly summarise this study, persons with long-term rheumatic pain face numerous and continuing emotional difficulties in their daily activities [25, 26]. In some persons with RA [25, 27] and PsA [26], experiencing long-term rheumatic pain can create feelings of helplessness [25], isolation and disempowerment [26, 27]. For persons with RA or PsA, the "problematic body" (also described by patients to be the "unfamiliar" and "unreliable body"); i.e., the non-performing or painful symptomatic body, typically causes daily life to become more difficult [26, 27]. Among persons with long-term rheumatic pain, personal suffering is often connected to fear of the long-term loss of several vital physical functions. As a threat to the self, or of an increase in general disability or pain in daily life activities. Attention is often given to a body part that is weakened by the pain or the part is not able to participate (or physically perform) in daily activities as it did before the occurrence of the illness [25–31]. Existential meanings have been reported; e.g., spending too much time alone and mistrusting the body and self (withdrawal from life's physical and social activities), fear of not being able to handle life with pain [26, 27], or being reduced to nothing as a result of the pain [28].

3 Emotional Meanings of Pain Associated with Rheumatism: The Painful and Untrustworthy Body

Long-term rheumatic pain, like other common types of chronic non-cancer pain, is a complex experience in which pain sensation, meaning and negative emotion interact to adversely impact practical aspects of daily life and functioning [32]. These common experiential factors are essential to understanding the experience of pain. This is because the intrusiveness or threat of pain, which is the common meaning of pain, can intensify the unpleasantness of the immediate pain sensation, which, in turn, can develop or intensify secondary meanings of pain, including concern about the implications of having pain and negative emotions or moods (e.g., frustration, anxiety and depression) [33–35].

How patients experience long-term pain requires a new orientation towards life and opportunities to talk about the pain experience. It must be recognised that patients with long-term pain, like most other people, are eager to retain their sense of dignity and self-respect [32]. Many negative feelings (e.g., grief, aggressiveness, fear, shame) are emotions closely related to participation restrictions in everyday life in RA [25, 27]. Anger and irritation, for example, can arise because the patient with pain is not able to perform as before in either domestic or employed work. Such feelings also occur in various social relationships, where the individual feels unable to continue valued social activities with others [25]. In patients with long-term rheumatic pain the threat of pain might be a threat to their personal values, attitudes or beliefs that are integral to personal identity across time, or social-existential in relation to the world that the patient shares with other human beings, or both [26, 27]. Anxiety, stress, depression and anger are commonly reported pain-related emotions or disorders in patients with RA or PsA [36–39]. Increased pain or inflammation in patients with PA or PsA usually coincides with depressive symptoms and greater functional disability [37–39].

Furthermore, many patients with RA or PsA experience unpredictable daily symptoms as a result of their rheumatic disease [1, 4, 40–42]. Familiar symptoms in RA can be painful and stiff joints in the morning, swelling and tenderness, fatigue or a low-grade fever. Arthritis typically begins in the small joints of the hands and feet that subsequently spreads to the larger joints, causing joint deformity and progressive physical disability. Symmetrical inflammatory polyarthritis is the primary clinical manifestation [1, 40, 41]. Psoriatic arthritis often begins in the distal joints, those farthest away from the core of the body, such as the joints in the fingers and toes. In PsA, symptoms include pain, swelling and tenderness (typically hand, foot or elbow), skin problems or brittle, flaking and peeling nails. PsA is frequently asymmetrical, affecting only one side of the body [4, 42].

These illness symptoms are often accompanied by negative emotions in patients with RA [25–27, 43] or PsA [26, 42] that contribute to reduced physical and psychosocial health-related quality of life. Persons with long-term rheumatic pain perceive their body as changing over time from a relatively healthy body to one showing a rapid decline in physical strength and functioning. This perception can lead some persons with RA or PsA to see their body as fragile or untrustworthy, which leads to a downward spiral of fear and avoidance of health-enhancing physical activities (HEPAs) [27, 43]. Studies have shown that heightened body awareness in some patients with RA or PsA pain, commonly triggered by different negative emotions, reveals a body that is untrustworthy to the patient [25–27]. Some patients described the experience metaphorically as though they were constantly trapped inside a feeble and disabled body. Some patients with long-term rheumatic pain described having feelings expressing fury, rage or grief over having a poorly functioning body [25–27]. This form of reactive body awareness [27] often involved a comparison with a younger or more healthy body.

Exclusion are described by patients when not being able to act like everyone else. For instance, when not being able to take long walk with the dog and the stroller. At work, exclusion was experienced when not being able to increase work pace, or when not being able to carry and use the stairs [25].

3.1 Fear-Avoidance Beliefs in Relation to Physical Activity or Social Interaction

Research findings show that a minority of persons with RA continue to perform HEPA and that psychosocial factors seem to be the most salient and consistent factors to explain variations in HEPA [44]. Fear-avoidance of physical activity and avoidance of exercise in relation to fear of increased rheumatic pain [45] can lead some patients to brood on their fragile physical and emotional state [25–27, 46]. Fear-avoidance beliefs of physical activity generally involve recurring experiences (episodes) of fear because of an expected increase in rheumatic pain [45], suffering from heightened pain and emotional distress during or after physical activity or fear associated with an expected increase in rheumatic disease activity or decreased immune system responses [26, 27, 43]. High fear-avoidance beliefs about physical activity in patients with RA are found in a study linked to being male, having a below average income, high level of pain, poor health, low health-related quality of life (HRQoL) and low self-efficacy to engage in regular physical activity [45].

Examples of other fear descriptions have been found in relation to "fumble fear", which led individuals to withdraw from activities because of the perceived fragility and uncertainty of the body. Fumble fear made it less possible to participate in those activities the individual used to enjoy. It is a recurrent emotion described in relation to situations where the patients were afraid of fumbling, falling or letting go of something or someone (such as fumbling when walking, while shopping or with friends) [25].

Other fears associated with long-term pain caused by RA or PsA include the fear of reduced social interaction with others, which occurs in some patient due to both persistent pain/suffering and consequent social isolation and withdrawal

[26, 27]. Living with long-term pain problems can inhibit individuals with pain from engaging in physical and social activities [25–29, 31]. During periods of solitude or episodes of sedentary behaviour, the individual's attention can be involuntarily focused on the rheumatic pain, the symptomatic body and the self, leading to a negatively toned body awareness and high fear-avoidance behaviours [43]. Persons with pain issues tend to see their pain as a symbol of disgrace or shame, which causes the individual to withdraw from social activities and contact with others. They feel confused, afraid for their future and vulnerable [47]. In a study with RA patients, pain experiences were found to range widely; from a reduction of symptoms and coping well with long-term pain, to being completely engulfed in the pain experience. This experience was followed by either a new body awareness, in the first patient group, or being entirely preoccupied by pain [31]. Shame or embarrassment have also been reported when participation restrictions became visible in public [25]. Some persons with long-term pain problems fear that they are not being believed (taken seriously) and therefore withdraw from interacting with others [47]. Avoidance, the most common behaviour in persons with RA to cope with pain, is associated with predominantly negative outcomes [36]. Another outcome associated with cognitive meanings of RA- or PsA-related pain [25, 26] includes avoidance of social interactions to protect oneself from society's perceived stigmatisation about long-term pain [26, 28, 47]. Patient emotions in relation to participation restrictions can give professionals a sense of what's important in life for specific individuals [25].

3.2 Cognitive Meanings of Rheumatic Pain

Cognitive meanings that focus on deliberation or rumination about the meaning of pain are frequently seen in patients with RA or PsA. For example, patients with rheumatic pain have concerns about their ability to live a normal, physically active life, or worries about the future implications of having rheumatic pain [25–28]. Some persons feel alienated from the outside world, see their debilitation as a barrier that prevents them from participating in physical activity, or believe they are becoming a burden to society and family [43]. These experiences are often accompanied by feelings of sadness or existential feelings of emptiness, isolation, hopelessness or helplessness, which are more about the person suffering from pain than the pain itself [25–29]. Illness fear is described as individual experiences related to worries for the future and possible increased disability [25].

Some persons may not understand or refuse to recognise that they have a long-term pain condition, its common symptoms and how it affects their coping [36], bodily functioning or body awareness [48, 49] of everyday life activities and events. Patients' beliefs and expectations about their pain have been identified as important disabling factors in long-term musculoskeletal pain [50]. Cognitive meanings can be, for example, a search for an explanation of the cause of the pain (or the illness) or comparing the current perceived self (with others) [47]. Consistent evidence has confirmed the role of fear-avoidance responses to pain, and pain-related disability (measured via self-report) [50].

Physical limitations in daily life, driven by negative or fearful (avoidance) thoughts that compare the present disabled body with the former, younger or healthier body [26, 27, 46–52] can exacerbate such concerns. People with RA or PsA, who dwell on the negative effects of pain, might develop other existential meanings, such as severe loneliness or helplessness [26, 27]. Understanding body awareness is thought to be useful in the management of long-term diseases [26, 27, 46, 49]. Body awareness and body reliance seem to have importance in the (cognitive) process of acceptance of the pained body [46, 49]. Experiences of the body, such as good balance and physical stability, are descried as basic experiences linked to the conception of wellbeing and control in daily life [29]. Existential meanings might lead to a negative spiral of worsening outcomes, including increased withdrawal from social or physical activities due to an intensified mistrust of the self or body, and fear or doubts about body signals (symptoms) [5, 26, 27, 47–52].

In addition, cognitive meanings of pain may manifest in uncertainty or doubt, following not being believed or taken seriously by others when reporting pain [47]. Some persons with long-term pain express a personal failure, believing there is no way forward for them. Integration of the aching body and sense of self-purpose seems impossible. In these patient experiences, pain emerges as something incomprehensible, unacceptable and unfair [49]. In contrast to those who completely avoid physical activity, some patients may be physically overactive, which might contribute to elevated pain intensity and joint stiffness [25, 26]. Other patients with RA pain may struggle to determine personally suitable physical activity forms, including activity intensity, volume or duration [52].

4 Long-Term Rheumatic Pain and Adaptive Body Awareness Towards Health Enhancing Physical Activity

Cognitive and emotional processes are crucial contributors to inter-individual differences in the perception and impact of rheumatic pain [39]. A person-centred approach [10] attempts to treat the mind and body as two parts of a single whole, rather than as separate entities. Such an approach can help patients with RA or PsA to find adaptive coping strategies to ensure the maintenance of health-promoting daily physical activities [26, 27]. Person-centred care is critical to understanding the meaning of long-term pain from a health-promotion perspective [10, 26, 53]. Focusing on the context, combined with verbal expression of emotions (patient narratives), can help discover vulnerable patients [25].

Persons with long-term pain problems underscore the importance of applying a person-centred approach in health care [26, 49, 54]. This approach can help to build not only the individual's inner strength or resilience to endure continued pain and pain-related suffering, but also provide the person with hope to maintain a physically active life in the future [26]. A sedentary lifestyle could lead to high stress levels and even more health complications [32, 55], such as increased risk for cardiovascular diseases and decreased mental wellbeing. Thus, health care professionals need to address, support and implement individually designed health-promoting physical activities in the patient's treatment plan [55]. This role closely corresponds with recommendations for the nurse's role in the management of long-term arthritis [11].

A person-centred approach places the person's views about his or her life situation and condition at the centre of care. Within this theoretical framework, patient narratives (using patient stories (subjective) to reflect on care) constitute an ideal starting point for person-centred care in the treatment of pain [10]. This approach stresses that an individual's identity is to some extent created through encounters with others [10]. Many patients describe a strong need to be confirmed and met with respect by health care personnel, which contribute to their sense of togetherness. Having a sense of attachment with caregivers can enhance the patient's physical, emotional or psychosocial health and wellbeing [56].

Implications for rehabilitation healthcare professionals should be aware that individuals with long-term pain conceptualize and hold different meanings of acceptance when starting rehabilitation. This should be considered and addressed in rehabilitation programmes. The meaning given to acceptance is related to the experience of the lived body and the sense of self, as well as to receiving legitimization and acceptance from others. Therefore, these aspects need to be considered during rehabilitation [49] for patients with RA and PsA. Health care professionals should serve to promote the patient's health and prevent illness while having regard for his or her autonomy. Health care professionals need to work to relieve patient suffering and restore self-dignity [53].

The term life-world, derived from the German *lebenswelt* as used in the philosophical discipline phenomenology, refers to the experiential qualities that appear to people in their conscious life experiences and the meanings of things in these experiences, including the meaning of objects, events, tools, the flow of time, the self and other people [57]. One phenomenological study described embodied identity in two categories: living in the body, and living in relation to others and society [47]. Another phenomenological study also identified this notion, being a meaningful part in the outside world is important for improving HEPA of the individual [26]. Persons with moderate to severe rheumatic pain also see themselves in a wider social context, reflecting a main-source of positive energy and self-efficacy to themselves and others [26]. Research has shown the importance of understanding emotions and needs through awareness of the body, which is understood here as the basis for self-confidence and trust in oneself, as well as the ability to take care of oneself both physically and mentally [47].

The term sensitivity implies an open (accepting) communicative approach in which listening to the patient's experiences is central to the care process and where autonomy is preserved as much as possible [56]. Low levels of communication and social support at the time of diagnosis consistently predict functional disability and pain [51]. Studies moreover show that health care professionals need to assist their patients to find new strategies that focus on meaning, acceptance and tolerance of their pain [46, 47]. Results indicate that patients with long-term pain can be found along a spectrum from accepting to rejecting the aching body. Body awareness and a trust in ones' body seem to be important on the path towards acceptance of the body as well as one's life situation as whole [46].

Using phenomenological research methods, health care clinicians (as researchers) can help patients to understand that body awareness is inseparable from the person's identity, given that it affects the health of the individual [47]. The concept of the body as a silent partner and informant can promote an appreciation of the body's own subjectivity that can be enhanced in nursing care through sensitivity, listening and creative coaching. Nurses can assist their patients to identify new and positive understandings of what their bodies mean to them after confusing or worrisome bodily changes have occurred [29, 48]. Bodily-existential challenges (e.g., reflection, coping with or accepting a transformed body) should be highlighted, as well as the importance of social support [26, 27, 49].

A wide range of competencies in rehabilitation clinics seem to be needed [49, 54]. Pain coping and social support, assessed very early in the disease process, can have a positive effect on life in long-term functional disability and rheumatic pain. Low levels of social support at the time of diagnosis consistently predicted functional disability and pain at 3 and 5-year follow-ups. Taken together, these results suggest that early interventions that focus on pain-related avoidance factors and social resources for patients at risk may beneficially influence long-term outcomes in RA [51]. Recommendations for physical activity in persons with rheumatic disease underline the need to reduce the risk of aggravated disease symptoms in connection with increased physical activity and develop an individually designed (person-centered) physical activity plan [32, 55].

As previously described, some people who experience rheumatic pain perceive their body to be fragile and unpredictable. Engaging in regular physical activity has been found to be important for general health and renewed bodily strength and functioning [25–27, 52, 55]. By educating patients to adopt a more favourable attitude towards the self, the body can at least be partially restored to a state of calm and balance [26, 27], which may improve their ability to cope with pain and enhance HRQoL. Pain always appears in a particular way. Perceived meaning is therefore extremely important in the way that pain is felt [58].

4.1 Let Me Be a Meaningful Part in the Outside World: An Example

To give a famous example from experiences of life with rheumatic pain, the French artist Pierre-Auguste Renoir (1841–1919) suffered from debilitating RA. Despite his severe pain, Renoir was able to maintain a high level of precision and efficiency during the later years in his life even after RA severely limited his mobility. One plausible reason why he was able to achieve this high level of competency is that he did not let his condition affect his passion for painting. He remained positive and applied a wide variety of coping mechanisms. Often, he used creativity to devise different ways to continue his passion in life [59].

Renoir's experience is illustrative because it shows that hopelessness, the hallmark of personal demoralisation, is associated with poor outcomes in physical and psychiatric illnesses [28]. Despite having long-term moderate to severe rheumatic pain, persons can experience more positive body awareness by engaging in personally meaningful life activities. One way to achieve this goal would be by living in a community with others and being accepted as part of the community, which means being treated with respect and dignity [43]. In addition, alternative treatments should be made available to effectively relieve pain and suffering, as described by patients [26, 27].

Communicating and sharing experiences with other people (such as working in groups and participating in various physical activities as often as possible) are powerful resources to have available for persons with rheumatic pain. This mode of thinking implies that persons strive to play an active part in society [26, 27]. An important aspect of the embodied self is the ability to interact with others and participate in the community [46]. Through different social networks (such as paid or volunteer work), some patients with RA or PsA can find their inner world become more meaningful, leading to a shift in focus from their own body and self to others (achieved through social interaction). Interacting with other people can lead to a sense of bodily calmness and tranquility of mind [26, 27]. Moreover, interaction with other people is important for cognition, development and wellbeing [22, 26, 27]. The sense of togetherness is also important because it makes people feel good about themselves and helps people to endure dark and difficult times [56]. Patients, however, do not always feel that the manner of the nursing staff is empathetic or sympathetic, which can lead to greater suffering and isolation [22].

Finally, research suggests that being socially connected and in a safe environment in which the individual is physically active reduces fear and decreases some components of the flight response [26]. Choosing the right activity and the right level of difficulty can be challenging [52]. Therefore, recommendations for physical activity in persons with rheumatic disease underline the need to reduce the risk of aggravated disease symptoms in connection with increased physical activity and develop an individually designed physical activity plan in alliance with the patient's multidisciplinary management team [32, 55]. Having the opportunity to participate in meaningful and purposeful daily real-world activities keeps the mind busy and can serve to switch off negative thinking and physical tension and is therefore central to a person's general health and wellbeing [43].

5 Conclusions

Long periods of solitude can lead to maintained fear-avoidant behaviour, followed by an inactive downward spiral that results in new negative existential meanings of loss, isolation or helplessness. Despite suffering from moderate to high self-rated rheumatic pain, patients point out their desire to participate in a wider social and active context without being subject to any form of stigmatisation from others. To patients with RA or PSA, this is a main source of positive energy and meaning in life. Personal factors and the social environment are important regarding the impact of pain on participation and sense of independence. However, during periods of solitude or sedentary behaviour, inner bodily attention can be directed towards the problematic body and self. Suffering from a long-term illness (and pain) can be perceived as a threat to the self.

Among persons with long-term rheumatic pain, suffering is often connected to fear for the long-term loss of several vital physical functions and of increased general disability in daily life activities. Attention is often directed to the symptomatic body, i.e., the perceived painful body. Because the meaning of threat increases immediate pain unpleasantness, it is important that persons with long-term rheumatic pain experience more positive body awareness by participating in personally meaningful life activities, including living in a community with others, being accepted and being cared for by others. Patients might also be given the opportunity to experience alternative clinical treatments to help relieve pain and suffering. Future research can examine high fear-avoidance beliefs towards physical activity in this patient group, and the common meanings of such beliefs. Other common meanings, such as helplessness (inactivity and passive coping) and escape/avoidant coping strategies (including denial and wishful thinking) should be further investigated. The concept of fear-avoidance beliefs towards physical activity is complex and a deeper understanding of what it means from the patient's perspective is needed.

References

1. Kilic L, Erden A, Bingham CO, Gossec L, Kalyoncu U. The reporting of patient-reported outcomes in studies of patients with rheumatoid arthritis: a systematic review of 250 articles. J Rheumatol. 2016;43(7):1300–5.
2. Riise T, Jacobsen BK, Gran JT. Incidence and prevalence of rheumatoid arthritis in the county of Troms, northern Norway. J Rheumatol. 2000;27:1386–9.
3. Scott DL, Wolfe F, Huizinga TW. Rheumatoid arthritis. Lancet. 2010;376:1094–108.
4. Myers WA, Gottlieb AB, Mease P. Psoriasis and psoriatic arthritis: clinical features and disease mechanisms. Clin Dermatol. 2006;24(5):438–47.
5. Gladman DD, Antoni C, Mease P, Clegg DO, Nash P. Psoriatic arthritis: epidemiology, clinical features, course, and outcome. Ann Rheum Dis. 2005;64(Suppl 2):ii14–7.
6. Mease PJ, Armstrong AW. Managing patients with psoriatic disease: the diagnosis and pharmacologic treatment of psoriatic arthritis in patients with psoriasis. Drugs. 2014;74:423–41.

7. Ogdie A, Yu Y, Haynes K, Love TJ, Mahila S, Jiang YH, et al. Risk of major cardiovascular events in patients with psoriatic arthritis, psoriasis and rheumatoid arthritis: a population-based cohort study. Ann Rheum Dis. 2015;74(2):326–32.
8. Finzel S, Englbrecht M, Engelke K, Stach C, Schett G. A comparative study of periarticular bone lesions in rheumatoid arthritis and psoriatic arthritis. Ann Rheum Dis. 2011;70(1):122–7.
9. Walsh DA, McWilliam DF. Mechanisms, impact and management of pain in rheumatoid arthritis. Nat Rev Rheumatol. 2014;10(10):581–92.
10. McCormack B, McCance T. Person-centred nursing: theory and practice. Chichester, West Sussex: Blackwell; 2010. 208 p.
11. van Eijk-Hustings Y, van Tubergen A, Boström C, Braychenko E, Buss B, Felix J. EULAR recommendations for the role of the nurse in the management of chronic inflammatory arthritis. Ann Rheum Dis. 2012;71:13–9.
12. International Association for the Study of Pain (IASP). Taxonomy [Internet]. 2017 [cited 13/04/2019]. Available from http://www.iasppain.org/Content/NavigationMenu/GeneralResourceLinks/PainDefinitions/
13. Head H. Studies in neurology. London: Oxford University Press; 1920. 352 p.
14. Landau W, Bishop GH. Pain from dermal, periosteal, and fascial endings and from inflammation: electrophysiological study employing differential nerve blocks. AMA Arch Neurol Psychiatry. 1953;69:490–504.
15. Price DD. Characteristics of second pain and flexion reflexes indicative of prolonged central summation. Exp Neurol. 1972;37:371–87.
16. Price DD, Hu JW, Dubner R, Gracely RH. Peripheral suppression of first pain and central summation of second pain evoked by noxious heat pulses. Pain. 1977;3:57–68.
17. Price DD, Barrell JJ. Inner experiences and neuroscience. Merging the two perspectives. Cambridge: MIT Press; 2012. 360 p.
18. Price DD, Harkins SW, Baker C. Sensory affective relationships among different types of clinical and experimental pain. Pain. 1987;28:291–9.
19. Harkins SW, Price DD, Braith J. Effects of extraversion and neuroticism on experimental pain, clinical pain, and illness behavior. Pain. 1989;36:209–18.
20. Staud R, Vierck CJ, Cannon RL, Mauderli AP, Price DD. Abnormal sensitization and temporal summation of second pain (wind-up) in patients with fibromyalgia syndrome. Pain. 2001;91:165–75.
21. Staud R, Craggs JG, Perlstein WM, Robinson ME, Price DD. Brain activity associated with slow temporal summation of C-fiber evoked pain in fibromyalgia patients and healthy controls. Eur J Pain. 2008;12(8):1078–89.
22. Hansson KS, Fridlund B, Brunt D, Hansson B, Rask M. The meaning of the experiences of persons with chronic pain in their encounters with the health service. Scand J Caring Sci. 2011;25(3):444–50.
23. Cassell EJ. The nature of suffering. London: Oxford University Press; 2002. 336 p.
24. Cohen M, Quinter J, van Rysewyk S. Reconsidering the international association for the study of pain definition of pain. Pain Rep. 2018;3(2):e634.
25. Östlund G, Björk M, Thyberg I, Thyberg M, Valtersson E, Stenström B, Sverker A. Emotions related to participation restrictions as experienced by patients with early rheumatoid arthritis: a qualitative interview study (the Swedish TIRA project). Clin Rheumatol. 2014;33(10):1403–13.
26. Lööf H, Johansson UB. A phenomenological study of body awareness and fear-avoidance beliefs towards physical activity in persons experiencing moderate to high rheumatic pain. J Clin Nurs. 2019;28(1–2):321–9.
27. Lööf H, Johansson UB, Welin Henriksson E, Lindblad S, Bullington J. Body awareness in persons diagnosed with rheumatoid arthritis. Int J Qual Stud Health Well Being. 2014;9:24670.
28. Armentor JL. Living with a contested, stigmatized illness: experiences of managing relationships among women with fibromyalgia. Qual Health Res. 2017;27(4):462–73.
29. Wilde MH. Embodied knowledge in chronic illness and injury. Nurs Inq. 2003;10(3):170–6.

30. Gyllensten AL, Skär L, Miller M, Gard G. Embodied identity-a deeper understanding of body awareness. Physiother Theory Pract. 2010;26(7):439–46.
31. Dragesund T, Råheim M. Norwegian psychomotor physiotherapy and patients with chronic pain: patients' perspective on body awareness. Physiother Theory Pract. 2008;24(4):243–54.
32. Methods of treating chronic pain. Stockholm: Swedish Council on Health Technology Assessment in Health Care (SBU); SBU report no 177/n.1&2 (in Swedish). 2006.
33. Sullvian MD. How chronic pain treatment falls short of patient-centered care [Internet]. 2017 [cited 13/04/2019]. Available from https://catalyst.nejm.org/chronic-pain-care-falls-short/
34. Ballantyne JC, Sullivan MD. Intensity of chronic pain—the wrong metric? N Engl J Med. 2015;373(22):2098–9.
35. Price DD, Harkinst SW. The affective-motivational dimension of pain. A two-stage model. APS J. 1992;1(4):229–39.
36. Ramjeet J, Smith J Adams M. The relationship between coping and psychological and physical adjustment in rheumatoid arthritis: a literature review. J Clin Nurs. 2008;17(11c):418–28.
37. Fifield J, Tennen H, Reisine S, McQuillan J. Depression and the long-term risk of pain, fatigue, and disability in patients with rheumatoid arthritis. Arthritis Rheum. 1998;41(10):1851–7.
38. Shih M, Hootman JM, Strine TW, Chapman DP, Brady TJJ. Serious psychological distress in U.S. adults with arthritis. J Gen Intern Med. 2006;21(11):1160–6.
39. Edwards RR, Cahalan C, Mensing G, Smith M, Haythornthwaite JA. Pain, catastrophizing, and depression in the rheumatic diseases. Nat Rev Rheumatol. 2011;7(4):216–24.
40. Stone AA, Broderick JE, Porter LS, Kaell AT. The experience of rheumatoid arthritis pain and fatigue: examining momentary reports and correlates over one week. Arthritis Rheum. 1997;10 (3):185–93.
41. Pollard LC, Choy EH, Gonzalez J, Khoshaba B, Scott DL. Fatigue in rheumatoid arthritis reflects pain, not disease activity. Rheumatology. 2006;45(7):885–9.
42. Lee S, Mendelsohn A, Sarnes E. The burden of psoriatic arthritis. A literature review from a global health system perspective. PT. 2010;35(12):680–9.
43. Lööf H. Pain, fatigue and fear-avoidance beliefs in relation to physical activity and body awareness in persons diagnosed with rheumatoid arthritis. Stockholm, Doktorsavhandling: Thesis. Karolinska Institutet; 2015. p. 1–82.
44. Demmelmaier I, Bergman P, Nordgren B, Jensen I, Opava CH. Current and maintained health-enhancing physical activity in rheumatoid arthritis: a cross-sectional study. Arthritis Care Res. 2013;65(7):1166–76.
45. Lööf H, Demmelmaier I, Welin Henriksson E, Lindblad S, Nordgren B, Opava CH, et al. Fear-avoidance beliefs about physical activity in adults with rheumatoid arthritis. Scand J Rheumatol. 2015;44(2):93–9.
46. Afrell M, Biguet G, Rudebeck CE. Living with a body in pain–between acceptance and denial. Scand J Caring Sci. 2007;21:291–6.
47. Osborn M, Smith JA. The personal experience of chronic benign lower back pain: an interpretative phenomenological analysis. Br J Health Psychol. 1998;3:65–83.
48. Tsay A, Allen TJ, Proske U, Giummarra MJ. Sensing the body in chronic pain: a review of psychophysical studies implicating altered body representation. Biobehav Rev. 2015;52:221–32.
49. Biguet G, Nilsson-Wikmar L, Bullington J, Flink B, Löfgren M. Meanings of 'acceptance' for patients with long-term pain when starting rehabilitation. Disabil Rehabil. 2016;38 (13):1257–67.
50. Hasenbring M, Verbunt JA. Fear-avoidance and endurance-related responses to pain: new models of behavior and their consequences for clinical practice. Clin J Pain. 2010;26 (9):747–53.
51. Evers AWM, Kraaimaat FW, Geenen R, Jacobs JWG, Bijlsma JWJ. Pain coping and social support as predictors of long-term functional disability and pain in early rheumatoid arthritis. Behav Res Ther. 2003;41(11):1295–310.

52. Ahlstrand I, Björk M, Thyberg I, Börsbo B, Falkmer T. Pain and daily activities in rheumatoid arthritis. Disabil Rehabil. 2012;34(15):1245–53.
53. International Council of Nurses (ICN). Svensk sjuksköterskeförening (SSF). ICN: Etiska Kod för Sjuksköterskor. [The ICN Code of Ethics for Nurses.] Stockholm: Svensk sjuksköterskeförening. 2018.
54. Kamper SJ, Apeldoorn AT, Chiarotto A, Smeets RJ, Ostelo RW, Guzman J, et al. Multidisciplinary biopsychosocial rehabilitation for chronic low back pain. Cochrane database of systematic reviews. BMJ. 2015;18:350.
55. Brodin N, Swärdh E. In: Physical activity in the prevention of treatment and disease, FYSS; Rheumatoid arthritis. 2017 [cited 13/04/2019]. Available from http://www.fyss.se/
56. Nygren Zotterman AN, Skär L, Olsson M, Söderberg S. Being in togetherness: meanings of encounters within primary healthcare setting for patients living with long-term illness. J Clin Nurs. 2016;25(19–20):2854–62.
57. Spiegelberg H. The phenomenological movement: a historical introduction. The Hague, The Netherlands. In: Kim H, Kollak I, editors. Nursing theories: conceptual and philosophical foundations. New York: Springer; 2006. 309 p.
58. van Rysewyk S. A call for study on the meanings of pain. In: van Rysewyk S, editor. Meanings of pain. Cham: Springer; 2016. p. 1–22.
59. Kowalski E, Chung KC. Impairment and disability: Renoir's adaptive coping strategies against rheumatoid arthritis. Hand (NY). 2012;7(4):357–63.

Chapter 7
The Importance of Pain Imagery in Women with Endometriosis-Associated Pain, and Wider Implications for Patients with Chronic Pain

Christopher J. Graham, Shona L. Brown, and Andrew W. Horne

Abstract Pain imagery is *"like having a picture in your head* [of your pain] *which may include things you can imagine seeing, hearing or feeling."* Pain imagery may offer a unique insight into a patient's pain experience. This chapter summarises findings from international pain imagery research in women with endometriosis-associated pain. Endometriosis is a chronic inflammatory condition associated with debilitating pain that affects 5–10% of women of reproductive age worldwide. Our international research has found that pain imagery is experienced by around half of women suffering from endometriosis-associated pain, and is associated with higher levels of catastrophising, depression, and anxiety. However, coping imagery is also reported, and prevalent, at 30%. Pain imagery in women with endometriosis falls into themes: sensory qualities of pain; loss of power or control; attack (by someone, "something," or self); pathology or anatomy envisaged; past or future catastrophe; pain as an object; and abstract images. Imagery content may therefore reveal the meanings of pain or endometriosis to these women. This chapter explores pain imagery content and its personal significance to patients, both for women with endometriosis-associated pain and for patients with other chronic pain conditions. The chapter concludes by discussing the clinical application of imagery, with example patient cases to contextualise the practicalities and therapeutic potential of imagery techniques.

Clinical Implications: Pain imagery was reported by half of women with endometriosis-associated pain in our international study and associated with higher levels of catastrophising, depression, and anxiety. Imagery content is extremely varied but can be categorised into themes, which may offer unique insights into each woman's pain experience. Coping imagery was prevalent at 30%. We believe

---

C. J. Graham (✉) · A. W. Horne
MRC Centre for Reproductive Health, University of Edinburgh, The Queen's Medical Research Institute, Edinburgh, Scotland
e-mail: christopher.graham@ed.ac.uk

S. L. Brown
Department of Clinical Psychology, NHS Lothian, Astley Ainslie Hospital, Edinburgh, Scotland

S. van Rysewyk (ed.), *Meanings of Pain*,
https://doi.org/10.1007/978-3-030-24154-4_7

imagery techniques may be particularly helpful for women with endometriosis-associated pain and discuss these techniques, which should be of interest to professionals involved in pain management.

Keywords Endometriosis · Persistent pelvic pain · Chronic pelvic pain · Pain imagery · Coping imagery · Imagery-based therapies

1 Pain and Chronic (Pelvic) Pain

Pain duration gives rise to two classifications. Pain that persists for at least either 3 months (for example, [1, 2]) or 6 months (for example, [3, 4]) is known as chronic pain, whereas pain that resolves quickly is known as acute pain.

A large-scale survey of chronic pain (defined as pain lasting 6 months or more) in 15 European countries and Israel ($N = 46{,}394$) found that 19% of adults surveyed experienced moderate to severe chronic pain limiting their occupational and social functioning [3]. In an exclusively UK-based community sample ($N = 3605$), high levels of distress and disability were associated with chronic pain (defined as pain lasting 3 months or more) [2]. Australia's seminal *National Pain Strategy* highlighted that pain is their nation's third most costly health problem and that chronic pain should be recognised as a disease in its own right [5]. Chronic pain is prevalent and damaging, to the sufferer and to society.

Chronic pelvic pain can be defined as *"intermittent or constant pain in the lower abdomen or pelvis of a woman of at least 6 months in duration, not occurring exclusively with menstruation or intercourse and not associated with pregnancy"* [6]. Endometriosis is the most common cause of chronic pelvic pain [7].

2 Endometriosis

Endometriosis is a chronic inflammatory oestrogen-dependent condition affecting women that can be associated with debilitating pain. It is defined by the presence of endometrial-like[1] tissue outside the uterus, commonly on the lining of the pelvis (the pelvic peritoneum) or on the ovaries [8].

Endometriosis affects up to 10% of women of reproductive age—an estimated 176 million women worldwide [9]. There is no non-invasive biomarker for endometriosis and diagnosis requires (laparoscopic) surgery. This leads to a typical diagnostic delay of 8–10 years [10, 11]. There is no cure for endometriosis. Endometriosis is managed surgically (excision/ablation of lesions) or medically by

[1]The endometrium is the lining of the uterus (womb) that undergoes cyclical regeneration as it thickens and is later shed during menstruation.

ovarian suppression. Symptom recurrence is common after surgery, and medical management is often associated with side effects and is not fertility sparing [8].

3 Mental Imagery

Mental imagery is defined as: *"cognitions, which take the form of sensory experiences (in any modality) in the absence of a direct percept"* [12]. This definition acknowledges the multisensory meaning of mental imagery. For example, someone who has had a bicycle accident may experience mental imagery in which they visualise their damaged bicycle, hear commentary from onlookers, and taste blood. Despite its multisensory scope, visual imagery has formed the focus of most research [13]. Mental imagery can be generated spontaneously or intentionally, recruiting similar brain regions as those involved in real perception (for example, [14, 15]). Similar to the effects of perceptual experience, it has also been shown that mental imagery can have physiological consequences: taste imagery increases salivation [16]; imagery of activity increases heart rate [17]; and more "active" imagery increases heart rate to a greater degree [18]; and imagery of white blood cells attacking germs produces measurable immune system change [19]. Mental imagery can affect the body as much as actual perceptual experience [14].

Mental imagery has been reported in a number of psychological conditions, including anxiety disorders [20] such as post-traumatic stress disorder [21] and social phobia [22], depression [23], bipolar disorder [24] and bulimia nervosa [25] (see Holmes and Mathews' review for more examples and discussion [26]).

Holmes and Mathews highlight that mental imagery is key in psychopathology because of its impact on emotion [26]: mental imagery is a potential emotional amplifier [27] and is reportedly more emotionally triggering than verbal thought [28]. Mental imagery can also maintain various psychological difficulties through avoidance or destructive behaviours [26]. Conversely, mental imagery is itself a therapeutic target [29, 30]—later explored in this chapter.

3.1 Pain Imagery

Pain imagery is: *"like having a picture in your head* [of your pain] *which may include things you can imagine seeing, hearing or feeling"* [31].

Berna and colleagues interviewed a small sample of women with chronic pelvic pain ($N = 10$), who all experienced pain imagery [32]. Each participant's "most significant" image (the one they felt was most relevant to their pain, or the one they had most often when in pain) was intrusive and negative. Intrusion is a key concept in understanding pain [33]: often signifying physical harm (or impending physical harm), intrusion may also signify an attack on "self," and being personally overwhelmed or consumed by pain [34], that is pain (and for many that experience it, pain imagery) is intrusive on both body and consciousness [33]. Eight women experienced "coping" imagery. For example, one woman reported the following: *"I put the pain into a box, but this is difficult as no box is big enough;" "I imagine that I grind an analgesic drug on my body, rubbing it in, or I imagine that I get an injection of a painkiller;" and "I see my husband. He gives me support."* Post-hoc, the coping imagery described by the eight women was categorised as: *allegoric treatment of an object symbolizing pain, imaginary treatment applied to the body,* and *supportive person.*

Philips reported a 78% prevalence of mental imagery in a mixed group of acute and chronic pain sufferers ($N = 59$) in an Occupational Rehabilitation Centre, following a work-related or motor vehicle accident [35]. Participants were asked to generate their "Index" image (the most powerful/distressing image they see when in pain). This led to worsened mood (increased anxiety, sadness and anger; decreased happiness and calm), and increased pain levels. Participants found it difficult to provide a predominant meaning of their Index image, and few had considered meaning prior to the open questioning at interview ("*...maybe it is due to this...or...maybe to that...*"). Participants were also presented with a discrete number of meaning categories to choose from, but this was also problematic as many participants provided a number of possible suggestions with overlapping themes. For example, one participant reported that her Index image (*"I see myself sitting alone in my wheelchair. I am old and unable to care for myself."*) meant that she would be a burden to her children in the future. Presented with various meaning categories, she chose five: *future catastrophe, physical disability, absence of control, unhappiness,* and *dependence.* In addition to the open-ended and categorical approaches used to elucidate image meaning, the investigator also classified participant responses: 23% gave negative self-appraisals (for example, *"I am a loser"*), 28.2% described future catastrophe, and 12.8% described past catastrophe (the accident or other past trauma).

Gillanders, Potter, and Morris discovered a pain imagery prevalence of 23% in their chronic pain sample ($N = 83$) [36]. Pain imagery was associated with significantly higher levels of anxiety, depression, and catastrophising. Participants' pain imagery content is reported as individual and idiosyncratic, and not readily categorised.

Gosden and colleagues found a pain imagery prevalence of 36% ($N = 105$) [31]. The imagery was frequent, vivid, and damaging (higher depression, anxiety and pain ratings) for most. More frequent imagery was associated with greater pain unpleasantness. The researchers categorised the imagery into themes relating to: anatomical representations (for example, *"Bones grinding together"*), pain as an attack (for example, *"Like my body is being attacked* [where problems are] *by someone with a voodoo doll or a little army inside me making sure I am in agony!"*), the sensory qualities of pain (for example, *"I have an image of an electric short circuit running down my legs!"*), pain as an object (pain encapsulated in object format, with a clear sense of shape and boundary, for example, *"My image looks like a large ball about the size of a tennis ball and it looks spongy and horrible"*), and pain as an abstract image (for example, *"I see faces in the sky"*).

Pain imagery may offer a unique insight into a patient's pain experience, as discussed by Berna in the first book of *Meanings of Pain* [37], and may provide a novel target for cognitive behavioural therapy [32] and other psychological therapies.

4 Endometriosis-Associated Pain Imagery

Our group at the University of Edinburgh has investigated pain imagery in women with endometriosis-associated pain in two progressive studies.

Our first, UK-based study [unpublished], suggested that pain imagery is experienced by 45% of women suffering from endometriosis-associated pain ($N = 378$) and is associated with higher levels of anxiety and depression. In this study, we piloted and adapted a pain imagery questionnaire used previously by Gosden [38] following interviews with women with endometriosis-associated pain imagery ($n = 6$) to create our Chronic Pain Mental Imagery Questionnaire. During interview, pain imagery and its trigger(s), associated affect(s), meaning(s) and impact on behavioural avoidance were explored (Table 7.1).

In our second, international study [in preparation], women aged 18 and over suffering from chronic pelvic pain (lasting 6 months or more) with a surgical diagnosis of endometriosis were invited to take part in an online questionnaire by six international endometriosis patient organisations. 785 responses from 43 countries met the inclusion criteria and were analysed.

The questionnaire, hosted on *Online Surveys* (www.onlinesurveys.ac.uk), was open from 7 November 2016 to 1 December 2016. The questionnaire included questions on participant demographics; a Chronic Pain Mental Imagery Questionnaire; the Short-Form McGill Pain Questionnaire 2 (SF-MPQ-2) [39]; the Pain Catastrophizing Scale (PCS) [40]; and the Depression, Anxiety and Positive Outlook Scale (DAPOS) [41].

Mann-Whitney U testing was used to explore differences between the imagery and non-imagery groups (independent t-tests were inappropriate as the Kolmogorov-Smirnov test revealed non-normality). Statistical analysis was undertaken using both Microsoft Excel for Mac 2011 and Statistical Package for the Social Sciences (Version 22). Qualitative data, from our Chronic Pain Mental Imagery Questionnaire, were thematically analysed as described by Braun and Clarke [42].

52% of women with endometriosis-associated pain reported experiencing pain imagery (409/785). Pain imagery prevalence was similar across nationalities. Age, pain frequency and duration, and average and worst pain intensity in a typical month were not significantly different between imagery and non-imagery groups. Most reported imagery was vivid (mainly either *"moderately clear and vivid"* (36%) or *"clear and reasonably vivid"* (30%), with 13% *"perfectly clear and as vivid as normal vision"*—these ratings were drawn from the Vividness of Visual Imagery Questionnaire [43]). Most women experienced imagery at least monthly, with 18% experiencing pain imagery daily. 94% reported that their imagery causes distress.

Table 7.1 Description of each woman's most common pain imagery ($n = 6$), with associated trigger(s), associated affect(s), meaning(s) and impact on behavioural avoidance

Imagery content	Trigger(s)	Associated affect(s)	Associated meaning(s)	Avoidance or functional impact
Weights pulling on womb and stabbing	Pain and walking movement	Stress, anxiety, frustration	No meaning offered	Avoidance of physical activity, stress and social contact
Hot poker	Intense pain and spasms	Upset, anger, distress	No meaning offered	Not applicable
Angry animal	Pain in the ovaries	Frustration	Frightening surgeries, infertility, unsuccessful IVF. *"A wee monster that's spoiled things."*	Makes her want to curl up and do nothing
Beast inside tearing to get out	Pain	Frustration, anger, hopelessness, sadness	Hopeless and helpless, as something has control of her	Avoidance of physical activity and social contact
Ugly piranhas gnawing	Feeling of being prodded	Incredulousness	*"They're ugly little [fish]... that's what the pain is like."* Wild: invaded by something that should not be there	Avoidance of social contact
Something inside growing bigger	Not applicable	Tearful, feels bad	Something is going wrong	Avoids all activities, stays at home

IVF In Vitro Fertilisation

Distress caused and the degree to which the imagery interferes with daily life were moderately correlated ($r_s = 0.673$, $p < 0.001$).

Pain imagery was associated with significantly higher catastrophising ($U = 67{,}231.0$, $p = 0.002$, $r = -0.109$), depression ($U = 62{,}877.5$, $p = <0.001$, $r = -0.160$), and anxiety ($U = 63{,}009.0$, $p < 0.001$, $r = -0.157$). Pain imagery was also associated with significantly higher scores across all dimensions of the SF-MPQ-2 (Total SF-MPQ-2 score: $U = 60{,}787.0$, $p < 0.001$, $r = -0.181$).

We found seven pain imagery themes: sensory qualities of pain, loss of power or control, attack (by someone, "something," or self), pathology or anatomy envisaged, past or future catastrophe, pain as an object, and abstract images.

30% of women reported coping imagery (124/409), which may include a "feeling," such as heat or flowing water; or "sounds" that helped women cope with their pain. Eight examples of coping imagery reported involved rescripting (changing distressing imagery into benign imagery, as discussed in *Clinical Implications* below). Two women described coping imagery as their main image:

- "I picture someone wringing out a dishcloth. I hold that image as long as the cramp lasts and let it go when the cramp passes. It helps me deal with the really bad ones."
- "A clear night sky, full of stars and constellations and a full moon. Sometimes there was clouds."

We found a pain imagery prevalence of 52%. In previous studies, a wide variation in pain imagery prevalence has been reported, from 23% in chronic pain patients [36] to 100% in women with chronic pelvic pain [32]. Pain imagery prevalence may vary in different pain conditions—this requires further clarification. Alternatively, the wide variation may be due to methodological differences between studies, with some researchers using questionnaires [31, 36] and others using interviews [32, 35]. Interview-based studies have reported the highest prevalence: 100% prevalence in women with chronic pelvic pain [32] and 78% in a mixed sample of acute and chronic pain sufferers [35]. This suggests that participants completing a self-report questionnaire may struggle to describe or understand their imagery, resulting in a lower reported prevalence.

Pain imagery was often reported as vivid by the women in our studies. Other researchers have been able to use the vividness of pain-related imagery to create artistic representations of pain and painful conditions (for example, [44–48]), and patients have recreated their own pain-related imagery through artistic means [49–51]. However, this imagery may be metaphorical (a way to communicate the patient's pain experience) as opposed to pain imagery (intrusive imagery one experiences when in pain).

The majority of women in our study found their imagery distressing, and it can interfere with daily life. Pain imagery was associated with higher levels of catastrophising, depression, and anxiety. The finding that pain imagery generally has a negative impact is consistent in pain imagery research [31, 32, 35, 36]. Although the differences in depression and anxiety scores between the imagery and non-imagery groups were highly significant, the effect size was small. For an individual woman, experiencing pain imagery may not signify a clinically significant increase in anxiety or depression. However, because an estimated half of women with endometriosis experience pain imagery, the morbidity associated with pain imagery may in fact be considerable. With regards to the increased levels of catastrophising found in the imagery group, Jamani and Clyde have proposed that mental imagery could itself be a form of catastrophising [52]. A study of patients undergoing dental procedures found that 37% catastrophised in their pain imagery ($N = 75$) [53].

We suggest the following potential meanings of the pain imagery themes we elucidated:

- *"Sensory qualities of pain"* may offer women a way to make sense of and qualify their pain experience;
- *"Loss of power or control"* may refer to endometriosis being an invasive condition, where "rogue" cells often take control of women's lives. As another explanation, many women wait years for a diagnosis with no control over their pain (in our international study, 27% reported that they had waited for more than 10 years for a diagnosis). Additionally, endometriosis is difficult to manage, and women often have little they can do to improve the condition themselves;
- *"Attack"* may simply represent the sensory quality of the pain (for example, stabbing imagery with sharp, stabbing pain), but could also hold deeper

meanings, such as the belief that their pain is a punishment or that they are a victim. "Attack by self" warrants further exploration: endometriosis could itself be described as an "attack by self," as endometrial-like tissue moves to where it should not be, causing inflammation and pain. Alternatively, could these women also have thoughts of self-harm?;

- *"Pathology or anatomy envisaged"* may offer insight into a woman's beliefs surrounding the cause and prognosis of their pain and endometriosis. If this imagery reveals a misunderstanding in the anatomy or pathophysiology of endometriosis or pelvic pain, clinicians could empower the patient with a greater understanding of their condition. It has been suggested that it may be beneficial to video record operations and have the surgeon discuss the video's content with the patient (enabling the woman to see the pathology and its removal) [54];
- *"Past or future catastrophe"* may highlight experiences or concerns that warrant extra attention and exploration. This imagery may reveal the woman's greatest fear or most distressing memory—a key discussion area;
- *"Pain as an object"* may signify that women feel their pain is "concrete." If these objects are "internalized" (where women see objects "within" themselves), imagery techniques, which "externalize" these objects, may be therapeutic; and
- *"Abstract images"* may reflect the status of endometriosis as a vaguely understood condition, or chronic pain as a difficult to conceive experience.

The imagery themes we found align closely with those in the pain imagery literature, which tend to be relatively consistent across patients with various causes of chronic pain [31, 32, 35, 36], as discussed by Berna [37]. Berna, Tracey and Holmes highlight that imagery can *"incorporate symbolic or real elements of the patient's individual pain experience"* [12]. These elements may prove key in pain management, as the meaning given to pain can have a powerful influence on its experienced intensity [55]. Indeed, meaning is *essential* to pain and the pain experience (for example, [33, 56, 57]).

Many examples of pain imagery reported by our participants and in the literature (for example, [32]) represent the pain itself, for example *"a knife stabbing."* Some may suggest that this imagery is voluntary and descriptive, so that sufferers can help others to understand their pain experience. Indeed, there is a separate literature that focuses on metaphors (voluntary descriptions used to communicate a patient's pain experience): see Stewart's chapter in the first book of *Meanings of Pain* for a review [58]. However, we agree with Berna, Tracey and Holmes [12] that this is not the case: patients live with their imagery far beyond the odd explanation to their clinicians, family or friends, and reported pain imagery is intrusive, distressing and often uncontrollable. Additionally, most participants have never talked about their imagery with anyone before [12]. In our questionnaire, we clearly stated: *"A mental image is **not a metaphor** for how the pain feels (it is not a way of describing your pain to others), but it is an image that pops into your mind when you are in pain."*

We found a substantial prevalence of coping imagery (30%), which may have been generated through psychological therapy or by women de novo as a form of self-management. We also discovered examples of naturally occurring rescripted imagery. In their in-depth interview-based study of women with chronic pelvic pain, Berna and colleagues found that eight out of the ten women with chronic pelvic pain interviewed reported coping imagery, with three proposed themes: *allegoric treatment of an object symbolizing pain, imaginary treatment applied to the body* and *supportive person* [32]. Our 124 examples of coping imagery suggest the following themes:

- *Positive, relaxing imagery*, for example white light, beach scene;
- *Rescripted imagery*, for example from *"ligaments contorting in a cartoon style"* they picture *"a warm hand smoothing over the contorting ligaments;"*
- *Memories*, for example, *"I picture myself with my husband on our honeymoon;"*
- *Supportive person*, for example, *"someone hugging me, giving me comfort, but I can't see who it is;"*
- *Imaginary treatment,* for example, *"a white or pale-yellow glowing light wrapping around my reproductive organs like ribbons until it envelops everything and reduces swelling;"* and
- *Taking "ownership" of their original pain imagery*, for example, *"Sometimes I can get a laugh out of picturing a little alien/demon inside me wreaking havoc."*

The main strengths of our study include the large sample size ($N = 785$) compared with previous pain imagery studies, which range from a sample size of 10 [32] to 105 [31], and the use of a previously piloted questionnaire with validated measures widely used in chronic pain research. However, convenience sampling may have introduced bias, and participants were largely from Westernised nations, and 90% were white. Due to the lack of diversity in our participants, we cannot truly generalise our findings to women with endometriosis worldwide. Self-report measures and self-declared characteristics, which were not clinically or externally validated, are open to bias and subjectivity. Co-morbidities (for example, depression or anxiety) and the use of concomitant medications (for example, analgesics, antidepressants or anxiolytics) were not explored. "Mental image," used in the questionnaire when describing pain imagery, may be misinterpreted by participants as a suggestion that their pain is "all in their head." This may have led to an artificially low pain imagery prevalence through questionnaire abandonment.

In summary, our international study aligns with our previous UK-based work suggesting that pain imagery is prevalent in women with endometriosis-associated pain and generally has a negative impact. This finding, in addition to the prevalence of coping imagery reported, suggests that imagery-based therapies (such as rescripting—see Case Study 2 below) may be an appropriate complementary treatment approach for women with endometriosis-associated pain.

5 Clinical Implications

5.1 Meanings of Pain

As Berna, Tracey and Holmes have described [12], intrusive cognitions experienced by those living with chronic pain provide a method to explore and understand the patient's unique pain experience. This does not just mean exploring the sensory qualities of pain, although this may be relevant, but also the meaning to that particular individual of the pain in terms of the impact on their emotions, function and ultimately quality of life. For professionals supporting individuals to adjust to life with any persistent health condition, working therapeutically with meaning appears key. For example, the meanings of pain can powerfully determine its associated affect [33], perceived intensity [55], and response to treatment [56].

It has been argued (for example, by Pearson and colleagues [13]) that psychological therapies have tended to focus on verbal cognitions and have largely neglected mental imagery. We suggest that researchers should redress this balance and that mental imagery should be an important consideration for clinicians working in pain management.

> One picture is worth a thousand words
> San Antonio Light, Page 6, 10 January 1918

Case Study 1 gives an example of exploring the meaning of pain imagery for a woman with endometriosis.

Case Study 1 Example of exploring the meaning of pain imagery

Clinician	"Some women who have endometriosis have told me that they sometimes get an image that pops into their head when they get their pain. By this, I don't mean a metaphor, which is a way to describe your pain, but an image that intrudes on your thoughts in a threatening and unpleasant way. Have you ever experienced such an image?"
Patient	"I've never spoken about this before, but yes, I have had a picture that comes with my pain."
Clinician	"Would you like to share what picture you see?"
Patient	"I picture a kitchen knife stabbing my ovaries, over and over, when I get my worst pain. I can see the knife and my insides clearly—just like in the image in the poster in my last surgeon's office."
Clinician	"That sounds so frightening and distressing."

(continued)

Patient "It is, and it makes me worry about my future."
Clinician "What do you mean by that?"
Patient "Well, I really want to have children, but each time I picture my ovaries being attacked, I worry about whether or not I ever will."

Clinician explores the patient's fertility fears and the theme of attack—the patient's previously unexplored imagery guides the clinic appointment.

Padfield and colleagues in their *face2face* project have explored the use of pain-related photographic images to allow patients to share their pain narrative [45]. Patients were given 54 pain images (previously generated in partnership with pain patients) and were asked to select those that had personal significance, which acted as the talking point of the consultation, forming a *"clinician-patient-image therapeutic triangle."* Padfield and colleagues argue that by using these images, patients can lead the dialogue, facilitating a more balanced interaction between patient and clinician [45]. Photographer Rosy Martin has commented: *"What these images offer is a narrative space for people to step into, the possibility of some kind of identification and empathy with the other...some kind of slippery surface for further narrative"* [44].

The way that pain is described is affected by both culture and language, which can be overcome using imagery that depicts different types of pain [48]. Therefore, the use of imagery in the exploration of pain may be more revealing than the use of language: *"Whatever pain achieves, it achieves in part through its unsharability, and it ensures this unsharability through its resistance to language..."* [59].

5.2 Imagery Techniques

Imagery therapy has been around for millennia [60]. Many imagery techniques exist for pain relief, as described in Pincus and Sheikh's book [61]. Hales and colleagues describe four key imagery techniques: imagery competing tasks, metacognitive imagery techniques, positive imagery techniques, and imagery rescripting [62].

Imagery competing strategies describe when concurrent visuospatial tasks are used to dampen down problematic imagery [62]. Holmes' group discovered that intrusive memories induced by experimental trauma (traumatic film footage) can be diminished by a competing visuospatial cognitive task (*Tetris* computer game play), even for established memories [63].

Metacognitive strategies describe when patients are shown that imagery simply involves mental representations (*"an image is just an image"*), rather than possessing emotionally significant meaning [62].

Positive imagery strategies (sometimes labelled "guided imagery" [64]) describe when the patient is helped to form and access soothing, pleasant or mood-enhancing imager [62]. This has long been a feature of psychological therapy, for example, as a form of relaxation *("imagine, using all of your senses, you are in a calm place...")*, which promotes a shift in attention from pain [65]. Most often, a script is used with a standard image (for example, a meadow) [37]. Alternatively, positive imagery can be built in partnership with the patient [30], which aims to distract, relax, increase optimism, and/or improve emotions [37].

There is encouraging (although inconclusive) systematic evidence that positive imagery strategies benefit patients with various causes of chronic pain (for example [66–68]), including in children [69]. Guided imagery has reduced mobility difficulties related to the painful condition [70] and improved self-reported capacity to cope with pain [71]. Guided imagery is recognised as a management option for chronic pelvic pain [72]. For more information on guided imagery, see Hall, Hall, Stradling, and Young's book [73].

Imagery rescripting (or "transformation") describes when negative or distressing imagery is transformed into benign imagery [62]. Spontaneous negative imagery is first identified, then the meaning and emotional impact of this imagery is explored by the clinician. This allows the patient to reflect on their imagery and, in partnership, the patient and clinician can form new imagery that counters the negative meaning and affective value of the original imagery [30]. This can be summarised as two key steps: (1) sensory re-description, and (2) cognitive re-appraisal [74]. Imagery rescripting aims to increase the sense of control individuals have over their imagery. For example, imagery rescripting has been used for survivors of childhood sexual abuse experiencing post-traumatic stress nightmares [75]. Rusch and colleagues include two case examples of imagery rescripting for the interested reader [76]. Imagery transformation can also take place under hypnosis [77].

Philips and Samson found that a short session of imagery rescripting can be enough to improve emotional status and reduce negative appraisals and pain [78]. The significant improvement in pain was independent of emotional change. However, long-term outcomes are unknown. This article includes three examples of participant rescripted imagery, for example: imagery changed from *"I see myself on all fours—like a dog but unable to move."* to *"I am at the start of a race...the gun goes off and the crowd cheers as I take off"* [78]. Participants found imagery rescripting easy and enjoyable.

Imagery transformation of experimentally-generated pain-related imagery (*"Let your pain take on a shape..."*)—as opposed to spontaneous pain imagery—can reduce pain levels [79] and change the meanings of pain and its perceived permanence [80]. Lewandowski, Good and Draucker compared imagery transformation (*"Put your pain object in your hands... How would you change the shape...the size...Now change the color...and its texture...Give it a different sound..."*) with routine treatment in chronic pain patients ($N = 42$), who described their pain broadly as *"never-ending," "relative," "explainable," "torment," "restrictive," or "changeable"* [80]. For patients in the transformation group ($n = 21$), pain became *"changeable,"* and the theme of pain as *"never-ending"* did not recur over the 4-day study period (and this theme did recur or was maintained in the control group).

Lewandowski and other colleagues thereafter found that the same imagery technique reduced pain intensity and disability over an 8-week period ($N = 27$) [81].

For an overview of evidence regarding imagery rescripting, see Arntz's review [82], and for more information on the imagery rescripting technique, see Winterowd, Beck and Gruener's chapter "Eliciting and modifying imagery" in *Cognitive therapy with chronic pain patients* [83]. For more information regarding mental imagery in cognitive therapy, see Renner and Holmes' chapter in Leahy's *Science and Practice in Cognitive Therapy: Foundations, Mechanisms, and Applications* [84].

Case Study 2 details an example of imagery rescripting and its relevance to endometriosis-associated pain management.

Case Study 2 Adhesions like piano strings: Imagery rescripting

Patient	"They have told me I have adhesions. I have this picture in my head of the adhesions being like a piano string being pulled."
Clinician	"That sounds like a really unpleasant image to be having."
Patient	"Yes. I get the picture when I get this tight, pulling sensation."
Clinician	"Do you have any worries about the adhesions?"
Patient	"Well, yes. I think about the fact that if you pull something too much, or too hard, it will eventually snap. What's going to happen to my insides if these adhesions snap?"
Clinician	"I'm wondering if it might be helpful to spend a bit of time talking about adhesions and about the fact that adhesions are scar tissue and very unlikely to snap. Do you think that would be helpful?"
Patient	"Yes, please."

Clinician completes some education to normalise occurrence of adhesions after pelvic surgery and de-catastrophise the experience of pulling sensations in the pelvis.

Clinician	"I'm wondering if there is a way you could manage this picture in your head of the piano strings being pulled. It sounds like a horrible image to be having. Other people can find it helpful to have a way of managing the image, perhaps by changing it into a less horrible or scary image. I'm wondering if there is any way you could change this image to be less threatening?"
Patient	"I'm not sure what you mean."
Clinician	"Well, one idea that is coming to my mind is that other people have sometimes found it helpful to find a way to relax and then change the image to a less scary one. So, for example, you might do some of the diaphragmatic breathing we practised last session but then with each breath imagine the piano string loosening and becoming less tense."

(continued)

Patient "Oh, I see what you mean. Yes, I could try that and see if it might help."

Clinician guides patient in completing a breathing exercise and encourages the patient to think about how to change the image so that it is less threatening. The clinician then asks for feedback and the patient agrees it would be worth experimenting with this at home to see if it might help.

To our knowledge, there are no systematic trials of the use of imagery in endometriosis management. There is only one small mixed-methods study ($N = 10$) of guided imagery therapy for women with endometriosis [85]. Women reported that this helped them to cope daily with their endometriosis symptoms and that their pain and anxiety improved with guided imagery, although neither outcome was statistically significant. The endometriosis-specific guided imagery script described may offer a useful basis for future interventional research [85]. For more information and examples of imagery techniques, see Singer's book *Imagery in Psychotherapy* [86] and Stopa's chapter "How to use imagery in cognitive-behavioural therapy" in *Imagery and the Threatened Self* [87].

5.3 Therapeutic Frameworks

Not all mental imagery is necessarily amenable to rescripting and this intervention alone may be insufficient when working clinically with imagery in pain management. Pain can be viewed by the patient as a barrier to proceeding with life and they may wish to put their life on hold until the pain is "cured." While we can understand the logic of this approach, this presents difficulties in the case of chronic pain. Mental imagery may be related to the belief of pain as a barrier to living life and this can be a useful area to explore clinically. For example, a patient may report *"I have the image of the pain as a little monster in my pelvis. It ruins everything."* On further discussion this patient may report: *"How can I possibly live a normal life with this monster. I need to get rid of it before can get back to living my life in the way I want."* In this case, the mental imagery allows exploration of the meaning associated with pain in terms of expectations of cure and a sense of pain taking this person away from the life they had before and want to get back to. Intervention may then focus on supporting adjustment to chronic pain and encourage the person to process the losses in life resulting from chronic pain but also how to engage with areas of life that are important to them, even in the presence of pain.

Acceptance and Commitment Therapy (ACT) (for in-depth coverage, see Hayes and Strosahl's guidebook [88]) is an approach focused on facilitating acceptance (that is a willingness to take what is given) and commitment to valued life areas. There is evidence that acceptance may have a mediating effect for change in physical functioning for patients with chronic pain [89], and that acceptance mediates the relationship between pain and physical dysfunction [90]. ACT uses discussion and metaphor to explore the workability of the patient's current strategies and facilitate development of more useful approaches in taking them towards the areas of life that are important to them. See Vowles and Thompson's chapter for an overview of ACT applied to pain management [91] or Dahl, Luciano and Wilson's book for in-depth coverage [92].

Journey metaphors are used in ACT. For example, we use the following bus metaphor: *The bus journey is a metaphor for valued life directions. For example, we may have the value of being a caring and supportive friend and plan to "drive" our bus in this direction (perhaps by meeting our friends regularly). However, chronic pain represents unwanted and controlling passengers on the bus. These passengers cause the person to stop driving the bus in the direction they want to, resulting in a sense of loss of control and self-reliance.* This metaphor acts as a springboard for problem-based learning and illuminating dialogue [58]. For example, discussion regarding the fact that although the passengers on the bus are extremely unpleasant, they are not harmful. It makes sense that the bus has stopped in the middle of the road or is going around in circles while the person argues with the horrible passengers. However, the patient can find ways to step back from these passengers, get back behind the steering wheel, and take the bus in the direction of the things that are important to them.

A key message in ACT is that attempts to get rid of pain may be part of the problem, and perhaps a different approach is needed. This is well portrayed by the *"Chinese finger trap experiment"* described by Stewart in the first book of *Meanings of Pain* [58]. If the patient is open to the idea that their current approach (for example, cure seeking) is not working, the intervention then focuses on defining important areas of the patient's life (their values) and taking steps towards these as defined in behavioural terms (for example, ways to be physically active if health is valued, but the individual is unable to engage in their previously enjoyed hobby of running).

As described in Case Study 1, imagery may be linked to concerns about fertility. Endometriosis is linked to subfertility [93], and its management can be contraceptive. Imagery may represent a woman's struggle to conceive or their concern that this is an area of life endometriosis will affect. These worries (and associated imagery) are linked to distress and represent a potential target for intervention. ACT metaphors can be used to help patients "defuse" or step back from their unhelpful thoughts. It may be that thoughts are biased, or, as in the fertility example, the future unclear and dwelling on these worries may increase distress and impact on functioning. Imagery-based exercises aim to help people see thoughts as thoughts, not necessarily truths (a metacognitive strategy). Techniques such as imagining putting the unhelpful thought in a box and sending it out to sea may help to gain a sense of being able to manage what our mind gives us.

Living with chronic pain is often associated with loss of identity. People can become isolated and perceive negative judgement from others. This can result in high rates of negative self-evaluation, shame and self-criticism

[94, 95]. Compassion-focused therapy (CFT) (see Gilbert's book [96]) was developed to address shame and self-criticism. Gilbert defines compassion as *"sensitivity to the suffering of self and others, with a deep commitment to alleviate it"* [96]. CFT includes use of imagery-based techniques aiming to develop individuals' ability to respond compassionately to their experience of chronic pain, and to develop their self-soothing abilities. There is also an explicit focus on developing coping strategies related to attention, with an evolutionary explanation of why our minds are directed to threat-based stimuli (such as pain) but that we have a responsibility to manage our minds and the focus of our attention. This is very relevant when working with imagery in pain management.

When working within a CFT framework, therapy attempts to direct attention and develop the emotion regulation system linked to self-soothing. This includes a focus on exercises aiming to facilitate the experience of receiving compassion from a *"compassionate other"* (a mental image of the "compassionate other," be it human or non-human, can be used [97]), and in imagining what it is like to be self-compassionate via guided exercises. These aim to help people increase their ability to get in touch with, and tolerate their distress, as well as take steps to alleviate this via a focus on compassionate responding. These can be powerful experiential exercises and can be useful in managing imagery linked to distress, in particular, attack and punishment imagery. CFT imagery can, in some individuals, generate a soothing affect system and diminish hypothalamic-pituitary-adrenal axis activity (that is activity of the body's core stress response system) [97]. Case Study 3 provides an example of compassion-focused therapy.

Case Study 3 *"Compassionate Other"* Exercise from Compassion-Focused Therapy

The clinician describes the rationale for completing an exercise in which the focus is to imagine receiving compassion from another. The clinician explains that this is an ideal "compassionate other" and could be someone from the person's life but may be better conceptualised as a figure from literature, history or a combination of known others to make the ideal "compassionate other." This other can also be a pet or an image from the natural world. The "compassionate other" encapsulates the qualities of warmth, non-judgement, strength (to bear the suffering the person is experiencing) and wisdom (to help them take steps to manage this suffering). In the current authors' experience, time would have been spent in previous sessions completing grounding, mindfulness, breathing and calm place exercises as well as discussing the concept of compassion with the patient with a progression to completing this exercise in an appointment.

The clinician then guides the patient with an initial focus on grounding (for example, "feel your feet on the floor, the points of contact of your body in the chair"), to a focus on mindful breathing (for example, "notice your breath as it

(continued)

enters and leaves your body"), to soothing rhythm breathing (for example, "try to have a rhythm to your breathing that is soothing for you, this might involve slowing your breathing down, having an evenness of the in and the out breath—you may like to count your breaths").

The clinician then invites the individual to use all of their senses to imagine they are in a calm place ("What can you see...hear...smell...touch in your calm place?") and then that their ideal "compassionate other" joins them in their calm place. The patient is invited to think about how they know this other is showing them compassion (for example "What do they say? What is their tone of voice, body language, and expression like? How does it feel to receive this compassion from another?") The clinician then brings the exercise to an end by inviting the patient to focus once again on their calm place, then their breathing, before returning to the "here and now," for example, with a focus of the feeling of their feet on the floor. The clinician will then debrief what the experience of completing the exercise was like.

Patient "It was tricky as my mind kept going to different people, but I did eventually settle on something."

Clinician "How did you know they were showing you compassion?"

Patient "Well, I imagined that my dog joined me in my calm place. He is always there for me, even when I'm having a bad pain day he seems to know and will just sit with me with his head on my knee until I get through it and the flare-up passes. He doesn't try to give me false promises as others do, like 'the pain will get better.' I'm starting to realise that 'better' can mean different things and my dog is there by my side helping me to cope and bear this pain."

The clinician and patient then discuss if this exercise would be worth repeating at home with the aim of developing the experience of receiving compassion from another as a coping strategy. This can be developed in a future session to include an imaginal exercise in which the person shows self-compassion and is asked to imagine how they know they are showing themselves compassion (for example, body language, what they say to themselves, tone of voice). Often people recognise that they are more compassionate towards others than they are towards themselves.

"Grounding" is a therapeutic technique that aims to help the person come into the "here and now," and offers a distraction from intrusive imagery. Kennerley has described a number of "grounding" techniques, from refocusing (for example, by focusing on breathing) to grounding words, phrases or images [98].

When working clinically with women with chronic pelvic pain we have found an integrated approach drawing on evidence-based interventions to be helpful. Table 7.2 provides examples of the therapeutic approaches described in this chapter

Table 7.2 Themes identified in endometriosis-associated pain imagery research, example imagery and possible clinical interventions

Theme	Example of imagery	Possible clinical interventions
Sensory qualities of pain	*"A fire burning"*	• Exploration of any worries or beliefs associated with this imagery • There may be a role for education. For example, discussing pain mechanisms to facilitate understanding of the burning experienced as characteristic of neuropathic pain as opposed to other potentially more sinister explanations the person may be concerned about • Imagery rescripting. For example, introducing diaphragmatic breathing and then with each breath imagining manipulating the image into a less emotive image, for example a sprinkler system putting out the fire
Loss of power or control	*"I often feel caged or trapped (by my pain) with bars made of barbed wire"*	• Intervention aiming to increase the individual's sense of being able to manage the image and pain. Imagery rescripting may aim to change the image in a step-by-step manner to imagine the barbed wire softening and bars bending to allow escape • Discussion regarding loss of power and if this is relevant in other areas of life. By exploring if there are any exceptions to this we are aiming to strengthen the person's sense of being able to cope with adversity and regain a sense of control
Attack	Attack by someone, "something" or self, for example: "*I picture a person holding my uterus. In this image they are squeezing it, twisting it and pulling it.*"	• Exploration of the possible meaning of this image to the individual and increase their insight into any links with other areas of their life • There may be potential to restructure the meaning associated with the image. For example, by what mechanism could chronic pain be a punishment for perceived wrong-doing? • Compassion-focused therapy techniques can be useful clinically, in particular *"compassionate other"* or *"compassionate self"* imagery exercises
Pathology or anatomy envisaged	*"I can imagine clumps of endo stuck to my ovary all angry and red"*	• Explore the pros and cons of focusing on this image. • Normalise that this image *will* be attention grabbing—this is not the person's fault but perhaps they have a responsibility to try to manage their mind and their mind's tendency to focus on the image • There may be a role for education regarding adhesions, fertility, pain physiology, and so on
Past or future catastrophe	*"I often have flashbacks...also what would happen if I was to die"*	• Acknowledge the emotion linked to the image • Exploration of "What if...[feared future catastrophe were to happen]?" How likely is the feared outcome? Or perhaps the person is underestimating their ability to cope? What might help them cope in this feared scenario?

		• Grounding approaches aiming to help the person come into the "here and now" (for example, by focusing on their breathing) when experiencing this image. Similarly, imagery competing strategies could be trialled
Pain as an object	*"A red ball with spikes"*	• Imagery rescripting. For example, the red spikes retracting with a deep breath. This responder had rescripted her own imagery: *"I imagine the red ball being covered in a thick pale blue liquid and it leaving my body into an ocean"* • Grounding techniques
Abstract images	*"A sea of rusty nails. Or just the colour green.* *That browny-green they use in cartoons to denote vomit"*	• Imagery rescripting. For example, the rusty nails being smoothed down by sandpaper or a similar method, or the green colour transforming into their favourite colour • Grounding techniques • Positive imagery strategies: *"imagine, using all of your senses, you are in a calm place..."*

as applied to examples of imagery reported by women with endometriosis-associated pain.

6 Summary and Conclusions

In summary, meaning is essential to the pain experience, and pain imagery may reveal hidden or unexplored meanings of pain. Our research has found that pain imagery is prevalent in the common condition endometriosis, and generally has a negative impact. We categorised reported pain imagery into themes: sensory qualities of pain, loss of power or control, attack, pathology or anatomy envisaged, past or future catastrophe, pain as an object, and abstract images. These themes and pain imagery content may reveal deeper meanings of pain to sufferers, providing clinicians with an effective and hitherto under-utilised (at least in endometriosis-associated pain) means to explore chronic pain, its personal significance, and perhaps provide novel (and personalised) clinical interventions (Table 7.2). Coping and self-rescripted imagery both exist, which suggests that a proportion of women with endometriosis-associated pain may be particularly amenable to imagery-based therapies.

In order to adequately explore pain imagery and its impact, a mixed-methods approach was adopted. Indeed, in order to effectively evaluate the meanings of pain, interview- or questionnaire-based qualitative methodology is essential. Qualitative methods encourage depth, detail and openness in participant responses—essential for researching personalised meanings of pain. However, qualitative methods are deemed by some to be less rigorous than quantitative methods, and often have fewer participants due to time and funding restraints. After piloting and seeking feedback on our pain imagery questionnaire at interview, we collected qualitative and quantitative data via online questionnaire, advertised online and via social media, which maximised response rate ($N = 785$, 43 countries).

We propose that women with endometriosis-associated pain may benefit from imagery-based therapies, and that women who experience pain imagery may benefit most from these (the presence of pre-existing pain imagery is a positive predictor of outcomes following imagery-based intervention [46]). Additionally, we believe routine assessment of women with endometriosis-associated pain would benefit from exploration of pain imagery. Imagery content may reveal more about women's pain experience, their understanding of endometriosis, and their beliefs and fears. We believe the next step is a clinical trial evaluating imagery-based techniques for the management of endometriosis-associated pain.

Allied research has shown that pain imagery commonly accompanies chronic pain, no matter its cause. We suggest that further research into pain imagery and the use of imagery-based techniques in pain management is required, both for women with endometriosis-associated pain and for patients with other chronic pain conditions. A reliable measure of pain imagery and its characteristics should be developed and validated.

Acknowledgments The authors are grateful to Dr. Simon van Rysewyk for his edits of this chapter; Dr. David Gillanders for offering comment on questionnaire design; Dr. Katy Vincent for expert review of our international research; the creators and distributors of the validated and reliable measures: Short-Form McGill Pain Questionnaire 2 ([39] access from Mapi Research Trust: https://eprovide.mapi-trust.org, adapted with permission), Pain Catastrophizing Scale ([40] now access is also from Mapi Research Trust: https://eprovide.mapi-trust.org), and Depression, Anxiety and Positive Outlook Scale [41]; all of the endometriosis patient organisations (Endometriosis UK, Endometriosis Foundation of America, Endometriosis.org, Endometriosis New Zealand, Endometriosis Australia, and The Endometriosis Network Canada) for advertising the online survey; and all of the participants who completed the questionnaire. The international pain imagery research was supported by an MRC Centre Grant (MR/N022556/1).

References

1. Elliott AM, Smith BH, Penny KI, Smith WC, Chambers WA. The epidemiology of chronic pain in the community. Lancet. 1999;354(9186):1248–52.
2. Smith BH, Elliott AM, Chambers WA, Smith WC, Hannaford PC, Penny K. The impact of chronic pain in the community. Fam Pract. 2001;18(3):292–9.
3. Breivik H, Collett B, Ventafridda V, Cohen R, Gallacher D. Survey of chronic pain in Europe: prevalence, impact on daily life, and treatment. Eur J Pain. 2006;10(4):287–333.
4. Johannes CB, Le TK, Zhou X, Johnston JA, Dworkin RH. The prevalence of chronic pain in United States adults: results of an internet-based survey. J Pain. 2010;11(11):1230–9.
5. National Pain Summit Initiative. National pain strategy: pain management for all Australians [Internet]. 2010 [cited 13/01/2019]. Available from https://www.painaustralia.org.au/static/uploads/files/national-pain-strategy-2011-wfvjawttsanq.pdf
6. Royal College of Obstetricians & Gynaecologists (RCOG). The initial management of chronic pelvic pain [Internet]. Green-top Guideline No. 41. 2012 [cited 13/01/2019]. Available from https://www.rcog.org.uk/globalassets/documents/guidelines/gtg_41.pdf
7. Hickey M, Ballard K, Farquhar C. Endometriosis. BMJ. 2014;348:g1752.
8. Zondervan KT, Becker CM, Koga K, Missmer SA, Taylor RN, Viganò P. Endometriosis. Nat Rev Dis Primers. 2018;4(1):9.
9. Adamson GD, Kennedy S, Hummelshoj L. Creating solutions in endometriosis: global collaboration through the world endometriosis research foundation. J Endometr. 2010;2(1):3–6.
10. Hadfield R, Mardon H, Barlow D, Kennedy S. Delay in the diagnosis of endometriosis: a survey of women from the USA and the UK. Hum Reprod. 1996;11(4):878–80.
11. Hudelist G, Fritzer N, Thomas A, Niehues C, Oppelt P, Haas D, et al. Diagnostic delay for endometriosis in Austria and Germany: causes and possible consequences. Hum Reprod. 2012;27(12):3412–6.
12. Berna C, Tracey I, Holmes EA. How a better understanding of spontaneous mental imagery linked to pain could enhance imagery-based therapy in chronic pain. J Exp Psychopathol. 2012;3(2):258–73.
13. Pearson J, Naselaris T, Holmes EA, Kosslyn SM. Mental imagery: functional mechanisms and clinical applications. Trends Cogn Sci. 2015;19(10):590–602.
14. Kosslyn SM, Ganis G, Thompson WL. Neural foundations of imagery. Nat Rev Neurosci. 2001;2(9):635–42.
15. McNorgan C. A meta-analytic review of multisensory imagery identifies the neural correlates of modality-specific and modality-general imagery. Front Hum Neurosci. 2012;6:285.
16. Barber TX, Chauncey HH, Winer RA. Effect of hypnotic and nonhypnotic suggestions on parotid gland response to gustatory stimuli. Psychosom Med. 1964;26:374–80.

17. Jones GE, Johnson HJ. Physiological responding during self generated imagery of contextually complete stimuli. Psychophysiology. 1978;15(5):439–46.
18. Jones GE, Johnson HJ. Heart rate and somatic concomitants of mental imagery. Psychophysiology. 1980;17(4):339–47.
19. Schneider J, Smith CW, Minning C, Whitcher S, Hermanson J. Guided imagery and immune system function in normal subjects: a summary of research findings. In: Kunzendorf RG, editor. Mental imagery. Boston, MA: Springer US; 1991. p. 179–91.
20. Hirsch CR, Holmes EA. Mental imagery in anxiety disorders. Psychiatry. 2007;6(4):161–5.
21. Hackmann A, Holmes EA. Reflecting on imagery: a clinical perspective and overview of the special issue of memory on mental imagery and memory in psychopathology. Memory. 2004;12(4):389–402.
22. Hackmann A, Clark DM, McManus F. Recurrent images and early memories in social phobia. Behav Res Ther. 2000;38(6):601–10.
23. Patel T, Brewin CR, Wheatley J, Wells A, Fisher P, Myers S. Intrusive images and memories in major depression. Behav Res Ther. 2007;45(11):2573–80.
24. Di Simplicio M, Renner F, Blackwell SE, Mitchell H, Stratford HJ, Watson P, et al. An investigation of mental imagery in bipolar disorder: exploring "the mind's eye". Bipolar Disord. 2016;18(8):669–83.
25. Somerville K, Cooper M, Hackmann A. Spontaneous imagery in women with bulimia nervosa: an investigation into content, characteristics and links to childhood memories. J Behav Ther Exp Psychiatry. 2007;38(4):435–46.
26. Holmes EA, Mathews A. Mental imagery in emotion and emotional disorders. Clin Psychol Rev. 2010;30(3):349–62.
27. Holmes EA, Geddes JR, Colom F, Goodwin GM. Mental imagery as an emotional amplifier: application to bipolar disorder. Behav Res Ther. 2008;46(12):1251–8.
28. Mathews A, Ridgeway V, Holmes EA. Feels like the real thing: imagery is both more realistic and emotional than verbal thought. Cogn Emot. 2013;27(2):217–29.
29. Holmes EA, Arntz A, Smucker MR. Imagery rescripting in cognitive behaviour therapy: images, treatment techniques and outcomes. J Behav Ther Exp Psychiatry. 2007;38 (4):297–305.
30. Hackmann A, Bennett-Levy J, Holmes EA. Oxford guide to imagery in cognitive therapy. Oxford: Oxford University Press; 2011. 233 p.
31. Gosden T, Morris PG, Ferreira NB, Grady C, Gillanders DT. Mental imagery in chronic pain: prevalence and characteristics. Eur J Pain. 2014;18(5):721–8.
32. Berna C, Vincent K, Moore J, Tracey I, Goodwin GM, Holmes EA. Presence of mental imagery associated with chronic pelvic pain: a pilot study. Pain Med. 2011;12(7):1086–93.
33. Price DD, Harkins SW. The affective-motivational dimension of pain: a two-stage model. J Pain. 1992;1(4):229–39.
34. van Rysewyk S. A call for study on the meanings of pain. In: van Rysewyk S, editor. Meanings of pain. Cham: Springer; 2016. p. 1–22.
35. Philips HC. Imagery and pain: the prevalence, characteristics, and potency of imagery associated with pain. Behav Cogn Psychother. 2011;39(5):523–40.
36. Gillanders D, Potter L, Morris PG. Pain related-visual imagery is associated with distress in chronic pain sufferers. Behav Cogn Psychother. 2012;40(5):577–89.
37. Berna C. Mental imagery in chronic pain: an access to meaning beyond words. In: van Rysewyk S, editor. Meanings of pain. Cham: Springer; 2016. p. 267–80.
38. Gosden T. Images of pain: exploration of the characteristics and functions of pain-related mental imagery in chronic pain. D Clin Psychol [Thesis] [cited 13/01/2019]. Edinburgh: University of Edinburgh; 2008. Available from Edinburgh Research Archive https://www.era.lib.ed.ac.uk/handle/1842/2742
39. Dworkin RH, Turk DC, Revicki DA, Harding G, Coyne KS, Peirce-Sandner S, et al. Development and initial validation of an expanded and revised version of the short-form McGill pain questionnaire (SF-MPQ-2). Pain. 2009;144(1–2):35–42.

40. Sullivan MJ, Bishop SR, Pivik J. The pain catastrophizing scale: development and validation. Psychol Assess. 1995;7(4):524–32.
41. Pincus T, Williams AC, Vogel S, Field A. The development and testing of the depression, anxiety, and positive outlook scale (DAPOS). Pain. 2004;109(1–2):181–8.
42. Braun V, Clarke V. Using thematic analysis in psychology. Qual Res Psychol. 2006;3 (2):77–101.
43. Marks DF. Visual imagery differences in the recall of pictures. Br J Psychol. 1973;64(1):17–24.
44. Padfield D, Janmohamed F, Zakrzewska JM, Pither C, Hurwitz B. A slippery surface... can photographic images of pain improve communication in pain consultations? Int J Surg. 2010;8 (2):144–50.
45. Padfield D, Chadwick T, Omand H. The body as image: image as body. Lancet. 2017;389 (10076):1290–1.
46. Carruthers HR, Miller V, Morris J, Evans R, Tarrier N, Whorwell PJ. Using art to help understand the imagery of irritable bowel syndrome and its response to hypnotherapy. Int J Clin Exp Hypn. 2009;57(2):162–73.
47. Carruthers HR, Morris J, Tarrier N, Whorwell PJ. Reactivity to images in health and irritable bowel syndrome. Aliment Pharmacol Ther. 2010;31(1):131–42.
48. Carruthers HR, Whorwell PJ. Letter: using pictures to improve communication between doctor and patient in functional gastrointestinal disorders. Aliment Pharmacol Ther. 2014;40 (11–12):1364.
49. Henare D, Hocking C, Smythe L. Chronic pain: gaining understanding through the use of art. Br J Occup Ther. 2003;66(11):511–8.
50. Phillips J, Ogden J, Copland C. Using drawings of pain-related images to understand the experience of chronic pain: a qualitative study. Br J Occup Ther. 2015;78(7):404–11.
51. Kirkham JA, Smith JA, Havsteen-Franklin D. Painting pain: an interpretative phenomenological analysis of representations of living with chronic pain. Health Psychol. 2015;34 (4):398–406.
52. Jamani N, Clyde Z. Treatment of pain-related fear in chronic (persistent) pain: the role of safety-seeking behaviour and imagery. Cogn Behav Ther. 2008;1(1):3–15.
53. Chaves JF, Brown JM. Spontaneous cognitive strategies for the control of clinical pain and stress. J Behav Med. 1987;10(3):263–76.
54. Slade P, Cordle C. Psychological aspects of the management of chronic pelvic pain. Curr Obstet Gynaecol. 2005;15(5):298–305.
55. Arntz A, Claassens L. The meaning of pain influences its experienced intensity. Pain. 2004;109 (1–2):20–5.
56. Cassell EJ. The phenomenon of suffering and its relationship to pain. In: Toombs SK, editor. Handbook of phenomenology and medicine. Dordrecht: Springer; 2001. p. 371–90.
57. Bayer TL, Baer PE, Early C. Situational and psychophysiological factors in psychologically induced pain. Pain. 1991;44(1):45–50.
58. Stewart M. Seeking order amidst the chaos: the role of metaphor within pain reconceptualisation. In: van Rysewyk S, editor. Meanings of pain. Cham: Springer; 2016. p. 325–53.
59. Scarry E. The body in pain: the making and the unmaking of the world. New York: Oxford University Press; 1985. 400 p.
60. Achterberg J. Imagery in healing: shamanism and modern medicine. Boston, MA: Shambhala Publications; 2002. 253 p.
61. Pincus D, Sheikh A. Imagery for pain relief: a scientifically grounded guidebook for clinicians. New York: Routledge; 2009. 311 p.
62. Hales SA, Di Simplicio M, Iyadurai L, Blackwell SE, Young K, Fairburn CG, et al. Imagery-focused cognitive therapy (ImCT) for mood instability and anxiety in a small sample of patients with bipolar disorder: a pilot clinical audit. Behav Cogn Psychother. 2018;46(6):706–25.

63. James EL, Bonsall MB, Hoppitt L, Tunbridge EM, Geddes JR, Milton AL, et al. Computer game play reduces intrusive memories of experimental trauma via reconsolidation-update mechanisms. Psychol Sci. 2015;26(8):1201–15.
64. Hart J. Guided imagery. Altern Complement Ther. 2008;14(6):295–9.
65. Philips H, Rachman S. The psychological management of pain. New York: Springer; 1996. 286 p.
66. Posadzki P, Ernst E. Guided imagery for musculoskeletal pain: a systematic review. Clin J Pain. 2011;27(7):648–53.
67. Posadzki P, Lewandowski W, Terry R, Ernst E, Stearns A. Guided imagery for non-musculoskeletal pain: a systematic review of randomized clinical trials. J Pain Symptom Manag. 2012;44(1):95–104.
68. Giacobbi PR, Stabler ME, Stewart J, Jaeschke AM, Siebert JL, Kelley GA. Guided imagery for arthritis and other rheumatic diseases: a systematic review of randomized controlled trials. Pain Manag Nurs. 2015;16(5):792–803.
69. Weydert JA, Shapiro DE, Acra SA, Monheim CJ, Chambers AS, Ball TM. Evaluation of guided imagery as treatment for recurrent abdominal pain in children: a randomized controlled trial. BMC Pediatr. 2006;6:29.
70. Baird CL, Sands L. A pilot study of the effectiveness of guided imagery with progressive muscle relaxation to reduce chronic pain and mobility difficulties of osteoarthritis. Pain Manag Nurs. 2004;5(3):97–104.
71. Ilacqua GE. Migraine headaches: coping efficacy of guided imagery training. Headache. 1994;34(2):99–102.
72. Udoji MA, Ness TJ. New directions in the treatment of pelvic pain. Pain Manag. 2013;3 (5):387–94.
73. Hall E, Hall C, Stradling P, Young D. Guided imagery: creative interventions in counselling & psychotherapy. London: SAGE; 2006. 143 p.
74. Brewin CR, Wheatley J, Patel T, Fearon P, Hackmann A, Wells A, et al. Imagery rescripting as a brief stand-alone treatment for depressed patients with intrusive memories. Behav Res Ther. 2009;47(7):569–76.
75. Smucker MR, Dancu C, Foa EB, Niederee JL. Imagery rescripting: a new treatment for survivors of childhood sexual abuse suffering from posttraumatic stress. J Cogn Psychother. 1995;9(1):3–17.
76. Rusch MD, Grunert BK, Mendelsohn RA, Smucker MR. Imagery rescripting for recurrent, distressing images. Cogn Behav Pract. 2000;7(2):173–82.
77. Barber J, Hilgard ER. Hypnosis and suggestion in the treatment of pain: a clinical guide. New York: W.W. Norton; 1996. 439 p.
78. Philips C, Samson D. The rescripting of pain images. Behav Cogn Psychother. 2012;40 (5):558–76.
79. Lewandowski WA. Patterning of pain and power with guided imagery. Nurs Sci Q. 2004;17 (3):233–41.
80. Lewandowski W, Good M, Draucker CB. Changes in the meaning of pain with the use of guided imagery. Pain Manag Nurs. 2005;6(2):58–67.
81. Lewandowski W, Jacobson A, Palmieri PA, Alexander T, Zeller R. Biological mechanisms related to the effectiveness of guided imagery for chronic pain. Biol Res Nurs. 2011;13 (4):364–75.
82. Arntz A. Imagery rescripting as a therapeutic technique: review of clinical trials, basic studies, and research agenda. J Exp Psychopathol. 2012;3(2):189–208.
83. Winterowd C, Beck AT, Gruener D. Cognitive therapy with chronic pain patients. New York: Springer; 2003. 392 p.
84. Leahy RL. Science and practice in cognitive therapy: foundations, mechanisms, and applications. New York: Guilford Publications; 2018. 418 p.
85. Dance AD. Guided imagery as a coping mechanism for endometriosis: a mixed-methods approach. Doctor of philosophy in integrative medicine and health sciences [Thesis] [cited 13/01/2019]. Oakland, California: Saybrook University; 2017. Available from https://search.proquest.com/openview/bf4c4b8056c00422703de97632d48f9e/

86. Singer JL. Imagery in psychotherapy. Washington, DC: American Psychological Association; 2006. 225 p.
87. Stopa L. Imagery and the threatened self: perspectives on mental imagery and the self in cognitive therapy. New York: Routledge; 2009. 282 p.
88. Hayes SC, Strosahl KD. A practical guide to acceptance and commitment therapy. New York: Springer US; 2004. 396 p.
89. Cederberg JT, Cernvall M, Dahl J, von Essen L, Ljungman G. Acceptance as a mediator for change in acceptance and commitment therapy for persons with chronic pain? Int J Behav Med. 2016;23(1):21–9.
90. Gillanders DT, Ferreira NB, Bose S, Esrich T. The relationship between acceptance, catastrophizing and illness representations in chronic pain. Eur J Pain. 2013;17(6):893–902.
91. Vowles KE, Thompson M. Acceptance and commitment therapy for chronic pain. In: McCracken L, editor. Mindfulness and acceptance in behavioral medicine: current theory and practice. Oakland, CA: New Harbinger Publications; 2011. p. 31–60.
92. Dahl J, Luciano C, Wilson K. Acceptance and commitment therapy for chronic pain. Reno, NV: New Harbinger Publications; 2010. 224 p.
93. D'Hooghe TM, Debrock S, Hill JA, Meuleman C. Endometriosis and subfertility: is the relationship resolved? Semin Reprod Med. 2003;21(2):243–54.
94. Purdie F, Morley S. Compassion and chronic pain. Pain. 2016;157(12):2625–7.
95. Smith JA, Osborn M. Pain as an assault on the self: an interpretative phenomenological analysis of the psychological impact of chronic benign low back pain. Psychol Health. 2007;22 (5):517–34.
96. Gilbert P. Compassion focused therapy: distinctive features. New York: Routledge; 2010. 248 p.
97. Rockliff H, Gilbert P, McEwan K, Lightman S, Glover D. A pilot exploration of heart rate variability and salivary cortisol responses to compassion-focused imagery. Clin Neuropsychiatry J Treat Eval. 2008;5(3):132–9.
98. Kennerley H. Cognitive therapy of dissociative symptoms associated with trauma. Br J Clin Psychol. 1996;35(3):325–40.

Chapter 8
Labour Pain

Laura Whitburn and Lester Jones

> *Where it is unavoidable, pain can be transformed into something useable, something which wakes us beyond the limits of the experience itself into a further grasp of the essentials of life and the possibilities within us...This insight illuminates much of the female condition, but in particular the experience of giving birth [1, p. 158].*

Abstract Contemporary thinking about pain suggests its ultimate function is more than just to indicate bodily injury, pathology or disease. This would seem especially important in the pain that a woman feels during labour and childbirth. The event of birthing a child is essentially a normal and vital physiological process but the pain women report can be extreme. In addition, it can be quite variable, and the variability cannot be explained by tissue-based factors alone. The variability extends not just to the intensity of the pain but also to its quality and behaviour. Equally variable is the ability for women to cope with the pain associated with childbirth. It can be anticipated that individuals will have differing capacities to cope, but the variability can also be a moment-to-moment proposition for the individual woman. In this chapter we will discuss the idea that the meaning of labour pain to the woman may be more important than its nature or intensity in determining the balance between coping and acopia and in defining her overall experience. In doing so, we will also highlight the limitations in current conceptions of pain that cannot yet fully account for unique occurrences of pain, such as the pain of labour and childbirth.

Capsule Summary: The concepts explored in this chapter emphasise the need to attend to the individual meaning that a woman ascribes to her pain experience during labour. We suggest that by conceptualising labour pain as a productive and

---

L. Whitburn (✉)
La Trobe University, Melbourne, Australia
e-mail: L.Whitburn@latrobe.edu.au

L. Jones
La Trobe University, Melbourne, Australia

Singapore Institute of Technology, Singapore, Singapore

S. van Rysewyk (ed.), *Meanings of Pain*,
https://doi.org/10.1007/978-3-030-24154-4_8

purposeful pain, prioritising individualised social support and attending to cognitive and emotional variables that shape a woman's perception of pain, women may be more likely to have positive experiences of labour pain and less need for pharmacological intervention.

Keywords Labour · Pain · Childbirth · Social support · Pain cognitions · Pain control · Maternity care

1 Introduction

The pain associated with labour is a unique and complex phenomenon. Whilst typical experiences of pain tend to be associated with injury or disease, labour pain emerges during a vital and highly positive event. In fact, as the pain intensity rises, the labour is seen to be progressing normally! [2, 3] This pain experience raises significant philosophical and theoretical questions due to its unique occurrence. One obvious question is: Why is a normal physiological process, one that is essential to human existence, associated with such intense pain? The unique context of this pain provokes two ideas: (1) the ultimate function of pain is more than just to indicate bodily injury, pathology or disease; and (2) labour pain may be better understood if considered as different from other types of pain.

A further complexity to this pain experience is its enormous variation between women or in the same woman on different occasions, independent of the physical demands of labour on her body. Labour pain is often described as the most challenging and intense pain experience a woman can undergo. However, reports of intensity vary significantly and descriptions of this pain range from excruciating through to pleasurable [4, 5]. Some women manage the pain very well, require minimal assistance and report positive experiences, whilst others do not cope well, experience great suffering and request intervention in order to avoid or alleviate the pain [6–8]. Curiously, women have even described labour pain as a paradoxical experience—one that is both excruciating but desirable because of its positive outcome: the birth of a child [5]. This variation in both intensity and descriptions suggests that the nature of labour pain is complex, and the experience of labour pain has determinants beyond those associated with the physiological state of the woman's labouring body.

A growing argument emphasises the meaning of pain as the determining factor in defining a woman's pain experience during childbirth. Women may interpret their pain as productive and purposeful and accept it as part of their labour experience, or as a threatening pain from which they wish to escape. Thus, the meaning a woman attributes to her pain affects her relationship to the pain and her ongoing responses to it. The individual interpretation of this pain experience is a complex evaluative process influenced by personal, social, contextual and physical inputs. Furthermore, an exploration of the concept of a *productive and purposeful* pain challenges current conceptions of pain and its function. In this chapter we will examine these ideas to better account for the individual differences in experiences reported by women, as well as to expand conceptions of pain more broadly.

1.1 A Brief Historical Perspective

Prior to the scientific revolution, conceptions of labour pain were mainly driven by religious and cultural beliefs. For example, the Judeo-Christian conception of labour pain was that it was Eve's punishment for her sins in the Garden of Eden: "I will make your pains in childbearing very severe; with painful labor you will give birth to children"—Genesis 3:16 [9]. Many women continue to call on their faith to give them strength during labour. In other cultures and religions, labour pain has many different meanings and functions. In traditional Japanese culture, birth is believed to be the work of the gods and overcoming labour pain is seen as an honour [10]. Chinese childbearing women report that it is shameful to scream during labour, and a proverb often used is, "If you wish to be the best person, you must suffer the bitterest of the bitter" [11]. More recent conceptions of labour pain reflect the medicalisation of birth, but many women still hold beliefs about labour pain drawn from cultural traditions.

1.2 Current Definition of Pain

> An unpleasant sensory and emotional experience associated with actual or potential tissue damage, or described in terms of such damage. —IASP definition of pain [12]

Definitions of pain have evolved to be more inclusive of the human pain experience. The longstanding International Association of the Study of Pain (IASP) definition of pain [12] (which should always be considered with its annotation) challenges a biomedical, exclusively tissue-based understanding of pain by emphasising the emotional component and incorporating the idea that perceived damage may be enough to explain pain.

> Pain is a distressing experience associated with actual or potential tissue damage with sensory, emotional, cognitive, and social components. —Williams and Craig's [13] proposed revised definition

Above is a recent attempt to provoke discussion for change to the IASP statement; primarily to emphasise the cognitive and social components of pain [13]. This inspired much commentary including a response from us about its capacity to (in) sufficiently account for labour pain [14] and subsequently there has been a further detailed analysis of the IASP definition [15].

In the context of labour pain, the IASP definition, including its annotation, is overly focussed on damage to tissues. The implication is that the primary function of pain is to indicate actual or potential tissue damage. Here is the dilemma: labour is a normal physiological process and so it is hard to justify that the function of pain in that context is to indicate damage. The definition also promotes that tissue damage—whether actual or perceived—is the main contributor to the pain experience. Again, this would seem inappropriate in a process such as labour that is tissue-challenging but not necessarily tissue-injurious [16].

At this point we would like to acknowledge that women can and do sustain tissue damage during childbirth, but the pain associated with labour, we argue, is often separate to the phenomenon of tissue injury.

1.3 What Is the Function of Pain?

The Australian and New Zealand College of Anaesthetists, Faculty of Pain Medicine (ANZCAFPM) promote a reorganisation of the term biopsychosocial to socio-psycho-biomedical, and this perhaps is a good starting point to capture the pain associated with labour and childbirth. However, *biomedical* may be seen to exclude the reactive, learned and pre-emptive *biology* associated with psychological and social contexts, so important in the human pain experience.

In line with the emphasis on social and psychological contexts, it has been argued that pain during labour may have more of a behavioural function, by encouraging the woman to act in a way to find a safe place to birth and to promote empathic and support behaviours in those around her [17, 18]. As labour continues, increases in pain are associated with progression [2, 16] and perhaps the ramifications of this, and maybe the function of this, is to focus the woman on the task of birthing. One benefit of the intensity of the sensation could be to disable any sophisticated cognitive processing, preventing the woman from *overthinking* and instead forcing her to *withdraw within* [19] and engage with innate and primal processes as her labour progresses.

Of course, these alternate ideas about the primary function of pain could simply be a convenient consequence of a tissue-based phenomenon, for those putting forward these more humanistic views. However, it is important to recognise that the expression of pain has evolved in a social context. It could be argued that there would not be expression of pain if it did not warn the social group of some danger and trigger a response that promoted survival of group members, or if it did not facilitate behaviour in nearby others to act in a way to protect and support the individual in pain. It is not too difficult to imagine that the perception and expression of pain are likely to have evolved together around this social function. In which case, perhaps the association of pain and tissue damage is simply a convenient consequence for those maintaining a biomedical view.

Indeed, to think of pain only as a sign of tissue damage reduces its function to one that is increasingly questionable. A large number of imaging studies on asymptomatic participants published since the 1990s would suggest structural variations occur with age and are not necessarily pathological or the source of pain [20–22]. Pain might draw us to explore tissue integrity but it is not sensitive or specific to that. It would seem wiser to recognise pain as a driver of species preservation behaviour influenced by the complexity of psycho-socio-neuro-immune-endocrine synergies and that this may be for either protective or (re)productive purposes. This would include behaviour in response to tissue disease and damage—or perhaps more correctly put—the response to the challenge to survival (immediate or latent) that tissue disease and damage are often associated with.

If accepted fully, this argument about the development of pain perception ignores the role of pain in self-preservation behaviour that might be independent of the social context. Nevertheless, it does promote a shift in thinking that may allow a more comprehensive understanding of all types of pain and pain behaviour, but especially pain associated with labour and childbirth.

2 Perspectives and Theories of Labour Pain

2.1 Women's Report of Pain Experience

The research into women's experiences of labour pain produces an additional concern about applying the current and proposed definitions of pain to labour and childbirth. The current definition uses the word "unpleasant" while the definition proposed more recently [13] uses "distressing." Studies report that contrary to the pain having a negative quality as these terms suggest, many women describe the pain experience in positive terms associated with empowerment, strength, happiness and even pleasure [5, 23]. Some women are even conflicted enough not to want to use the term "pain" because the pervading negative connotation does not match their birthing experience, which is overwhelmingly positive [23]. It would seem that notions of working with the pain, commonly promoted by midwifery practitioners, aligns well with embracing a positive, more physiological interpretation of pain, and taking away the pain, commonly promoted by medical practitioners, aligns more with a negative, pathological interpretation of pain. The latter obviously is also more aligned with current pain definitions.

It would also seem apparent that the approach to pain promoted by a woman's carers needs to align to her concepts of pain and her experiences can be influenced by her ability to maintain her conceptual framework during her labour. A recent review of the literature of labour pain [24] emphasised the concept of the *individual meaning* of a woman's pain experience and this may in fact be its defining feature. It has been demonstrated across numerous studies involving women from various cultural backgrounds and birthing in various models of care, that a woman's pain experience is shaped by the personal meaning that she ascribes to it. The meaning of the pain is influenced by factors including personal beliefs, the context of the pain, cognitive attributes of the woman, and the immediate social and broader socio-cultural environment in which she is birthing. Women who ultimately described their labour pain as "productive and purposeful" tended to demonstrate a greater capacity to cope. Alternatively, other women described their experience of labour pain as a "threatening" pain, and this was often associated with a diminished capacity to cope and greater call for pain intervention.

To understand how women develop and sustain a meaning for their pain, we must first consider the broader socio-cultural and philosophical perspectives of pain. These overarching pain "beliefs" will no doubt form the foundation for a woman's own perspectives and understanding of her pain experience.

2.2 Current Labour Pain Theories

Two opposing theoretical views regarding labour pain and how it should be best managed exist currently. One can be described as a "biomedical" or "medicalised" view, and the other a "midwifery" or "working with pain" view, based on their differing philosophical perspectives of the function of labour pain and how it should be managed. The biomedical view of pain is that it is a sign that things are not right. Therefore, if effective pain relief is available, then it is unnecessary for any woman to experience labour pain [25]. This view advocates for the availability and use of pharmacological interventions to eradicate pain. The working with pain view sees labour pain as a normal part of labour and birth that can be used constructively [26]. This view emphasises the use of (non-pharmacological) resources to support the labouring woman to cope with the sensations of labour. Neither view believes that a woman should suffer during labour and birth. However, the biomedical view assumes that if a woman is in pain then she must be suffering [27], whilst the working with pain view separates pain from suffering and focuses on supporting the woman to cope with the pain of labour [28].

Over the past 50 years in Western societies, the medicalised view of labour pain has dominated. Within this view, labour pain is conceptualised using a tissue-based model that focuses on peripheral contributions to the woman's pain experience: labour pain is described as an "excellent model of acute pain;" that is, one that is clearly attributed to noxious stimulation [29]. Much literature on labour pain within this model emphasises that nociceptive input is the reason for a woman's pain and leaves little space for consideration of non-tissue-based influences [16, 30, 31]. Accordingly, pharmacological management of labour pain that targets the nociceptive input is prioritised. For example, in Wall & Melzack's Textbook of Pain (fifth ed.) over 12 pages is dedicated to describing pharmacological methods of managing labour pain, whilst less than one page discusses non-pharmacological methods. As stated on page 793: "The modern theory of pain management in labour and delivery points out that pain should and must be relieved effectively" [32]. The focus on the eradication of pain in labour is further illustrated by a statement made by the American College of Obstetrics and Gynecology in 2017: "Labor causes severe pain for many women. There is no other circumstance in which it is considered acceptable for an individual to experience untreated severe pain that is amenable to safe intervention while the individual is under a physician's care" [25, p. 766]. In an Australian study, a critical analysis of hospital documents provided to women described how the use of epidural analgesia during labour is framed as safe whilst the use of water immersion during labour is framed as risky, despite these claims not being supported by evidence [33]. It is clear from examples such as these how pervasive the medical model is. An assumption is made regarding how women understand and relate to their pain during labour, and subsequently how it will be managed, with little space for the possibility of it to be perceived as a productive and purposeful pain.

The opposing working with pain view sees labour pain as central to the process of labour and birth. Within this view, labour pain is described as "functional" pain in that it is "physiological pain felt in a healthy body working well, but at levels of high intensity, beyond usual comfort levels" [34]. The term "functional discomfort" is also suggested as an alternative to labour pain, in order to further differentiate it from pathological pains and the negative connotations associated with the term [35]. Within the working with pain view, it is believed that, given the right environment and circumstances, women possess the capacity to cope with the pain of normal labour.

A key feature of this thinking is that the pain experience is beneficial and various functions of labour pain may include:

- labour pain forces a woman to stop and divert her attention to her body, to recognise that she is about to give birth
- labour pain triggers a woman to summon support
- the challenge of labour pain marks the significance of the occasion—birthing a child
- the discomfort of labour heightens the joy of receiving a baby

At a biochemical level, research has identified that pain plays a vital role in triggering a cascade of neurohormones that optimise the labour process, such as oxytocin [36]. In addition, the production of the body's natural pain-relieving opiates—beta-endorphins—demonstrates that human physiology was designed to attenuate the nociceptive input generated by the woman's labouring body, to facilitate her coping.

It is important to note that a distinction is made between "normal" labour pain, which is that associated with the physiological process of normal labour, and "abnormal" pain, which may be associated with a complication such as labour dystocia or damage to tissues. The working with pain view recognises that abnormal pain may warrant pharmacological intervention. However, during normal labour, the view prioritises non-pharmacological intervention to support the labouring woman. This is justified by the growing body of evidence demonstrating negative effects of pharmacological interventions on hormonally-mediated mechanisms that support and drive labour, breastfeeding and maternal-infant attachment, as well as on mothers' and babies' health and outcomes (See Leap and Anderson [26] and Whitburn et al. [24] for summaries). In promoting normal birth, the view focuses on supporting the labouring woman to engage with, and work through any pain associated with normal labour, rather than trying to take it away. Importantly, the working with pain view emphasises the role of the woman's support people in helping her cope with her pain: The attitudes and actions of her support people will have a powerful influence on her own perceptions of her pain and ability to cope.

It is clear that each view assumes a different meaning for the pain associated with labour. The medicalised view does not differentiate labour pain from pains associated with pathology, injury, disease or over-applied adaptive changes to the nervous system. The meaning of labour pain is simply associated with nociceptive input from tissue damage, and its management (i.e., to relieve all pain) suggests its redundancy in the process of labour and birth. On the other hand, the working with pain view recognises that labour pain has a different context and function to other pains. Labour pain is functional pain in that it is associated with desirable (if extreme) adaptive tissue changes that occur within a normal physiological event. The working with pain view also acknowledges possible philosophical, social, and personal implications of this pain experience that, from an evolutionary perspective, are important to consider.

Ultimately, we may describe a continuum in which at one end labour pain is viewed as an unnecessary by-product of the labour and birth process that can and should be avoided. At the other end labour pain is viewed as a central component to the transformative process of becoming a mother and has several specific functions. Across the continuum are likely to be mixed, uncertain or ambivalent feelings relating to the role of pain in labour. Recognising these two alternate views regarding labour pain allows us to appreciate the likely social influences that women face prior to, and during, labour and birth. Undoubtedly, these attitudes will subsequently shape the woman's personal meaning for her pain during labour and may set the stage for her interpretations of labour pain as a productive and purposeful pain, or as a threatening pain (Fig. 8.1).

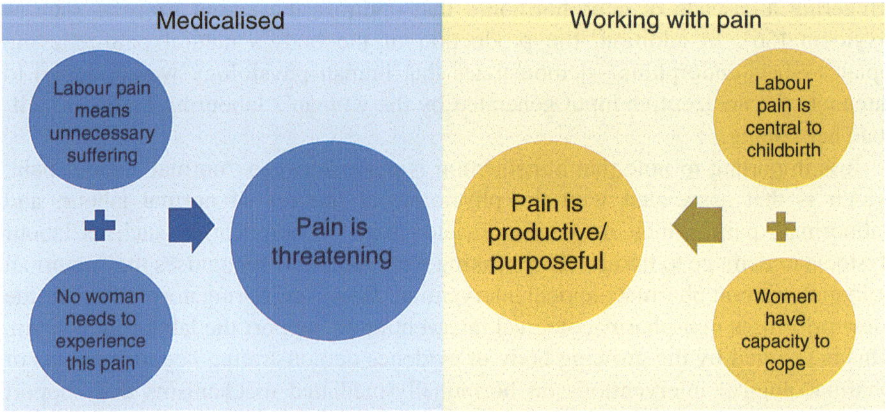

Fig. 8.1 Opposing theoretical views regarding labour pain and its management. The medicalised or biomedical viewpoint: labour pain is associated with suffering and is unnecessary and something to be avoided; the working with pain viewpoint: labour pain plays a central role in labour and birth and given the right environment and support during a normal labour women possess the capacity to cope. This work is licensed under the Creative Commons Attribution-NonCommercial-ShareAlike 4.0 International License. To view a copy of this license, visit http://creativecommons.org/licenses/by-nc-sa/4.0/

3 Factors that Shape the Personal Meaning of Labour Pain

3.1 Personal Beliefs

Deep personal beliefs are rooted in cultural and social perspectives and are also shaped by personal prior experiences. Although the biomedical view dominates in many Western birthing contexts, women will be influenced by their own set of personal experiences, cultural backgrounds, and may hold their own ideas and beliefs, which may go beyond the biomedical tissue-based model. In a study we conducted involving primiparous women, participants were asked, pre-birth, what they thought the function of pain during labour may be [37]. Women's thoughtful responses revealed various theories, including:

Pain as a signal of the normal progress of labour.

> *Maybe it's there (the pain) as a sign that everything is going the way it should be going.* Participant 2203 [37]

The challenge of the pain matches the significance of the event.

> *It's a pretty amazing thing (labour and birth) that's happening so if it was easy then it wouldn't. . .really match up to what you're going through.* Participant 2111 [37]

The challenge of the pain triggering a sense of accomplishment.

> *If somebody wants to climb a mountain it's never going to be easy, it's never going to be painless. But once they've finished they know they've achieved because they've gone through the pain.* Participant 2113 [37]

Pain as a trigger of maternal-child attachment.

> *It could be a strong bonding point with your child, that you've done all this for them. Not in a selfish way, but in. . . that I've gone through this to have you in my life.* Participant 2104 [37]

These responses by women anticipating their first birthing experience demonstrate that despite the dominance of the biomedical model of labour pain, many women seem aware of its limitations in making sense of this unique occurrence of pain.

The personal meaning of pain also has another reference: the woman's prior pain history. Women participating in our research had experienced pain associated with various bony injuries, chronic back pain, kidney stones, a mandible infection, wisdom tooth removal, endometriosis and one even reported being tasered. These experiences can affect the woman's belief about labour pain, her self-efficacy for labour and her approach to managing labour pain. We can access some of the personal meaning about labour pain for women with reference to other painful life experiences by exploring two quotes from our research.

> *It (labour pain) was definitely the worst (physical) pain I've ever experienced. But then, there are worse pains when you're really sick. When I was younger, I was really sick and the intensity of my abdomen pain was scarier because it was an unknown, I was thinking, 'What's going on with me, am I going to die?' Whereas when you're in labour I knew I was in labour and wasn't going to die. So there are worse pains, but labour's pretty bad too.* Participant 8 [38]

> *... I don't like to talk about it as pain like I don't think of it as being in pain, it was a really intense physical experience but I never really thought of it as, yeah, being painful, it just ... you know it took a lot of concentration to get through it and that sort of thing but yeah...* Participant 6 [23]

These two women provide examples of pain being physically intense but not threatening compared to pain associated with illness or injury. In the latter case we also get some insight into the cognitive demand required to cope with the intensity.

Post-birth data reveal an additional interpretation of pain to a labouring woman: Women report using their labour pain to "track" their labour.

> *...you had it in your mind the whole time that the contractions were good even though they were painful it was good because it was sort of tracking your progression.* Participant 2106 [18]

In such cases the meaning of pain to the woman is a sign of the productive work she is doing in labour; that progress is being made. The ability to be consistent and persistent in this interpretation of her body's internal signalling relies on supportive messages from her external environment and, one assumes, a certain amount of resilience.

3.2 *Cognitive Attributes*

By cognitive attributes, we mean the woman's resilience, her persistence with and ability to focus on a task despite adversity and her ability to interpret signals from interoceptors. Over the past two decades research has revealed important clues regarding the role of cognitive attributes in the experience of labour pain. To fully appreciate how many of these cognitive variables influence a woman's pain, we will consider how they contribute to the meaning of the pain to the labouring woman.

3.2.1 State of Mind

> *They assessed me and told me that I would need a forceps delivery so they were going to give me an epidural...So then I thought, 'Oh well, the epidural is going to take care of the pain and the contractions now so I can stop focussing. It's going to be easy now'. And as soon as I lost my focus I started getting pain. I became more focussed on the room and the people around me. I was focussing on all the outside stuff instead of focussing on what was going on inside me. I had more of a normal everyday mind.* Participant 12 [23]

A woman's state of mind during labour influences her relationship with the pain and interpretation of it. In a study we conducted in 2014 [23], the women's retelling of labour indicated a shift between two states of mind during their labour: a "mindful acceptance" state and a "distracted and distraught" state. The mindful acceptance state was characterised by women remaining focused in the present moment, on their bodies and on their sensory experiences, without reacting to the experience or judging themselves. When in this state, women appeared to be in tune with their bodies. Importantly, their experience of pain was accepted as one component of their overall experience. This state of mind appeared to have a powerful effect on women coping well with the pain, through this quality of acceptance. It may be that a mindful acceptance state attenuates the pain experience by preventing pain from re-entering an active threat-response system as an additional threatening input. Several other studies have also identified that an attitude of acceptance and "going with the flow" helps women interpret the pain as less threatening and enables them to work with it [39, 40].

A distracted and distraught state was characterised by women not focussing on the present moment or their bodies. Instead, their thoughts included reactive responses (particularly worrying about the pain) and critical judgement of their capacity to cope. A sense of helplessness highlighted this state and resulted in a negative relationship to the pain. Although helplessness may be considered useful in the context of labour—i.e., a recognition of not being able to cope alone that triggers behaviours that summon support—sustained catastrophic thinking may undermine an individual's sense of coping. The distracted and distraught state featured key elements of pain catastrophising, described as an exaggerated negative mindset in relation to an actual or anticipated pain experience. Catastrophising has also been linked in other studies to measures of labour pain intensity, use of pain interventions during labour, and the length of postpartum physical recovery [41, 42].

3.2.2 Distractions

Women in our study reported shifting between the two states throughout their labour. They could be pulled out of a mindful acceptance state by distractions in their environment, or by an internally generated loss of focus. This included the sounds and features of the space they were labouring in (e.g., bright lights, the sounds of monitors or clocks on the wall), who was around them and what those present in the room were doing, or their own thought processes triggered by this environmental *noise*.

> *I was focused on not having seen my daughter for 2 days . . . I got distracted and out of my zone.* Participant 8 [23]

An unexpected finding was how some women described that their focus was often drawn to a concern for how their partner was coping. It became apparent from this data, that there was a strong interrelationship between a labouring woman's pain experience and her physical and social environment, and that her own thought processes, including those related to the care of others, could draw her away from a mind-state beneficial for the progress of labour.

3.2.3 Other Life Events

One of the key influences on the development of the cognitive attributes in responses to pain may be the outcomes of prior pain experiences, as outlined in a previous section, and other life learning about pain including parental behaviours [43, 44].

> *My mum was a pretty tough cookie so she . . . I don't know, she was kind of not hesitant but she was kind of like 'Oh you'll be okay, you'll be fine, you'll get over it' kind of thing. And my sister for example broke her arm and my mum didn't think much of it until she complained about it for like quite a few hours and then Mum's like 'Okay, we better get this sorted.' So yeah, Mum and Dad are quite strong and tough. So I think that kind of got passed on to us.*
> Participant 2 (unpublished data)

Yet another factor that potentially shapes women's labour experience are the stories and descriptions provided by childbirth educators and other women, and the increasingly accessible images and recordings available online. Most women do not get to witness a live birth before they have their own experience and so must build a version of what it might be like, drawn from personal accounts of others or edited versions presented in the media.

3.2.4 Self-Efficacy for Labour

The imagery of labour presented in the media and online and the personal storytelling, especially by those she views as similar to her, has the potential to strongly influence a woman's self-efficacy. Self-efficacy relates to the belief in one's ability to accomplish a task. In the context of labour, it is a woman's belief in her ability to labour successfully, despite the associated intense sensory and emotional experiences, including pain. Prior self-efficacy for labour has been found to influence a woman's labour pain experience [45]. Higher levels of self-efficacy are associated with persistence despite difficulty and, in labour, reflect that a woman feels she has the necessary cognitive and behavioural resources to manage the pain and so is less likely to rely on passive pain interventions to cope.

A recent study investigating distress and the use of epidural analgesia found that women who were more distressed during pregnancy were more likely to use epidural as their sole tool for pain management [46]. While self-efficacy was not measured, it is possible a distressed woman would self-evaluate her resources and ability to cope as lower than a non-distressed woman. A longitudinal study of primiparous women did measure self-efficacy [45]. While the study found that self-efficacy did not influence pain tolerance (measured as the percentage of time during labour without pain intervention), higher levels did change women's evaluation of the intensity of pain and how distressing the pain was. These findings have recently been supported and extended in a study of more than 200 women using pre- and post-birth measures of self-efficacy. Additionally, this study reported that women with higher levels of self-efficacy were more likely to use coping strategies during the labour than those with low levels [47].

Self-efficacy has been shown to be shaped by numerous factors. One factor that is particularly critical during labour is that of verbal persuasion. While encouraging and supportive comments can increase self-efficacy, comments interpreted by the woman that she is not coping potentially have the opposite effect. In one of many examples from a qualitative study examining 50 women's labour stories, one woman said: "I was asked eleven times if I wanted drugs...it tears away at your self-confidence..." [48].

It is possible to draw in on the meaning of pain here. The self-evaluation of the ability to labour successfully, is likely to be different for a woman who is accepting of the pain associated with labour and who views it predominantly as non-threatening, compared to a woman who considers labour pain threatening. It could be speculated the woman with the more accepting, non-threatening view of pain, might consider she needs fewer or more self-regulated resources to cope, and so would have a higher self-efficacy for labour. Importantly, the woman's caregivers can provide implicit or explicit cues to her regarding her capacity to cope, or not, and in doing so can influence both her self-evaluation of coping and the meaning of the pain.

3.2.5 Anxiety Sensitivity

Anxiety has long been considered an influence on pain but the role of general anxiety on labour pain seems uncertain. On the other hand, anxiety sensitivity, defined as the belief that anxiety-related symptoms are themselves dangerous or threatening, has been shown to be a strong predictor of labour pain [49, 50]. It could be expected that a woman with high levels of anxiety sensitivity would be hypervigilant for bodily experiences during her labour (e.g., the physical experience of a uterine contraction) and attribute these to negative outcomes (e.g., the subsequent experience of pain), and to interpret them as more dangerous (i.e., the pain is more threatening).

3.2.6 Attachment Pattern

A woman's attachment pattern prior to labour has also been demonstrated to influence her experience of labour pain. Attachment is conceptually thought of as the tendency of a person to establish an emotional bond to attachment figures for safety and security [51]. According to attachment theory, childbirth is a significant life event that should activate the attachment system, thus calling upon a woman's attachment tendencies when engaging with her caregivers for support and assistance. Anxious and avoidant attachment patterns have been found to be associated with more severe pain reports and to be predictive of analgesia use [51, 52].

One of the primary purposes of attachment patterns is thought to be the regulation of negative affect. During childbirth, women may use attachment behaviours to manage their emotions and threatening experiences, and subsequently increase their sense of safety. Anxious and avoidant attachment patterns both represent suboptimal cognitions, emotions and attachment behaviours in relation to caregivers, and therefore may heighten the perceived sense of the threat of labour, and thus result in a more threatening pain experience.

Whilst the positive effect of support during labour is well documented, particularly in relation to a woman's ability to reframe her pain and capacity to cope, the emerging data on attachment patterns demonstrate the complexities of how support may be differently perceived by different women.

3.3 Coping and Suffering and the Meaning of Labour Pain

I remember thinking 'this hurts, but it also feels awesome!' Participant 15 [23]

In developed countries where labour pain is often conceptualised as a negative pain, pain and suffering are often inextricably linked—a woman experiencing pain during labour is presumed to be suffering. It is claimed that women should not be made to suffer through labour pain and will need to be "rescued" through the implementation of pain interventions. Pain and suffering, however, are separate experiences, and whilst they often co-exist in situations of (particularly extreme) pathological pains, in relation to labour pain this may not always be the case. Consistent findings in the literature demonstrate that women who experience labour pain as a productive and purposeful pain, associated with positive emotions and cognitions, do not describe a sense of suffering. Suffering is often associated with women who feel alone or unsupported during their labour. Chuahorm et al. [6] describes the experiences of Thai women for whom support people are not allowed in the hospital labour room. Women described a sense of helplessness exacerbated by a sense of being alone. Similar findings were reported by Wang [7] regarding women giving birth in Shanghai, China. One participant explained:

> *When I was in pain, I would yell and no one would pay attention. Then by the time it hurt even more severely, I wanted to cry, but not even one tear would come out. Really, at the time I thought I wanted to die. . .Everyone [the nurse-midwives] wanted you to give birth yourself. Any they would just chat, talk to each other, and make jokes. And it was just me, alone, suffering—no one paid attention. At that time, I lost hope because there was not one person to comfort me. I felt like I didn't want to give birth anymore.* Participant Dongmei [7]

It is important to recognise that pain is not sufficient for suffering and instead it is the individual's unique interpretation of their experience, including the perceived impact on physical and emotional wellbeing, meaning and coping resources that determines whether they experience suffering in relation to their pain [27]. As Turk and Wilson [27] explain, "Viewing suffering as an inevitable consequence of pain may unwittingly initiate and reinforce suffering." It may therefore be that the current approach to viewing and managing labour pain in many Western societies inadvertently contributes to women's suffering.

3.4 Pain Context

As soon as I found out I'd need a caesarean section it felt more painful because I knew that it wasn't working towards giving birth. Participant 10 [23]

Interacting with these cognitive attributes of the woman, is the pain context. The context of the woman's pain experience will shape the meaning ascribed to it. Even though all labouring women are proceeding through the same process, which is working towards the birth of a child, this context may be interpreted differently by different women.

For some women, the pain of labour is a signal of labour progressing, is accepted as a normal part of the experience, and for some is even embraced as an opportunity for growth and achievement. The social environment can facilitate this contextual understanding. Caregivers who are known, trusted and calm can facilitate this positive interpretation of the context. A woman's caregivers can help steer her away from pain catastrophising and help her remain in a focussed state of mindful acceptance. Through implicit and explicit actions and words, the context of the pain is represented as leading towards a positive outcome: the birth of a child.

If a woman interprets the context of the pain as not working towards a goal, she is more likely to interpret the pain as threatening. For some women the rate of progression, or the intensity of the pain, does not match their expectations, and is not linked to progression through labour. Thus, the meaning of the pain is that it is a threat to her or her baby's well-being. A prior fear of the pain of labour, or low self-efficacy for labour, can further prime a woman for a negative evaluation of the context of the pain. Caregivers can have a powerful influence over a woman's response to the context. A lack of support can make a woman feel unsafe, heightening the sense of pain as threatening. Alternatively, caregivers who interfere with a woman's focus, or influence her interpretation of the pain through verbal and non-verbal cues, can further increase the pain's threat-value. In our evaluation of women's experiences of labour pain, we found that simply reporting the findings of a cervical dilation assessment could have detrimental effects on a woman's pain evaluation and sense of capacity to cope.

When they told me I was 3 cm... that's probably the main thing out of my whole labour that really got me. I started crying 'cause I was just so upset because, like, you hear you have to be this many centimetres ... But I reckon if they were to tell me that no, look, you are 8 cm, this is the pain at 8 cm, I would have been like alright, I'm managing with the gas then. Participant 2201 [18]

This quote also highlights another important contextual feature: the woman's emotional state. It would seem that losing her focus accompanied by a change in emotional context, may challenge a woman's resilience significantly.

Women's understanding of the context of the pain influences the pain's meaning. A productive and purposeful pain is associated with labour progression, is accompanied by positive cognitions and emotions, and with a supportive and sensitive social environment. A non-productive pain is one that is interpreted as not leading towards the goal, is not embraced as a useful component of the labour, and these messages may be implied by the actions or words of people in the social environment.

3.5 Social Environment

In many of the examples above, we see how the woman's social environment influences each variable, thus shaping her pain experience. Humans are social beings: Our brains are relational organs that drive us to connect with others. In doing so we contextualise, form an understanding of, and identify meanings for, our experiences. Pain is one such experience that is determined by an appraisal of an individual's needs at a time to survive and thrive in the physical and social environment. Pain during labour has strong social uses in driving a labouring woman to seek, and remain engaged with, caregivers. Even maladaptive pain cognitions such as catastrophising may be functioning to enhance the labouring woman's urge to seek help. The emerging role of the endogenous opioid system in socialisation [53] may further reinforce a link between pain and social bonding with carers, above and beyond what is currently realised.

Ultimately, the social context of the woman's pain during labour gives that pain meaning, which then contributes to its place in her labour and birth "story." The people present during a woman's labour are somewhat predetermined by choices she made, or others made, about the safety of her and her baby. However, there are often no guarantees that preferred staff will be there, especially for the duration of labour. Similarly, the preferred personal support people may not always be available or allowed to be present, for example due to restriction of numbers. This is important as it would seem apparent that caregivers and support people have a significant impact on a woman's pain experience. This chapter is not the place to explore or expand on care provision. However, models of midwifery care that provide continuity of care in small teams or via one-to-one midwifery care [54] and culturally informed initiatives such as Birthing on Country [55] can help to provide a supportive social environment that would appear to have many benefits, including for the labour pain experience [56].

4 Conclusion

Labour pain is a significant component of the birthing experience of women across the world. However, not all women experience labour pain in the same way. A defining variable in a woman's experience appears to be her interpretation of its meaning. What is this pain telling me? Research suggests that women can view pain as a positive sign of progression of labour, or as a sign of damage and even threat to their life. Importantly, a woman who may go into labour with a strong belief one way or the other can undergo challenges to that belief which may change the course of the experience. As the woman finds her resources to cope depleted or enhanced, the assistance she seeks and the capacity to persevere will likely change. This may be a moment-to-moment proposition and it may transform the birthing process. Our research suggests that it is important for women to think of labour pain as part of a natural physiological process (i.e., purposeful) and a sign of progression towards the birth of her child (i.e., productive). Logically, a woman supported by carers promoting a more physiological approach to pain may maintain a belief that pain is productive and purposeful and would be more likely to persevere with the effort of childbirth and show greater tolerance to the process. Conversely, a woman who has a belief that she or her child are at risk of harm is unlikely to persevere and is more likely to request and agree to medical interventions. Giving birthing women the confidence to acknowledge their pain experience as a sign of progression and to support them to respond by working with the pain, reinforces a meaning of labour pain that is distinct and unique.

In this chapter we have attempted to provide a review of the literature that demonstrates the important relationship between a woman's experience of labour pain and its meaning. We have drawn from our own work, including author Whitburn's doctoral thesis, as well as a broad range of theoretical and empirical literature from numerous contexts, countries and models of health care. However, the complexity of this experience could never be comprehensively explored in one book chapter. Therefore, there will be facets that we have not covered, or only brushed on. This is partly due to the limited available research that explores the concept of the meaning of labour pain.

We hope that future studies further explore the personal attributes and sociocultural dynamics that shape a woman's pain experience during labour. Due to the subjective nature of pain these ideas must be studied through robust qualitative inquiry. We also hope that the unique nature and context of labour pain helps to expand and improve conceptions and definitions of the human experience of pain and its function.

Note Informed consent was obtained from all individual participants included in the study.

References

1. Rich A. Of woman born: motherhood as experience and institution. London: WW Norton; 1995. 274 p.
2. Corli O, Grossi E, Roma G, Battagliarin G. Correlation between subjective labour pain and uterine contractions: a clinical study. Pain. 1986;26(1):53–60.
3. Lowe NK. Critical predictors of sensory and affective pain during four phases of labor. J Psychosom Obstet Gynaecol. 1991;12(3):193–208.
4. Melzack R, Kinch R, Dobkin P, Lebrun M, Taenzer P. Severity of labour pain: influence of physical as well as psychologic variables. Can Med Assoc J. 1984;130(5):579–84.
5. Lundgren I, Dahlberg K. Women's experience of pain during childbirth. Midwifery. 1998;14 (2):105–10.
6. Chuahorm U, Sripichyakarn K, Tungpunkom P, Klunklin A, Kennedy P. Fear and suffering during childbirth among Thai women. Thai J Nurs Res. 2007;11(1):49–61.
7. Wang E. Requests for cesarean deliveries: the politics of labor pain and pain relief in Shanghai, China. Soc Sci Med. 2017;173:1–8.
8. Beigi NMA, Broumandfar K, Bahadoran P, Abedi HA. Women's experience of pain during childbirth. Iran J Nurs Midwifery Res. 2010;15(2):77.
9. The Holy Bible: new international version. London: Hodder & Stoughton; 2015.
10. Doering K, Patterson J, Griffiths CR. Japanese women's experiences of pharmacological pain relief in New Zealand. Women Birth. 2014;27(2):121–5.
11. Callister LC, Khalaf I, Semenic S, Kartchner R, Vehvilainen-Julkunen K. The pain of childbirth: perceptions of culturally diverse women. Pain Manag Nurs. 2003;4(4):145.
12. Merskey H. Pain terms: a list with definitions and notes on usage. Recommended by the IASP Subcommittee on Taxonomy. Pain. 1979;6:249–52.
13. Williams ACC, Craig KD. Updating the definition of pain. Pain. 2016;157(11):2420–3.
14. Whitburn LY, Jones LE, Davey MA, Small R. Supporting the updated definition of pain. But what about labour pain? [Comment]. Pain. 2017;158(5):990–1.
15. Cohen M, Quintner J, van Rysewyk S. Reconsidering the international association for the study of pain definition of pain. Pain Reports. 2018;3(2):e634.
16. Rowlands S, Permezel M. Physiology of pain in labour. Baillieres Clin Obstet Gynaecol. 1998;12(3):347–62.
17. Visser EJ, Davies S. Expanding Melzack's pain neuromatrix. The threat matrix: a super-system for managing polymodal threats. Pain Pract. 2010;10(2):163.
18. Whitburn LY, Jones LE, Davey MA, Small R. The meaning of labour pain: how the social environment and other contextual factors shape women's experiences. BMC Pregnancy Childbirth. 2017;17:157.
19. Olza I, Leahy-Warren P, Benyamini Y, Kazmierczak M, Karlsdottir SI, Spyridou A, et al. Women's psychological experiences of physiological childbirth: a meta-synthesis. BMJ Open. 2018;8(10):e020347.
20. Reilly P, Macleod I, Macfarlane R, Windley J, Emery R. Dead men and radiologists don't lie: a review of cadaveric and radiological studies of rotator cuff tear prevalence. Ann R Coll Surg Engl. 2006;88(2):116–21.
21. Englund M, Guermazi A, Gale D, Hunter DJ, Aliabadi P, Clancy M, et al. Incidental meniscal findings on knee MRI in middle-aged and elderly persons. N Engl J Med. 2008;359 (11):1108–15.
22. Frank JM, Harris JD, Erickson BJ, Slikker W III, Bush-Joseph CA, Salata MJ, et al. Prevalence of femoroacetabular impingement imaging findings in asymptomatic volunteers: a systematic review. Arthroscopy: J Arthrosc Relat Surg. 2015;31(6):1199–204.
23. Whitburn LY, Jones LE, Davey MA, Small R. Women's experiences of labour pain and the role of the mind: an exploratory study. Midwifery. 2014;30(9):1029–35.

24. Whitburn LY, Jones LE, Davey MA, McDonald S. The nature of labour pain: an updated review of the literature. Women Birth. 2019;32(1):28–38. https://doi.org/10.1016/j.wombi.2018.03.004.
25. Obstetric Analgesia and Anesthesia. Practice bulletin no 177. American college of obstetricians and gynecologists. Obstet Gynecol. 2017;129:e73–89.
26. Leap N, Anderson T. The role of pain in normal birth and the empowerment of women. In: Downe S, editor. Normal childbirth: evidence and debate. Edinburgh: Churchill Livingstone; 2008. p. 29–47.
27. Turk DC, Wilson HD. Pain, suffering, pain-related suffering—are these constructs inextricably linked? Clin J Pain. 2009;25(5):353–5.
28. Leap N, Dodwell M, Newburn M. Working with pain in labour: an overview of evidence. NCT New Digest. 2010;49:22–6.
29. Melzack R. Labour pain as a model of acute pain. Pain. 1993;53(2):117–20.
30. Hess P. Pain in the obstetric patient. In: Elliott JA, Smith HS, editors. Handbook of acute pain management: informa healthcare; 2010. p. 276–91.
31. Brownridge P. The nature and consequences of childbirth pain. Eur J Obstet Gynecol Reprod Biol. 1995;59:S9–S15.
32. Pan PH. Obstetric pain. In: McMahon S, Koltzenburg M, Tracey I, Turk DC, editors. Wall & Melzack's textbook of pain. 6th ed. Philadelphia, PA: Elsevier Health Sciences; 2013. p. 772–93.
33. Newnham EC, McKellar LV, Pincombe JI. Documenting risk: a comparison of policy and information pamphlets for using epidural or water in labour. Women Birth. 2015;28(3):221–7.
34. Dempsey R. Embracing the intensity – celebrating birth as a peak bodily performance. Birth matters. 1998:4–6, 20.
35. Sanders R. Functional discomfort and a shift in midwifery paradigm. Women Birth. 2015;28(3): e87–91.
36. Buckley SJ. Hormonal physiology of childbearing: evidence and implications for women, babies, and maternity care. Washington, DC: Childbirth Connection Programs, National Partnership for Women & Families; 2015. 225 p.
37. Whitburn LY, Jones LE, Davey MA, Small R. First-time mothers in Victoria – a pilot cohort study.
38. Whitburn LY, Jones LE, Davey MA, Small R. Expectations, fears and pain experience: identifying internal and external threats to perceived safety for women in two birthing environments. BMC Pregnancy Childbirth. 2017;17:157.
39. Karlsdottir SI, Halldorsdottir S, Lundgren I. The third paradigm in labour pain preparation and management: the childbearing woman's paradigm. Scand J Caring Sci. 2014;28(2):315–27.
40. Johnson TR, Callister LC, Freeborn DS, Beckstrand RL, Huender K. Dutch women's perceptions of childbirth in the Netherlands. MCN Am J Matern Child Nurs. 2007;32(3):170–7.
41. Flink IK, Mroczek MZ, Sullivan MJL, Linton SJ. Pain in childbirth and postpartum recovery—the role of catastrophizing. Eur J Pain. 2009;13(3):312–6.
42. Veringa I, Buitendijk S, de Miranda E, de Wolf S, Spinhoven P. Pain cognitions as predictors of the request for pain relief during the first stage of labor: a prospective study. J Psychosom Obstet Gynaecol. 2011;32(3):119–25.
43. Kraljevic S, Banozic A, Maric A, Cosic A, Sapunar D, Puljak L. Parents' pain catastrophizing is related to pain catastrophizing of their adult children. Int J Behav Med. 2012;19(1):115–9.
44. Fillingim RB, Edwards RR, Powell T. Sex-dependent effects of reported familial pain history on recent pain complaints and experimental pain responses. Pain. 2000;86(1–2):87–94.
45. Berentson-Shaw J, Scott KM, Jose PE. Do self-efficacy beliefs predict the primiparous labour and birth experience? A longitudinal study. J Reprod Infant Psychol. 2009;27(4):357–73.
46. Jonsdottir SS, Steingrimsdottir T, Thome M, Oskarsson GK, Lydsdottir LB, Olafsdottir H, et al. Pain management and medical interventions during childbirth among perinatal distressed women and women dissatisfied in their partner relationship: a prospective cohort study. Midwifery. 2019;69:1–9.

47. Sánchez-Cunqueiro MJ, Comeche MI, Docampo D. On the relation of self-efficacy and coping with the experience of childbirth. J Nurs Educ Pract. 2018;8(6):48.
48. Rafalovich A. Pain is the club: identity and membership in the natural childbirth community. Qual Sociol Rev. 2016;12(3):100–16.
49. Lang AJ, Sorrell JT, Rodgers CS, Lebeck MM. Anxiety sensitivity as a predictor of labor pain. Eur J Pain. 2006;10(3):263–70.
50. Curzik D, Jokic-Begic N. Anxiety sensitivity and anxiety as correlates of expected, experienced and recalled labor pain. J Psychosom Obstet Gynaecol. 2011;32(4):198–203.
51. Costa-Martins JM, Pereira M, Martins H, Moura-Ramos M, Coelho R, Tavares J. The role of maternal attachment in the experience of labor pain: a prospective study. Psychosom Med. 2014;76(3):221–8.
52. Quinn K, Spiby H, Slade P. A longitudinal study exploring the role of adult attachment in relation to perceptions of pain in labour, childbirth memory and acute traumatic stress responses. J Reprod Infant Psychol. 2015;33(3):256–67.
53. Ballantyne JC, Sullivan MD. Discovery of endogenous opioid systems: what it has meant for the clinician's understanding of pain and its treatment. Pain. 2017;158(12):2290–300.
54. Forster DA, McLachlan HL, Davey MA, Biro MA, Farrell T, Gold L, et al. Continuity of care by a primary midwife (caseload midwifery) increases women's satisfaction with antenatal, intrapartum and postpartum care: results from the COSMOS randomised controlled trial. BMC Pregnancy Childbirth. 2016;16(1):28.
55. Kildea S, Hickey S, Nelson C, Currie J, Carson A, Reynolds M, et al. Birthing on country (in our community): a case study of engaging stakeholders and developing a best-practice Indigenous maternity service in an urban setting. Aust Health Rev. 2018;42(2):230–8.
56. Leap N, Sandall J, Buckland S, Huber U. Journey to confidence: women's experiences of pain in labour and relational continuity of care. J Midwifery Womens Health. 2010;55(3):234–42.

Chapter 9
Living with Complex Regional Pain Syndrome: Understanding the Battle

Colleen Johnston-Devin, Florin Oprescu, and Marion Gray

Abstract Living with complex regional pain syndrome (CRPS) can be described as similar to living with any other chronic pain condition, but with extra complications. Many health professionals have never heard of the condition and it is even less known in the general community. There is a diversity of presentations, no objective medical test for diagnosis, and it remains a diagnosis of exclusion based on clinical signs and symptoms. The pathophysiology is not fully understood and there is no dedicated treatment. There are only low-quality data supporting any treatment approach. Although there is no known cure it may go into remission. It may also reappear. The condition may occur spontaneously. Although the mechanisms are not well understood, it can spread to other parts of the body.

This chapter is based on a PhD study entitled "Battling Complex Regional Pain Syndrome: A Phenomenological Study." The aim is to describe and discuss the meaning of living with CRPS. Using a heuristic hermeneutic phenomenological approach to investigate the phenomenon of living with CRPS, 17 patients and four health professionals from 6 different countries were interviewed. Textual material from internet blogs and a book containing patient stories were also included as data sources. As the first author has been diagnosed with the condition, she was interviewed as well. Serving multiple purposes, the interview transcript was used as data while allowing for explication of preunderstandings and assumptions. This is an important step in phenomenological research. Findings revealed that living with CRPS is a daily battle. Within the battle analogy, readers can recognise aspects of the themes and relate to them on some level. Within this battle, there are many smaller, but important, fights. These fights are the themes which are:

- Dealing with an unknown enemy;
- Building an armoury against a moving target;
- Battles within the war;
- Developing battle plans with allies; and
- Warrior or prisoner of war.

C. Johnston-Devin (✉) · F. Oprescu · M. Gray
School of Health and Sport Sciences, University of the Sunshine Coast, Sippy Downs, QLD, Australia
e-mail: Cjohnst2@usc.edu.au

S. van Rysewyk (ed.), *Meanings of Pain*,
https://doi.org/10.1007/978-3-030-24154-4_9

The enemy is CRPS. Patients are the soldiers or warriors fighting it. From fighting for a diagnosis and coping with strange symptoms and disbelief, patients struggle with a condition that has no known cure and constantly changes. They battle against moving a painful body part too much and causing a flare or moving it too little and losing use of it. They face judgement from others, including health professionals, who do not understand CRPS. Support can be difficult to find. Patients must become resilient and must rise to the challenge with a warrior like attitude. Those who don't may feel like a prisoner of war.

This chapter provides vignettes composed from patient interviews. The vignettes could facilitate a better understanding of lived experiences from the patients' perspective.

Clinical Implications: Health professionals (HPs) have little understanding of the smaller battles faced by patients living with CRPS and patient goals do not necessarily correlate to those of the HP. It is anticipated that with better understanding of the lived experience, HPs will deliver more empathic patient focussed care. This could result in the clarification of realistic patient goals and needs as opposed to HP expectations of adherence and compliance to therapeutic interventions and could foster more collaborative and supportive self-management strategies for patients.

Keywords Complex regional pain syndrome · Pain · Qualitative research · Phenomenology · Interviews · Lived experience · Insider research · Patient experience

1 Introduction

The rare, little known condition, complex regional pain syndrome (CRPS) is difficult to diagnose, difficult to treat, and has the dubious reputation of being the most painful condition known to man as measured by the McGill Pain Index[1] [1, 2]. First observed in the sixteenth century, it was fully described for the first time during the American Civil War in 1864 and while awareness and understanding of the condition has grown immensely since then, there are many people, including health professionals who have never heard of CRPS, and many facets of the disease remain unknown or misunderstood [3].

In the past, lack of agreement regarding diagnostic criteria and outcome measures has caused difficulties such as delayed or missed diagnosis, a lack of comparison studies and inability to empirically quantify rates of recovery or permanent disability [1, 4, 5]. Adding to the confusion, despite the name CRPS being coined in 1993, it is still sometimes referred to by its outdated name of reflex sympathetic dystrophy (RSD), particularly in the United States [1] and online in many Facebook support groups. It is also frequently referred to as "the suicide disease" in the online CRPS community.

[1]Online patient support groups often include an infographic of this index which shows CRPS to be more painful than childbirth or limb amputation. For example: http://www.rsdhope.org/mcgill-pain-index%2D%2D-where-is-crps-pain-ranked.html.

While there have been many research efforts into diagnosis, treatment and understanding of the pathophysiology of CRPS, little qualitative research has been performed. As a result, there is little published scientific literature on the experiences of people living with this condition. Unless they have the condition themselves, few people understand the day to day lives of CRPS sufferers.

2 What Is CRPS?

CRPS may occur spontaneously but usually occurs following known tissue injury, fracture being the most common [1, 6, 7]. The condition is quite rare and between 0.5–2% of patients with injury or trauma develop CRPS [7]. Current understanding of the pathophysiological mechanisms involved in the initial development and the transition from acute to chronic CRPS remains unclear and recent hypotheses include differentiation of subtypes such as: predominance of inflammation, central neuroplasticity, psychological and autoimmune [8], and that CRPS may be better understood as four components of altered function; tissue trauma, abnormal pain processing, autonomic imbalance and alteration in the immune system [7].

Continuing, severe pain disproportionate to the inciting event appears to be an indication of emerging CRPS rather than a delay in normal healing [1, 6]. The diagnostic criteria for CRPS has been refined over many years and was last updated by the IASP in 2012 [1, 3]. Entitled the Budapest Criteria, clinical diagnosis is based on signs and symptoms as shown in Table 9.1.

CRPS presents with a variety of symptoms such as skin colour and temperature changes, oedema, allodynia, increased hair growth, tremors and dystonia, but the common symptom is extreme, ongoing pain disproportionate to the precipitating event [7]. Patients may also experience altered body perception, motor function reduction and “neglect-like” limb behaviour [1, 6, 7]. Unlike neglect in stroke, pain is critical in the development of the underuse, decreased movement attempts and fear avoidance behaviour in the development of “neglect like” symptoms in CRPS and many patients must concentrate on the affected limb in order to use it [8, 9].

Diagnosis can take years despite the initial florid signs, and referral to a specialist pain centre for treatment is often delayed despite suggestions that early diagnosis and treatment leads to better outcomes for the patient [3]. CRPS is known to change over time, the pain mostly remains, but the peripheral signs fade, and many patients under treatment improve within the first 12 months [1, 7, 10]. Improvement however, does not necessarily mean recovery, with only one third of patients considering themselves fully recovered six years after disease onset [10].

CRPS may spread with and without secondary trauma to another site in the body contralaterally (mirror image), ipsilaterally (such as hand to foot) and diagonally [1, 11, 12]. Although CRPS rarely develops concomitantly in two limbs, spread to another limb occurs in ~7% of patients [10]. The debilitating consequences of having CRPS lead to a higher suicide risk and long standing CRPS can lead to malignant swelling, ulceration and infection and is associated with a poorer quality

Table 9.1 Budapest diagnostic criteria for CRPS [6, p. 2]

1. Continuing pain, which is disproportionate to any inciting event
2. Must report at least one symptom in *three of the four* following categories: • *Sensory:* reports of hyperesthesia and/or allodynia • *Vasomotor:* reports of temperature asymmetry and/or skin colour changes and/or skin colour asymmetry • *Sudomotor/oedema:* reports of oedema and/or sweating changes and/or sweating asymmetry • *Motor/trophic:* reports of decreased range of motion and/or motor dysfunction (weakness, tremor, dystonia) and/or trophic changes (hair, nail, skin)
3. Must display at least one sign at time of evaluation in *two or more* of the following categories: • *Sensory:* evidence of hyperalgesia (to pinprick) and/or allodynia (to light touch and/or deep somatic pressure and/or joint movement) • *Vasomotor:* evidence of temperature asymmetry and/or skin colour changes and/or asymmetry • *Sudomotor/oedema:* evidence of oedema and/or sweating changes and/or sweating asymmetry • *Motor/trophic:* evidence of decreased range of motion and/or motor dysfunction (weakness, tremor, dystonia) and/or trophic changes (hair, nail, skin)
4. There is no other diagnosis that better explains the signs and symptoms

of life in comparison with people living with other chronic pain conditions [1, 10, 13].

3 Phenomenology as a Methodology for Researching the CRPS Lived Experience

Qualitative research is not often considered to be part of evidence-based medicine, is often ascribed low level evidence, and is not usually recommended to inform practice or policy [14, 15]. Arguments to consider the uniqueness of each individual person's experience and to value their individual knowledge of the lifeworld (the world of immediate experience, or the world as we live it [16, 17]) is gaining momentum in the published literature because qualitative evidence is increasingly considered as important and indispensable for practice and policy [14, 15, 18, 19]. Gaining a greater understanding of the factors affecting the quality of life of patients can help health professionals to understand the various influences on health and illness experiences and therefore provide compassionate person-centred care [19].

The inclusion of the CRPS patient in research endeavours has historically been at the discretion of the researcher. CRPS patient representatives have been included in projects developing outcome measures [4] and co-creating information leaflets [20] and patient conferences have been conducted with support groups including Burning Nights (UK), Reflex Sympathetic Dystrophy Syndrome Association (RSDSA, USA), and CRPS Forum (USA).

There are examples in the CRPS literature of lived experience being investigated, but in each case, it is a specific aspect of lived experience that has been researched such as the exploration of participants lived experiences of CRPS to understand their specific information needs [21]. Despite these initiatives, there remains a lack of research looking purely at the lived experience of the condition for the purpose of understanding the patient experience. Qualitative research is an avenue that could provide a better understanding of CRPS [22–24]. A better understanding of lived experiences of this condition may improve clinical management by inducing comprehension and empathy from clinicians [25].

Phenomenology is the study of essences; a qualitative research method concerned with the lifeworld, or world of lived experience, and the question of what is the essence, nature or meaning of something [16, 26]. It has been used in CRPS research in studies conducted by researchers such as Rodham and McCabe [27], who utilised interpretative phenomenological analysis (IPA) when examining how an online message board was used by people living with CRPS, and when examining the experience of transitioning from a hospital rehabilitation program to home [28]. This study used phenomenology as described by van Manen [16] who uses the term *human science* interchangeably with phenomenology and hermeneutics and describes an approach to human science research which he believes shows "a semiotic employment of the methods of phenomenology and hermeneutics" [16, p. 1].

For van Manen [16], phenomenology aims to transform a description of lived experience into a textual representation of its essence and is an attempt to explicate meanings as we live them in our everyday existence. Essence is the true nature or meaning of an experience; the description of which shows the lived quality and significance of the experience in a deep and complete manner [16]. His approach to phenomenology was chosen due to it being relevant to those researching social science to transform lived experience descriptions into a word-based depiction of its essence [16].

Hermeneutics is also known as *interpretative phenomenology* and the researcher is considered inseparable from their preunderstandings and assumptions [29]. Interpreting data through the process of the hermeneutic circle, the researcher and the participant experiences come together to let things speak for themselves [16, 29].

As a research method, phenomenology is considered to be free from prescriptive methodology [30] and as stated by van Manen [16], "the method of phenomenology and hermeneutics is that there is no method" [16, p. 29]. Therefore, elements of heuristic inquiry were incorporated, as the spontaneous creation of methods evoking experiential meaning are encouraged within heuristics [31]. Embracing self-experience rather than detaching from it, heuristics retains the person's essence and the research question is deeply personal [26, 32]. Within heuristics and phenomenology, researchers are required to become immersed in the question, which aligns with the concept of the utilisation of the experience of the first author in this research in order to gain a complete picture of the phenomenon of living with CRPS [16, 32, 33].

As CRPS treatment is palliative rather than curative and the goals of health care professionals are pain reduction and restoration or improvement of function [6, 13, 34], a study of the meaning of living with CRPS was required so that clinicians can better understand the lives of their patients and in so doing, improve the quality of care they provide [25]. Thus, the aim of the study was to understand the phenomenon of chronic pain in the lifeworld of people living with CRPS. It is expected that with greater awareness and deeper understanding of the phenomenon and the lives of people living with CRPS, the knowledge base of health practitioners (HP), patients, carers and the general population will be extended. It is anticipated that empathy for the experience of living with CRPS will be increased as people gain an appreciation of the daily struggles patients often face [25, 35]. The research question underpinning this qualitative study was: *What is the experience of living with chronic pain in the lifeworld of complex regional pain syndrome?*

4 Data Collection

To answer the research question, purposive sampling was used to recruit participants who had been diagnosed with CRPS.[2] After extensive online advertising, word of mouth and snowballing, 17 people living with CRPS were interviewed face to face, on Skype or telephone. The conversational interviews lasted an average of 51 minutes and were recorded and transcribed either by the first author or a professional transcriber. The first author was interviewed after the pilot interviews by the second author. This interview allowed for the explication of preunderstandings and assumptions and provided another data source. Although all other participants chose a pseudonym, the first author did not. This strategy ensured transparency of the researcher's voice, safeguarded against unintentional bias and assisted co-researchers to understand and follow the audit trail.

Although not suffering the condition themselves, HPs play a large part in the lives of patients through their clinical and research endeavours. They have influence over the health/illness trajectory for patients and also influence policy and procedures regarding the management of the patient living with CRPS and other chronic pain conditions. Therefore, HP participants were included to discuss emerging themes from the patient interviews and to generate deeper insights into the phenomenon.

Health practitioners were recruited following the International Association for the Study of Pain Special Interest Group in CRPS (IASP SIG CRPS) Conference in Cork, Ireland in 2017. Four health professionals who are considered leading world CRPS experts were interviewed by Skype or Skype telephone for an average of 42 minutes. As HP participants are easily identifiable, demographic data was collected to ensure their anonymity could be maintained.

[2]Ethics approval was granted by University of the Sunshine Coast Human Research Ethics Committee (S13577).

4.1 Data Analysis

Exploration of the lifeworld of CRPS from the different perspectives occurred through a reflective process and five themes which were enmeshed, yet distinct from each other, were identified. Interview transcripts were read multiple times, and those transcripts not transcribed by the researcher were read more often to ensure the same level of familiarity with the data. Following hermeneutic reflection,[3] five themes emerged from the patient participant data. Textual data from one book [36] and the first five websites [37–41] discovered upon seeking blogs about living with CRPS were consulted as the next stage of thematic analysis to provide evidence for the emergent themes. The emergent themes were found supported by themes identified in the book and in the blogs and they were discussed and verified during the health practitioner interviews.

5 Findings from Heuristic Hermeneutic Phenomenological Research into CRPS

The patient participants were a mix of white Anglo-Saxon males (n = 3) and females (n = 14), who lived in Australia, England, Wales, Singapore or USA, and were aged from 22 to 65 years old. They had lived with CRPS between four months and 18 years and their time to formal diagnosis ranged from three weeks to nine years. Health practitioners resided and worked in three countries and had been working with CRPS patients for an average of 23.5 years. Each HP participant had either previously or currently worked clinically and in research. As further detail of participants risks their anonymity, it will not be provided here.

Each person interviewed described a constant daily fight or battle, and people in the online CRPS community referred to themselves as warriors fighting the condition. Phenomenological writing aims to evoke a recognisable experience and be immediate yet prompting reflection [42]. Therefore, the essence or meaning of the phenomenon and themes derived from this research have been named to be deliberately evocative and recognisable. The overarching essence of the phenomenon was found to be a battle. Living with CRPS is a battle. The patients are akin to soldiers or warriors and the enemy they are fighting is CRPS. Health practitioners may be considered the enemy or allies. The themes were named in keeping with the battle analogy. Within the battle, there are separate fights which make up the following themes:

[3]Hermeneutic reflection involves studying the descriptions of lived experience to make meaning and discover what the experience is really like. Existentials to guide reflection on how humans experience the world are lived space (spatiality), lived body (corporeality), lived time (temporality) and lived human relation (relationality). Readers are referred to van Manen [16], Chap. 4, "Hermeneutic Phenomenological Reflection," for further information.

1. Dealing with an unknown enemy;
2. Building an armoury against a moving target;
3. Battles within the war;
4. Developing battle plans with allies; and
5. Warrior or prisoner of war.

Rather than discuss the battle here, this chapter will utilise vignettes derived from the patient interviews so that the meaning that CRPS has on the lives of these people will become apparent through their voices. It must be kept in mind that CRPS is a complex condition and these findings can be taken to be transferable only to the populations represented by the demographics of the participants.

5.1 Dealing with an Unknown Enemy

Delayed diagnosis is common amongst people with CRPS and pre-diagnosis, patients will often act on incorrect or poor advice due to their poor health literacy and a lack of knowledge from the health care professional [1, 34]. Participants reported that they knew something was wrong but were often accused of exaggerating or faking their pain. Delayed diagnosis and disbelief lead to emotional distress and unnecessary suffering in these participants.

There is also a psychological response to the development of symptoms which is often not acknowledged by health professionals. Patients feel confused because the injury should have healed and will try almost anything to stop the pain. Participants described losing all self-confidence and not being able to trust their bodies. They often looked to the internet for solutions and spent large amounts of time and money visiting multiple mainstream and alternate health practitioners. Those with low health literacy are likely to unwittingly put their health at risk. There are reports of patients rubbing sandpaper on their affected limb in an attempt to desensitise it.

Jackie was 55 and had CRPS for four months when interviewed. She broke her wrist and was in continuous pain for the next three months getting her wrist replastered multiple times. Jackie continued to see a private physio after she was given public appointments because she had been told she required intensive physio and she did not think weekly appointments were enough. She had physio four times a week and the public and private physiotherapists were unaware of the other. The private physio suggested she had CRPS. This was confirmed by a pain management specialist who did not explain the condition or its treatment to Jackie. Jackie did not like how the medication (gabapentin) made her feel, so she refused to take it and she also refused tramadol and amitriptyline believing they were toxins.

Jackie then commenced alternate therapies including acupuncture, osteopathy and electric differential treatment. She underwent a three-hour myofascial treatment which she described as very, very painful and the practitioner told her that she was feeling all her emotions. Jackie also put frankincense and myrrh on her arm and also a wintergreen concoction on her hand. She developed a rash which that practitioner explained as the gabapentin coming out of her body. The pain in her hand from seven and a half hours of chakra cleansing was explained as toxins leaving her body and when she had meditation with crystal gong healing that practitioner said she was carrying the emotional burdens of other people. She said she was usually an organised person but had brain fog and worried she was rambling.

Martin was 32 at the time of the interview and had lived with CRPS for five years. He developed leg pain a week after being in a car accident. His initial diagnosis was whiplash and referred back pain. He saw the foot clinic at his local hospital and was eventually referred to a public orthopaedic specialist. An appointment was not available for six months and Martin's wife managed to get a private appointment with the same doctor with less waiting time. This doctor suspected CRPS.

Martin saw that doctor through the public system and a cycle of X-rays, MRI's and some very little treatment commenced. A specialist CRPS program had been mentioned, but the orthopaedic doctor and the GP did not know how to get Martin an appointment there. Martin's father-in-law rang the specialist centre and a few months later Martin had an appointment. The centre confirmed the CRPS diagnosis and directed Martin to resources and accurate information.

Martin initially expected the inpatient course at the specialist centre would cure him. He then realised he was being trained in how to live with CRPS. He had been using a single crutch which was too small to help him walk for 18 months before attending the program and had developed a carpal tunnel in his left wrist. Martin was taught to always use correctly sized two crutches, but the damage had already been done. He then developed another carpal tunnel in his right wrist and CRPS in his left wrist. He was looking at getting a wheelchair for those times he was unable to mobilise with crutches.

Martin was unable to continue in his job and it had taken six months to gain further employment with another company, but he was made redundant within six months. Nine months later he found another job which did not require anything physical. He found that the transition from being a physically active person to requiring aids in the house frustrating and had refused hand rails on the stairs because he did not want physical signs of his disability everywhere. He had rails in the bathroom and a perching stool in the kitchen. He had a custom-made pram so that he could push his eight-month-old baby in it. At the time of the interview, Martin wanted to raise awareness for CRPS and was considering participating in a marathon.

Sharon was 46 when interviewed and had lived with CRPS for five years. She lived on the outskirts of a capital city with her 18-year-old son. After injuring her wrist, Sharon was in plaster for ten weeks. She had been working in two jobs and had to give both of them up after hurting her wrist. She described feeling like acid had been poured down her cast and knew something was wrong, but no one believed her. CRPS was suspected by a hand specialist but the doctor at the public pain clinic she was referred to didn't believe her. Sharon said she was bullied by the doctor who told her she had arthritis and a psychiatric problem.

Lyrica was too expensive at the time, so Sharon took pain killers and did a pain program where she learnt art therapy, meditation, graded motor imagery and did hydrotherapy. After asking for a second opinion for two and a half years, Sharon finally saw another doctor who officially diagnosed her with CRPS. Nerve blocks did not work so Sharon started ketamine infusions which made a big difference to her pain levels. The combination of medication and a sedentary lifestyle caused Sharon to gain 40 of 50 kilograms which caused additional problems.

At the time of interview, Sharon had swelling and redness from her fingers to between her elbow and shoulder on the right side. Her right shoulder and right side of her face were also red and swollen. She required carers every day to help her wash and dress and they would come back at lunchtime to make lunch and dinner and assist her to the toilet. Getting assistance at her age was extremely difficult. After having a fall and having no use of either arm, Sharon was told she wasn't disabled enough for a disability pension until she said she had depression. That process took six months.

After developing bowel and bladder issues, and being unable to self-care, depression did become a problem and Sharon's elderly mother came to help at times. Sharon did not think it appropriate for her son to assist her with hygiene care. She felt she had lost all her independence once she couldn't get herself to the bathroom or change her underwear. The carers though, did not always turn up and Sharon couldn't go out on her birthday two years in a row because she was unable to get dressed. Some carers did not understand the magnitude of the pain if they touched certain parts of Sharon's arm. Major pain episodes could cause her to be bedridden for months. She developed seizure like attacks because showering was sometimes so painful. The water felt like needles piercing her skin.

Hospital stays for ketamine infusions were also traumatic and doctors and nurses disbelieving Sharon's pain were common. She said that being a patient in the surgical wards were better because the nurses were too busy to bully her about the opiate prescriptions written by her pain specialist.

Sharon required community services to drive her to the hospital for appointments. Her mother accompanied her when possible as Sharon was not confident being out on her own. One time, she was unable to walk from the drop off point to the clinic on her own and had gone home without getting to her appointment on the next floor. The community car would not drive Sharon's mother home as she was not a client and although she had early stage dementia, she was forced to get a bus home after accompanying Sharon to hospital for a ketamine infusion. Sharon is now permanently in a wheelchair which has enabled her to leave the house for leisure activities and not just for treatment.

5.2 Building an Armoury Against a Moving Target

CRPS is known to change over time [1, 7, 10, 43]. Patients can experience an improvement in their symptoms or may experience a flare due to further injury, doing too much, illness or anything else that puts their bodies under stress. Medications can sometimes cease to have an effect, or the side effects become unbearable and patients are forced to explore different treatment options. As CRPS is known to spread, it causes additional problems with an extended affected area, and the potential for spread causes fear in many people. For some, the fear is irrational and there are reports online of patients being too scared to shave their legs in case they cut themselves and cause a spread.

Alice was 30 at the time of her interview. Five years before that she was working overseas when she injured her knee and had to return home to a regional town. Within the first two months, Alice couldn't walk, and a physio suggested she might have CRPS. She then saw multiple doctors who disbelieved her, and it spread to her entire leg before she was diagnosed with CRPS by a sports doctor nine months after the injury.

Alice had tried many medications and treatments including nerve blocks and spinal cord stimulation but said that nothing had worked. Within the first year CRPS had spread to both legs and hips, then to her spine. Around the third year she had CRPS in both arms and hands and she was losing teeth due it being in her mouth. She stated that it had recently spread to her lungs, stomach and digestive system when she was interviewed.

Alice lived at home with her parents and was on a disability pension. The family had been forced to move to a house that could accommodate her disability and wheelchair. There was a financial burden on her father for medical bills for Alice and her mother was chronically ill and unable to work.

Showers were painful for Alice, so she showered only once or twice a week. She wore incontinence pads and tried to move every 20–30 min because pressure on her body caused pain. She was usually able to cook the family meals and tried to go on the train to the city to meet her small group of friends every fortnight. She was hoping to get a licence and hand controls for a car so that she could have some more independence. Another goal for Alice was to lose the 40 kilograms she had put on from medication and had lost about 16 kilograms when interviewed.

Alice tried to spread awareness during the CRPS awareness month of November. She had sold orange ribbons and t-shirts at markets and used every opportunity to speak to people about her condition. She stated people still disbelieve her condition despite being in a wheelchair. Many people, including health professionals, doubt her inability to take any medication.

Mel was 41 when interviewed and lived in a regional town. Nine years before the interview an incident at work caused torn cartilage and a Bankart Lesion in her shoulder and all her treatment had to be approved by WorkCover. After a physiotherapist suggested CRPS, Mel was referred to an orthopaedic surgeon and a pain clinic where the diagnosis was confirmed. This was four and a half years after the injury. She left the pain clinic after dry needling to her shoulder caused immense pain and saw another pain specialist who recommended ketamine infusions and a spinal cord stimulator.

WorkCover refused permission to undertake these procedures, so Mel tried alternative therapies including kinesiology and acupuncture. At the court case, Mel was accused of being a drug addict and faking her condition despite the diagnosis being confirmed by two pain management specialists. This caused financial problems so bad that Mel was considering declaring bankruptcy at the time of her interview. She had been unable to work since her injury and had a dependant teenager at home who said he didn't believe in her pain because the judge didn't either.

Mel found it difficult to perform tasks with her non-dominant hand and had altered her routine to cope with her disability. She sold her manual car and bought an automatic with a steering wheel knob. She shopped daily because she was unable to push a shopping trolley. She used frozen vegetables, ate a lot of mince and made meals in in the slow cooker as much as possible. She used an electric toothbrush and had pump shampoo and conditioners which she could managed one handed. Mel could not wear a bra and none of her clothing had zips or buttons. She said her life was going really well until CRPS took away all her hopes and dreams.

5.3 Battles Within the War

The onset of CRPS results in the patient having a body that does not do as much as it used to. Finding that balance of maintaining function whilst not overdoing things is extremely difficult when the body is not consistent in its abilities. Patient participants described the difficulty to not overdo things on days they are feeling well and consequently often 'suffered' the day after being particularly active. Kinesiophobia, an irrational debilitating fear of physical movement, was described by many participants, particularly in the initial stages of their condition.

Patients often require extended sleep and day naps following big days. This is frustrating particularly for those who previously led full lives and must adapt to doing less each day and stopping before things become painful. Despite the necessity for some daytime naps, night-time sleep is often a problem for patients. Patients on CRPS Facebook sites often discuss "painsomnia" which means not sleeping due to pain.

Besides coping with the loss of their former body and life, patients also face financial issues. There is a cost associated with health care which is often not considered by HPs. The cost of appointments, including travel and parking when combined with an inability to work at previous levels and exhaustion of sick leave entitlements increases quite quickly. The impact of being unable to complete tasks is not often appreciated by HPs and those prioritising medical care over the meaning of the experience on the patient do not display compassion or empathy.

Laura was 29 when interviewed and had lived with CRPS for 15 years. She sustained a meniscal tear to her right knee when she was 14 and was initially told by physiotherapists and surgeons that she was a hysterical teenager and a hypochondriac and needed a psychiatric referral for saying her pain had never gone away. At this time, she lived close to the city with her parents and younger brother and her condition caused problems between Laura and her brother for many years. After another meniscal tear following several falls when she was 20, the surgeon diagnosed her with CRPS and sent her to a specialist physiotherapist and a pain specialist.

After developing dystonia in her foot, Laura was sent to another pain specialist who commenced Botox which was extremely painful. She then tried a spinal cord stimulator and Baclofen pump. Laura was a university student at this time and was attempting to do her clinical placement on crutches. It was winter, and she couldn't wear long pants or proper shoes until after the implant. She was getting very depressed. Around this time, Laura spoke to her doctor about self-harm and all the knives and medication in the house were put out of her reach.

A CRPS specialist physiotherapist told Laura that she should drop out of university because she would never finish her degree and she would never work in her field. Laura graduated in the top 10 of her class of 80. After graduating and getting a job, Laura was mobilising with crutches and developed swelling and pain in her left foot and a spinal stimulator was inserted into her left leg. She also developed gastroparesis and spent one month in hospital and had a nasogastric tube for about six months. She never went back to work and was coping with many more co-morbidities which may or may not be attributed to CRPS.

Laura was working in a hobby/business which gave her something to concentrate on and keep her occupied but as she was on a disability pension, her financial future caused her to worry. She was living with her parents and relied on them to drive her to appointments and was not sure how long that could last as they were ageing. The specialist Laura was seeing was over an hour's drive away, so taxis were not a viable option for the future.

She did not remember life without CRPS, saying it was just part of her life. She had a big toolbox to manage her condition which included people such as her parents, doctors, her physio and some friends from a CRPS support group, and also medication, mindfulness, pets, and visualisation therapy. She used a wheelchair outside the house and tried to manage with crutches and furniture at home. She made a point of going shopping with her parents every Saturday morning, so she was out of the house for a few hours. She had trouble pacing herself and easily overdid things when she was feeling well. She had a CRPS blog internet site of her own but was not able to keep it up to date.

Hannah was 22 when interviewed and had lived with CRPS for eight years. She was 14 when she tore the tissue between her thumb and index finger in her dominant hand at school, but she was not diagnosed for about two years. She taught herself to write with her non-dominant hand so that she could complete her schooling. She initially had no treatment and then poor treatment and described her CRPS as deteriorating in her right and spreading into her left hand. At the time of her injury, Hannah lived with her parents and brother. Her brother had autism and displayed repetitive behaviour in the form of air drumming (similar to playing an air guitar) which included a hitting a pretend drum whilst making a repetitive drumming sound. The vibrations set off pain in Hannah's arm and she tried not to be in the same room as him when she was at home.

Hannah was living in the city with a friend while completing an undergraduate degree. Motivation to continue her studies to PhD level came from a university supervisor who disclosed a diagnosis of fibromyalgia and encouraged Hannah to continue her studies despite living with chronic pain. Hannah was completing her own CRPS research as part of her degree and hoped to have a career in educational psychology.

Because she pushed through and adapted to living with CRPS at a young age, Hannah was not consciously aware of most of the strategies she used to cope with her physical limitations unless her hand was 'really bad'. Her anxiety and depression were getting worse which she attributed to stress from university combined with CRPS. She spoke about not being able to do the things that a typical 22-year-old would do such as working in retail while studying. She did not work, and Hannah's parents had helped with financial issues when required. She stated that prescriptions in her country are free. Her pain was usually well controlled, but Hannah suffered from brain fog as a result of medication.

Hannah described herself as lucky in that although her condition had worsened over the past few years, she could still manage it and did not remember what it was like to not have CRPS. She considered she still had a decent quality of life despite having CRPS.

5.4 Developing Battle Plans with Allies

Patients require support to deal with living with CRPS, but support does not necessarily come when or where it is expected. While there are many dedicated HPs with extensive CRPS knowledge, some have never heard of the condition and suspect their patients are fabricating many of their symptoms. Patients often look online for information and support and here they find allies who have had similar experiences. Group members who had to advocate for better care often consulted each other for recommendations of knowledgeable health care providers.

Many of the patient participants had formed strong friendships with people in online support groups but also acknowledged the amount of catastrophising, misinformation and competition to be the worst case or have the most pain existed in some groups. Some of the HPs interviewed discussed strategies they used to counter the misinformation online and recognised that multidisciplinary teams provide the best care and support for patients. Patient participants agreed.

Carolyn was 65 and had lived for 18 years with CRPS when interviewed. She lived with her husband in a rural village and her CRPS started with a pain in her elbow. It took three and a half years for her to be diagnosed with CRPS. A second injury following this caused CRPS in her leg. She had recently retired to spend time with her grandson.

Carolyn said her biggest challenge was walking, and she had an electric scooter but said that the hardest thing to overcome was losing her independence. She had been driving with a steering wheel knob but had 'wobbly' sessions, so it was no longer safe. She also required someone to cut up her food and needed help to get dressed.

Her family had developed strategies over the years to make life easier. They adhered a tin lid to their kitchen bench, so she could butter her own toast and they walked on her affected side in crowds so that no one could bump her. Carolyn felt that if she sat around at home all day she would be bored. The pain management clinic she attended had taught her techniques so that she could interact with her grandchildren when they were born, and she had learned to knit after she retired. Carolyn said she might turn into a cabbage if she didn't keep doing things. She did not want to walk around with a label saying she is disabled; she just wanted to try to fight it.

Dianne was 50 at the time of her interview and had CRPS for four and a half years. She injured her shoulder during a Pilates class. During the first 12 months she saw a GP, physiotherapist, a neurologist, a vascular surgeon, an orthopaedic surgeon and had stopped using her arm due to excruciating pain. Dianne went to another physiotherapist who mentioned CRPS and told her to carry on doing things and just forget about it and the CRPS will melt away. Dianne's GP refused to refer to her to a specialist CRPS centre she had heard about incorrectly reasoning that it was too expensive. Another GP referred Dianne to a CRPS doctor and she waited a long time for the appointment. She was told by this doctor that because she'd had it for three years, there was nothing that could be done, and that Dianne would just have to live with the pain. During this time, Dianne lost her job, could no longer drive, and as a health care professional, was frustrated knowing that her condition was very treatable if she had been diagnosed and treated earlier.

Dianne got a referral to the specialist centre she had originally asked about. She had two admissions to this centre where she received medication, education and therapy. She was taught to meditate and taught how to start re-using her right hand again. The specialist centre helped Dianne to adjust her routines. She started wearing a bra again for increasing lengths of time when leaving the house and told how she was wearing tops two sizes too big so that the material touching her inner arm, elbow, wrist and palm did not hurt. Before that, she had been wearing nothing on her top half at home because instinctively she wanted to avoid pain. She was slowly overcoming disassociation of her affected arm.

Dianne had a masters level education and had earned more than her husband prior to leaving work. She felt guilty that she was not earning an income because she could no longer help her young adult children financially. She also could not help her elderly parents. One of her goals was to be able to wear a variety of clothing and feel like herself again. Her other goals were to gain a part time or casual job where she could work from home and to be able to participate in a yoga class.

Karen was 55 and had CRPS for seven years. She lived with her husband and teenage daughter on the outskirts of a capital city. She suffered a spiral fracture to her foot on public transport and it took three years for her to be diagnosed with CRPS. She saw a physio and had two surgeries during this time. The pain specialist who diagnosed her forgot to write to the insurance company for permission for ketamine treatment three months in a row, so Karen organised to see another pain specialist who apart from referring her to a psychologist, did nothing besides writing prescriptions for medication. In this time, Karen went from limping to using a walking stick to using crutches full time and had a physical deformity to her foot.

Karen sought advice from an online support group and was recommended another pain specialist. There was a long waiting period to see the new specialist and he was on the other side of the city. The insurance was going to pay for her treatment, but the claim took 12 months and was extremely stressful both mentally and physically for a small financial gain.

She worked full time for years with CRPS and gave up on the advice of her GP as she was getting too tired to drive home safely after work. Karen used to be very social and did not go out much anymore due to the difficulty mobilising, the pain and the exhaustion afterwards.

There was a big impact on Karen's family following her injury and her husband had to do most of the housework and care of their teenage children, which Karen felt guilty about. She was looking forward to getting a power assisted wheelchair so that she would have some more independence.

5.5 *Warrior or Prisoner of War?*

Many people with a chronic illness refer to themselves as warriors, rising to the challenge of their condition and demonstrating courage and resilience in their efforts to be in control of their illness. Others succumb to the enemy and feel they have no control of their condition and become prisoners in their own home. Most participants described moments of being both. Using the "fake it until you make it" attitude, participants spoke about putting on a mask to hide their true levels of pain to protect both themselves and others. These patients are not faking their illness but rather are faking being well. Psychological support is important here. The HP participants agreed with the terminology, stating that having a warrior attitude helped patients activate resilience factors, which led to better health outcomes.

Many patients feel that their lives would be easier if more people in the general community were aware of the existence of CRPS. Many of the patient participants had started support groups, internet blogs and joined campaigns such as Colour the World Orange in attempts to raise awareness. Some participants were studying CRPS to answer some of their own questions and some participants delivered talks to local community groups and health students. CRPS is often named 'the suicide disease' online and while some of the participants disclosed suicide ideation, none divulged actual attempts.

Fred who was 41 when interviewed, had CRPS for 11 years and lived in a capital city with his wife and four children. He suffered a crush injury and ligament damage to his arm at work but as he does not like them, he didn't see a doctor for six months when he became unable to drive. The surgeon neglected to tell him he had inserted two pins into his arm and also didn't tell the physios he worked with. This resulted in Fred using a TENS machine which caused further harm. CRPS was diagnosed two years after that but Fred was told that CRPS is just excessive pain. Fred did not know that the other symptoms he was experiencing were part of CRPS and he thought he was going crazy. When he mentioned that the pain was spreading into his hand and he was unable to make a fist, he was told that it was "silly" by a pain specialist.

A pain clinic wanted to perform a spinal block, but Fred has a needle phobia and fainted at the first attempt. He was told that unless he had morphine injected into his spine he couldn't have mirror box treatment. Therefore, Fred had no treatment, and the CRPS spread throughout his arm and to his stomach and neck causing vomiting and headaches. He told of extreme temperature changes in his body from being boiling hot to being so cold he shakes so violently that he has torn chest muscles.

Fred's problems with WorkCover lasted 11 years have affected the family finances. With four children to provide for, Fred's' wife is his carer, but she also worked four days a week outside the home. As Fred was often unable to leave the house, he missed a lot of family events, including Christmas day at his mother's. When his two youngest children were between two and six years of age, Fred was afraid of them hurting him with their exuberance. He hid from them, confining himself to his bedroom for years.

Fred's work had also been his hobby and he missed that more than anything else. He used to enjoy fishing but being unable to wind his own rod, he gave it up. Fred was spending his days watching TV and playing computer games on his own, which he never did before the injury. His friends had abandoned him completely once he could no longer drive due to a combination of brain fog and pain. He had to reduce his opiate dose due to being "“spaced out” and found he tired easily and had trouble maintaining conversations. Most medications had had little or no effect on his pain levels and he described his pain level as 9.549012 out of 10 on his current medication stating that it had reduced his pain.

Fred wanted to see a pain specialist again but said he need a break from fighting WorkCover to pay for it. He had held his arm in the one position for so long that his elbow no longer straightened, and his shoulder had little movement. After researching CRPS on the internet, Fred discovered that his odd symptoms of chronic fatigue, rashes and increased sweating were part of CRPS and he said it felt good to know that he wasn't alone in how he was feeling.

Fred decided to help other people in similar positions and was planning to make YouTube videos featuring people with CRPS (including someone who has gone into remission) so that others in his position could learn. He had already made a short video explaining CRPS in his own words so that non-medical people with brain fog would be able to understand it. This was a project he could do when he was feeling well.

Rosemary was 64 at the time of her interview and had lived with CRPS for eight years. She lived in a rural area with her husband who was her carer. Her CRPS started after developing pain in her foot, which was eventually diagnosed as a neuroma. She saw many health professionals both private and public during the next five years. Rosemary had three surgeries on her foot performed by three different surgeons and her pain levels were not acknowledged until her last surgery. She saw two pain management specialists and eventually gave up on doctors after feeling she was not being taken seriously. It was taking her three hours to get to the city and at least two hours to travel to larger towns for treatment and the travel was aggravating her pain.

Five years after the pain started, Rosemary saw another pain management specialist who diagnosed CRPS and suggested spinal cord stimulation. Four months later she was still waiting to be booked in and decided to give up on the public system. After asking for suggestions on an internet forum, Rosemary made an appointment with a new specialist in the city. Six years after the initial surgery, Rosemary received a spinal cord stimulator trial and an implant four months after that.

She changed to a new private pain clinic in a regional town that was two hours away, saving her one hour travel each way. They started her on ketamine infusions and she saw a physiotherapist who taught her how to pace herself by limiting the time weight bearing to avoid flaring her foot.

Rosemary had contemplated suicide in that first five years and credits the local psychologist she was seeing for helping her to find a way to live with her pain. She said she was fortunate that she had private health insurance and no mortgage to cause major financial concerns. After giving up work due to pain, Rosemary said she found a new identity as an artist and converted a carport into a studio in her backyard. She was able to work on her art and unless the pain was really severe to direct all her attention to creating the artwork, describing it as an almost meditative state.

Rosemary had a number of mobility aids. At home she used crutches and a knee walker which she sourced from America. When leaving the house, she preferred a portable scooter which she described as much easier to manage in public than her wheelchair. Rosemary didn't go out very often, which she felt guilty about because her husband and her had planned to travel when they retired. They liked to take drives and do sketches together but every outing had the potential to cause pain.

Rosemary directed me to her CRPS website during her interview. She made the site after experiencing difficulty getting home help and wanted people to know that her condition and physical limitations can change on a daily basis.

6 Involving Participants: Discovering the Battles

Three patient participants developed the condition as teenagers. All are very intelligent and well educated as evidenced by the university study they have completed. Living with CRPS though means that Laura works in a hobby business making and selling soaps and teddy bears, instead of working in her profession where she would be earning a high income. Both Laura and Alice live with their elderly parents which poses the question of where will they live when their parents are no longer able to provide care? After completing her initial degree, Jasmine has continued to study at post-graduate level and is now married. Her diagnosis raises questions for her about potential issues related to pregnancy and child rearing.

Hannah's' story highlights a common side effect of CRPS and/or CRPS medication, brain fog, which is often discounted by health professionals as not significant. People have novel ideas for coping with brain fog in their everyday lives, such as making shopping lists and setting reminder alarms. For students like Hannah, Laura, Jasmine and Colleen though, brain fog can mean the difference between passing and failing. This can be the difference between working in a rewarding career and holding down a job to survive.

CRPS caused issues with siblings for both Laura and Hannah. Hannah has difficulty being in close proximity to her brother and stated that this impacted on family life when they were both younger and as adults, family gatherings remain difficult at times. Hyperesthesia can be challenging when situations involving crowds such as shopping, movies or travelling on public transport are a part of everyday life. Family life was impacted with all participants and guilt was expressed many times due to the burden placed on family members.

7 Implications of This Research

In order for appropriate patient focussed care to be delivered, the lifeworld of those living with CRPS must be understood. Using a framework of the themes as a guide for posing questions, clinicians can ascertain and discuss what the needs might be for each patient. Through dialogue with the patient, a determination of the meaning that CRPS has for that person can be established and plans can be implemented which address needs according to the priority of the patient. Specific CRPS patient education to both patients and health professionals will ensure that both are working from a larger knowledge base of the battles experienced. Health professionals must be aware that CRPS exists to begin with and must feel comfortable to refer suspected cases to specialists in pain medicine. Health professionals must be aware that differences exist in patients. For example, some patients fear being touched without warning, some are battling a fear of spread, and others are not believed at home. Improving the pain literacy of patients will help them to better understand their condition and alleviate some of the distress caused by accessing misinformation and from not understanding medical terminology.

Note Informed consent was obtained from all individual participants included in the study.

References

1. Bruehl S. Complex regional pain syndrome state of the art review. BMJ. 2015;350:h2730.
2. Tahmoush AJ. Causalgia: redefinition as a clinical pain syndrome. Pain. 1981;10(2):187–97.
3. Kocz R. History and epidemiology. In: Nader ND, Visnjevac O, editors. Complex regional pain syndrome: past, present and future. New York: Nova Science; 2015. p. 1–18.
4. Grieve S, Perez RSGM, Birklein F, Brunner F, Bruehl S, Harden RN, et al. Recommendations for a first core outcome measurement set for complex regional pain syndrome clinical studies (COMPACT). Pain. 2017;158(6):1083–90.
5. Harden RN, Maihofner SGMC, Abousaad JE, Vatine BJJ, Kirsling BA, Perez BR, et al. A prospective, multisite, international validation of the complex regional pain syndrome severity score. Pain. 2017;158(8):1430–6.

6. Birklein F, Dimova V. Complex regional pain syndrome-up-to-date. Pain Rep. 2017;2(e624): 1–8.
7. Russo M, Georgius P, Santarelli DM. A new hypothesis for the pathophysiology of complex regional pain syndrome. Med Hypotheses. 2018;119:41–53.
8. Birklein F, Schlereth T. Complex regional pain syndrome – significant progress in understanding. Pain. 2015;156:S94–S103.
9. Punt TD, Cooper L, Hey M, Johnson MI. Neglect-like symptoms in complex regional pain syndrome: learned nonuse by another name? Pain. 2013;154:200–3.
10. Goebel A. Management of adult patients with long-standing complex regional pain syndrome. Pain Manag. 2013;3(2):137–46.
11. van Rijn M, Marinus J, Putter H, Bosselaar SJ, Moseley GL, van Hilten J. Spreading of complex regional pain syndrome: not a random process. J Neural Transm. 2011;118(9):1301–9.
12. Goebel A. Complex regional pain syndrome in adults. Rheumatology. 2011;50:1739–50.
13. van Velzen GAJ, Perez RSGM, van Gestel MA, Huygen FJ, van Kleef M, van Eijs F, et al. Health-related quality of life in 975 patients with complex regional pain syndrome type 1. Pain. 2014;155:629–34.
14. Sale JE, Thielke S. Qualitative research is a fundamental scientific process. J Clin Epidemiol. 2018;102:129–33.
15. van Wijngaarden E, Meide H, Dahlberg K. Researching health care as a meaningful practice: toward a nondualistic view on evidence for qualitative research. Qual Health Res. 2017;27(11): 1738–47.
16. van Manen M. Researching lived experience: human science for an action sensitive pedagogy. New York: State University of New York Press; 1990.
17. Crotty M. Phenomenology and nursing research. South Melbourne, VIC: Churchill Livingstone; 1996. 202 p.
18. Florczak K. Adding to the truth of the matter: the case for qualitative research. Nurs Sci Q. 2017;30(4):296–9.
19. Gelling L. Qualitative research. Nurs Stand. 2015;29(30):43–7.
20. Rodham K, Gavin J, Coulson N, Watts L. Co-creation of information leaflets to meet the support needs of people living with complex regional pain syndrome (CRPS) through innovative. Inform Health Soc Care. 2016;41(3):325–39.
21. Grieve S, Adams J, McCabe C. 'What I really needed was the truth'. Exploring the information needs of people with complex regional pain syndrome. Musculoskeletal Care. 2016;14:15–25.
22. Johnston CM, Oprescu FI, Gray M. Building the evidence for CRPS research from a lived experience perspective. Scand J Pain. 2015;9:30–7.
23. Butler S. Qualitative research in complex regional pain syndrome (CRPS). Scand J Pain. 2015; 9:62–3.
24. Harris PS, Bedini LA, Etnier JL. An examination of leisure and quality of life in individuals with complex regional pain syndrome (CRPS). Annual in Therapeutic Recreation. 2016;23:1–11.
25. van Rysewyk S. A call for study on the meanings of pain. In: van Rysewyk S, editor. Meanings of pain. Cham: Springer; 2016. p. 1–22.
26. Patton MQ. Qualitative research and evaluative methods. 4th ed. Thousand Oaks: SAGE; 2015. 832 p.
27. Rodham K, McCabe CS, Blake D. Seeking support: an intrepretative phenomenological analysis of an internet message board for people with complex regional pain syndrome. Psychol Health. 2009;24(6):619–34.
28. Rodham K, Boxell E, McCabe C, Cockburn M, Waller E. Transitioning from a hospital rehabilitation programme to home: exploring the experiences of people with complex regional pain syndrome. Psychol Health. 2012;27(10):1150–65.
29. Converse M. Philosophy of phenomenology: how understanding aids research. Nurse Res. 2012;20(1):28–32.
30. Dowling M, Cooney A. Research approaches related to phenomenology: negotiating a complex landscape. Nurse Res. 2012;20(2):21–7.

31. Douglass B, Moustakas C. Heuristic inquiry: the internal search to know. J Humanist Psychol. 1985;25(3):39–55.
32. Moustakas C. Heuristic research: design, methodology, and applications. Thousand Oaks, CA; 2011. 130 p.
33. Johnston CM, Wallis M, Oprescu FI, Gray M. Methodological considerations related to nurse researchers using their own experience of a phenomenon within phenomenology. J Adv Nurs. 2017;73(3):574–84.
34. Harden RN, Oaklander AL, Burton AW, Perez RSGM, Richardson K, Swan M, et al. Complex regional pain syndrome: practical diagnostic and treatment guidelines. Pain Med. 2013;14: 180–229.
35. Meugnot A, Jackson PL. The contribution of new technological breakthroughs to the neuro-scientific research of pain communication. In: van Rysewyk S, editor. Meanings of pain. Cham: Springer; 2016. p. 87–106.
36. Rodham K. Learning to cope with CRPS/RSD: putting life first and pain second. London: Singing Dragon; 2015. 144 p.
37. Burning Nights CRPS. Our founder [Internet]. 2019 [cited 13/04/2019]. Available from https://www.burningnightscrps.org/ourstory/our-founder/
38. Princess. The tower [Internet]. 2019 [cited 13/04/2019]. Available from http://princessinthetower.org/
39. Kelly. About me [Internet]. 2019 [cited 13/04/2019]. Available from https://theinvisiblewarrior.com/about/
40. Genevieve. About me [Internet]. 2019 [cited 13/04/2019]. Available from http://www.shipwithnosails.com/about-me/my-story/
41. Jessica [Internet]. 2019 [cited 13/04/2019]. Available from http://hopeforjessica.blogspot.com.au/
42. van Manen M. Phenomenology online: a resource for phenomenological inquiry [Internet]. 2011 [cited 13/04/2019]. Available from http://www.phenomenologyonline.com/inquiry/methodology/
43. Casale R, Atzeni F, Masala IF, Sarzi-Puttini P. The words of pain in complex regional pain syndrome. Best Pract Res Clin Rheumatol. 2015;29:71–6.

Chapter 10
Cancer Pain and Coping

Sara E. Appleyard and Chris Clarke

Abstract Receiving a diagnosis of cancer can be devastating. Cancer continues to be one of the most feared diagnoses, and experiencing pain is a major fear for people diagnosed with cancer. Cancer pain is complex in aetiology and can be acute or chronic and can be caused by various compression, ischaemic, neuropathic or inflammatory processes. Many people with cancer will experience excruciating pain, which is often underreported and undertreated. The reasons for this are complex and include various factors including fears and beliefs held by patients. Cognitive factors are important modulators of pain and the appraisals, meanings and beliefs that people have in relation to illness, cancer, and pain, have implications in regard to help-seeking behaviours and the coping strategies people adopt. Cancer pain can impact a range of psycho-social factors across its course. Cancer pain relates to higher rates of psychological distress, anxiety and low mood, and the perception of pain intensity is, in turn, influenced by psychological factors. Cancer pain can negatively affect psychological health, and psycho-social factors can affect the pain experience. It has been suggested that people with a life-limiting illness experience 'total pain', and this can encompass psychological, social, practical, spiritual domains. Research demonstrates that cancer is predominantly an illness affecting older people, yet there is a higher risk of under-treated pain in this age group and there is a paucity of research into the subjective experiences of older people managing their cancer pain. Many older people are required to self-manage cancer pain at home as outpatients, due to drivers to keep people out of inpatient care, such as the high healthcare costs of inpatient treatment, and patients wanting to manage their illness, and die, at home. Our own research into this area found that the self-management of cancer pain involves a sequential and temporal process, which centres on perceptions of control. We describe how the older people in our study experienced a perceived loss of control, followed by a sense of gaining control over

S. E. Appleyard (✉)
Department of Clinical Psychology for Older People, Humber Teaching NHS Foundation Trust, Hull, UK
e-mail: sara.appleyard@nhs.net

C. Clarke
Department of Psychological Health & Wellbeing, University of Hull, Hull, UK

S. van Rysewyk (ed.), *Meanings of Pain*,
https://doi.org/10.1007/978-3-030-24154-4_10

pain through various experience of certain internal and external factors. The assessment and treatment of physical pain should be done in combination with assessment and treatment of psycho-social and spiritual pain, and interventions for cancer pain need to focus on increasing positive affect and reducing helplessness. No person should suffer with poorly controlled pain and we argue for the need for further research in this area to ensure adequate treatment for all.

Clinical Implications: Assessment and treatment of physical pain in people with cancer needs to be conducted in combination with assessment and treatment of psycho-social issues and spiritual pain. Psychological and behavioural approaches have strong evidence supporting their efficacy for reducing cancer pain. Interventions should target helplessness and focus on increasing positive affect through positive psychological states including fighting spirit and resilience. Person-centred interventions that focus on helping people with the search for meaning may help those with cancer pain derive positive benefits.

Keywords Cancer · Oncology · Pain · Older person · Coping · Meaning · Appraisal · Self-management · Systematic review · Interpretative Phenomenological Analysis (IPA)

1 Introduction

There are few of us whose lives have not been touched by cancer. Receiving a diagnosis of cancer for oneself, or for a family member, friend or colleague, can be devastating. Cancer continues to be one of the most feared diagnoses [1], despite increasing rates of survival when diagnosed early [2]. One of the main fears for people diagnosed with cancer is whether they will experience pain [3] and these fears appear well founded, as over one third of cancer patients describe their cancer pain as distressing or intolerable [4]. It is a tragedy in today's society that many people experience excruciating cancer pain which is often underreported and undertreated [5]. We will discuss some of the reasons why throughout this chapter.

Both personally and clinically, the first author has been witness to multiple experiences of cancer, and cancer pain has been a central theme across those experiences. In our work as clinicians, we are interested in the subjective nature of the pain experience from the perspective of the patient, and reflexivity is a vital component of research into this area. As researchers, we must therefore acknowledge our own assumptions about the meanings of cancer pain. The analysis of qualitative data in particular is sensitive to interplay between those assumptions and the individual experiences we encounter, and it is from this stance that we present our research findings.

From an academic and clinical viewpoint, we have been particularly interested in how people cope with cancer pain. In this chapter, we aim to outline the coping strategies people may employ in relation to their cancer pain, along with the underpinning meanings of pain that might affect the coping strategies which people adopt. In order to achieve this aim we will draw upon our own research in this field as well as existing empirical findings and conceptual accounts. We will present a brief overview of cancer as a disease within an epidemiological context to give the reader an introduction to the scale of the problem. Meanings and appraisals of cancer pain will then be discussed with reference to current research in the field. We present a summary of the findings of our own research examining whether there are indicators of individual differences in the selection of coping strategies for cancer pain, for which we conducted a systematic literature review. We will also present and discuss the findings from our own original research on the self-management of cancer pain, which focused on one of the largest groups of people to be affected by cancer: older people. Finally, we will discuss the clinical implications from our own and others' research to present the clinician with evidence-based treatment options to promote the successful psychological management of cancer pain.

2 What Is Cancer?

Cancer is a disease of the cells of the body. It is an umbrella term that describes many different types of disease that can occur due to multiple causes and in a variety of primary sites e.g., breast, prostate, lung etc. [6] Cancer is often described in terms of its stage (severity) and whether or not the cancer has metastasised (spread from its primary site to secondary sites), which occurs when abnormal cells from malignant tumours grow larger than their original boundaries. The causes of cancer are understood to be an interplay between individual genetic and external factors in the form of various carcinogens: physical carcinogens e.g., radiation; chemical carcinogens e.g., tobacco smoke; and biological carcinogens e.g., infections such as hepatitis B [6].

Cancer is not a new disease; some of the earliest recorded examples appear within Egyptian mummies and manuscripts dating from as early as 1600 BC where it appears that surface tumours were surgically removed in much the same way as in modern times [7]. Later, the Greek physician Hippocrates (460–370 BC) described malignant tumours as visually resembling crabs in terms of their spreading outline (from the Greek word for crab "karkinos" we get the English word "cancer"). Medical cancer treatments have rapidly developed throughout the twentieth century to include chemotherapy, radiotherapy, and immunotherapy, with our knowledge of the role of gene mutations in tumour growth and development occurring as a result of the discovery of DNA [7]. However, a focus on psychological treatments for cancer pain and distress dates from much more recent times.

3 The Scale of the Problem

Cancer affects many millions of people across all cultures, countries and levels of socio-economic status. Whilst recent advances in treatment have positively impacted survival rates, cancer remains one of the leading causes of death worldwide [6]. Recent estimates suggest that cancer cases are rising, in part due to increasing populations, but also in response to lifestyle factors that have a negative health impact such as obesity, tobacco smoking and physical inactivity. In 2012, 32.6 million people were living with cancer (within 5 years of diagnosis) and 14.1 million new cancer cases were diagnosed [8]. Recent estimates suggest there will be 9.6 million deaths from cancer in 2018 [6]. There is variation across world populations in regard to incidence rates, with male incidence varying from 79 per 100,000 in Western Africa to 365 per 100,000 in Australia/New Zealand; female incidence rates vary from 103 per 100,000 in South-Central Asia to 295 per 100,000 in Northern America [8]. Men are more likely to have a cancer diagnosis, variations in gender mean that the incidence rate for males is close to 25% higher, with overall rates at 205 per 100,000 whilst for females, in contrast, this accounts for 165 per 100,000 [8]. The most common cancers are lung and breast (accounting for 2.09 million cases) with lung cancer accounting for 1.76 million deaths per year [6].

In the UK, data for the years 2014–2016 indicate that around 450 people die every day from cancer (164,000 per year), accounting for just over a quarter (28%) of all deaths in the UK [9]. Age is a major factor in developing cancer [6]. According to Cancer Research UK [9], just over half of all UK deaths from cancer are in the over 75 years age group (53%) with mortality rates highest in the oldest-old age group (aged 90 years and above).

At the same time, cancer survival rates have increased over recent years. In 2010–2011, 50% of people diagnosed with cancer in England and Wales were still alive after 10 years or more [9]. There is an age effect, however, and survival rates in the UK are generally highest for those aged under 40 years [9]. Globally, the picture is much worse and dependent on available health resources for assessment and treatment, which for millions is non-existent.

4 Cancer Pain

As well as being one of the most feared symptoms, cancer pain is a reality for many of those diagnosed with the disease. Figures suggest that half of all cancer patients experience pain during the treatment phase and this figure increases to around two-thirds of people with advanced metastatic cancer [4]. This is despite advances in medical pain management such as the use of opioids based on guidance from the World Health Organisation's analgesic ladder, which should provide adequate pain relief for around 95% of cancer patients [6]. Oral morphine alone can relieve pain in around 85% of people with cancer pain; however, across the global population, the

World Health Organisation [6] reports that 80% of cancer patients have no access to opiates.

Cancer pain is complex in aetiology and can be acute or chronic. Pain can be caused by various compression, ischaemic, neuropathic and inflammatory processes across single or multiple sites in the body, depending on the spread of the cancer [5]. In addition, cancer treatments such as chemotherapy, surgery and radiotherapy can cause persistent chronic pain [5].

Cancer pain is underreported and undertreated even in countries with high standards of cancer care [5]. The reasons for this are complex and not fully understood, but there a range of recognised factors which affect the reporting of pain including fears and beliefs held by patients [10], which we explore further in this chapter.

Cancer pain can impact on a range of psycho-social factors across its course, from diagnosis, to treatment, to survivorship and end of life. Cancer pain is related to psychological distress, anxiety and low mood [11] and the perception of pain intensity is influenced by psychological factors. The interplay between cognitive, sensory, emotional, behavioural, environmental and social factors, is highly individual, complex and bi-directional in relation to pain, impacting on wellbeing and spiritual health. For example, cancer pain can negatively affect psychological health, and psycho-social factors can affect the pain experience [1]. Cognitive factors in particular have been highlighted as important modulators of pain and relate to the appraisals, meanings and resultant coping strategies employed to manage pain [5]. Personal assumptions regarding the aetiology, course, and treatability of pain, form a part of an appraisal process by which people come to understand the nature of their pain experience, which in turn can affect healthcare seeking behaviours [5].

5 Appraisals, Meanings and Beliefs

Whilst the experience of pain is a uniquely subjective experience [12], cancer pain is particularly sensitive to individual interpretation and psychological influences [5]. How people cope with pain depends on how pain is appraised and what meanings people attribute to their pain. For example, if pain is appraised as "threat" this may be interpreted as a sign that the cancer has spread and is no longer treatable. This may lead the person to believe there is nothing more that can be done and therefore avoid seeking their physician. This would negatively impact on mood, pain experience, relationships and quality of life. In contrast, if pain is appraised as "challenge" this may lead to more proactive help-seeking behaviours to manage pain, positively impacting on mood, relationships and wellbeing.

Leventhal et al. [13] originally described how appraisal of perceived threat can influence responses and adjustment to illness. They identified particular components of appraisal that are important in the meaning of personal illness, and which will influence coping strategies:

1. Cause (beliefs about what may have led to the illness)
2. Identity (beliefs about the aetiology of the disease)
3. Timeline (beliefs about the personal timescale of the illness)
4. Consequences (beliefs about the short- and long-term effects of the illness)

Primary appraisals involve an evaluation of the benefit or harm of a particular person-environment interaction, and secondary appraisals involves the evaluation of what can be implemented to affect the impact of this; for example, what could be done to increase personal benefit or reduce potential harm [14]. Research has identified three primary appraisals that people make when faced with illness: challenge (anticipation of growth); threat (the potential for harm/loss; centrality (perceived significance on wellbeing), and one type of secondary appraisal: resources [15]. Lee et al. [16] found that all of the studies in their systematic literature review alluded to the threat posed by cancer and this appears to be a common meaning.

The meanings people attribute to cancer, pain, and pain experienced during a cancer illness can be highly individual with a range of influences, including individual histories, family experiences, cultural understandings and expectations, and attitudes towards illness and wellness. Meanings attached to cancer were found to be an important factor in regard to self-esteem, optimism and reduced psychological distress [16] and can be divided under four major categories, presented in Table 10.1.

Table 10.1 The construct of meaning in cancer research: thematic findings from a systematic literature review [16] using categories proposed by Park and Folkman [17]

Meaning category	Definition and themes
Global meaning	Definition: Existential beliefs providing a framework from which to understand cancer in the context of past and future life experiences and expectations; beliefs that contribute to creating a sense of life purpose Themes: perceiving the world as less controllable and more random; perceived loss of control, self-esteem and self-worth; the need to preserve a sense of continuity between past, present and future; the struggle to reconcile differences between past views of self and life and current reality; religious and cultural attitudes towards the meaning of cancer
Appraised meaning	Definition: The perceived level of threat or challenge that cancer poses based on how life goals are affected Themes: increased awareness of own mortality; the perceived potential for personal growth that the cancer experience offers
Search for meaning	Definition: Cognitive coping strategies used to reduce the discrepancy between previously held global meaning and appraised meaning Themes: reprioritising values; deciding to accept loss, uncertainty and vulnerability; deciding to focus on life rather than cancer; speculating on the aetiology of the cancer and its impact
Meaning as outcome	Definition: The outcome of the search for meaning including positive outcomes, enhanced social resources, philosophical changes and the perception of personal growth Themes: increased compassion towards others; a renewed commitment to oneself; discovering a sense of fulfilment despite uncertainty

This framework suggests that a diagnosis of cancer can disrupt one's sense of global life meaning sufficiently to drive a new search for meaning in the context of living with cancer, and may relate to how people derive meaning, and a sense of personal control, in the face of living and coping with cancer pain. The beliefs that people have in relation to their pain are also an important factor in influencing how people cognitively and behaviourally manage their pain. For example, believing that a reduction in physical activity and rest is beneficial for pain may lead people to reduce rewarding activity, leading to lowered mood and negatively impacting self-esteem and confidence [5]. Unhelpful beliefs that can act barriers to successful cancer pain management have been identified [10] and include:

- Fear of addiction to medication
- Fear of medication tolerance
- Fear of unmanageable medication side effects
- Fatalistic beliefs about the inevitability of pain
- The desire to be a "good patient" and not complain
- Reluctance to distract the doctor from treating the cancer
- Fear that more pain indicates a worsening of the cancer
- Fear of injections

Certain cancer diagnoses are also more likely to attract social stigma; for example, in lung cancer there is a societal perception of the disease being self-inflicted due to its association with smoking tobacco. Awareness of social stigma surrounding a particular cancer diagnosis could lead people to form negative beliefs and meanings that in turn adversely influence well-being and coping responses, such as poor social support seeking [18].

6 Coping

It has long been understood that personal meaning, psycho-social problems and coping strategies are critical factors in the experience of cancer [19]. In the course of our research into cancer pain, we became increasingly interested in understanding the role of coping. Despite a wealth of research into the psychological aspects of coping, there continues to be varied complex conceptualisation of coping, which lack a unified definition [20]. However, within health psychology, illness models coping is described as a dynamic process where people utilise behavioural and/or cognitive responses as a way of managing the discrepancies that exist between perceived external demands (including pain) and individual resources [21].

Chronic pain research has demonstrated relationships between coping, beliefs, and adjustment to pain and it appears that control is a central issue, as those who believe they can control their pain tend to experience a reduction in perception of pain severity and increased functionality [22]. Self-efficacy (one's belief in a personal ability to accomplish goals or tasks) has been shown to moderate the effects of stress on quality of life in research focused on patients receiving palliative care

[23]. In contrast, helplessness is related to emotional distress and has been shown to be related to increased risk of disease recurrence and death [24].

Lazarus et al. [21] categorised coping strategies into two groups: emotion-focused coping strategies, including distraction and avoidance, which are active on the emotional consequences of the stressor, and problem-focused coping strategies such as social support and seeking information, which are actively directed towards the problem itself. A meta-analysis examining how primary appraisals in relation to cancer (threat; challenge; harm/loss) influence coping (problem-focused; emotion focused) and coping effort (approach; avoidance) found that appraisal of threat was related to problem-focused coping, appraisal of harm/loss was related to avoidance coping and appraisal as challenge was related to problem-focused and approach coping [25].

Coping strategies have been understood in regard to their stability. Situational approaches suggest that people choose various strategies dependent on the stressor, whereby a dispositional approach suggests stable patterns of coping despite the variability of the stressor [26]. Research by Roberts et al. [27], who specifically studied appraisals and coping in outpatients with advanced cancer, suggested however that coping has to be reactive to rapidly changing events, fluctuating symptoms, deterioration over time and uncertain timescales; "living around" and "living with," revisiting, reframing and redeveloping coping strategies. They describe this as an evolving process with ongoing potential for distress as advanced cancer includes multiple losses moving towards death as the final stage. They argue that understanding primary appraisals as "threat, harm or challenge" may be too simplistic as classifications and that appraisals in real life may be more complex or encompass multiple appraisals. Research has also shown that people with cancer who appraise their illness as severe may employ multiple coping strategies [28].

Positive personal traits such as optimism and "fighting spirit" may have benefits in terms of appraisals, as people may view adverse events in terms of challenge rather than threat. This may positively impact on physical health [29] although this remains contentious in terms of affecting disease progression [30]. Previous research has linked pain outcomes with coping strategies; for example, catastrophising has been associated with increased pain [31, 32]. However, there is a paucity of literature in regard to coping with cancer pain and, in particular, to what extent relationships exist between self-initiated pain coping strategies and particular patient variables. Poorly controlled pain increases psychological distress and, in combination with maladaptive coping strategies, has implications in terms of inpatient stays and long-term disability, which in turn has negative cost implications for already stretched healthcare systems [5].

7 A Systematic Literature Review: Coping Strategies and Cancer Pain. Who Chooses What?

We were interested in exploring coping strategies further. Examining the literature base, it soon became apparent that previous research has been directed towards examining the use of various coping strategies and whether they are related to positive or negative outcomes (e.g., pain intensity). However, we found an area in which there had been little exploration: whether there are relationships between clinical or patient variables and particular cognitive or behavioural coping strategies for managing cancer pain. Therefore, we were interested in discovering whether there were any predictors of who would choose particular coping strategies. A systematic literature review was selected as a method of investigating and analysing previous research to understand and synthesise existing data and establish whether there are individual differences in the selection of coping strategies and what factors may influence or underpin such differences.

The studies included in our review [33] were required to describe the use of cognitive or behavioural coping strategies as well as describing demographic, psychological or clinical characteristics of participants in their studies. Psychology, health and medical electronic databases were searched, and we did not exclude on grounds of methodology to ensure wide coverage of the literature. The review therefore included a range of qualitative, quantitative and mixed-methods studies. We concentrated our review on community-based patients and excluded inpatients. Outpatients are the group with the highest prevalence of cancer pain [34] and for a number of factors, including environment, medical management and availability of resources, the needs and experiences of inpatients are qualitatively different enough to warrant their exclusion from this review.

Fifteen international studies reached our inclusion criteria, the majority of which had been conducted in the USA (10 studies), with the remainder conducted in the UK, Canada and Taiwan. We used narrative synthesis [35] to analyse the findings of included studies; due to their heterogeneity meta-analysis was not viable. The studies included in our review explored a number of variables in relation to coping with cancer pain, which we describe in more detail below.

7.1 Demographic Variables

In the demographic variable category, studies reported on the effects of age and ethnicity. Whilst Bennett et al. [36] did not report any significant differences in the selection of coping strategies across older and younger patient groups, Gagliese et al. [37] found that older patients utilised accommodation strategies, including goal and activity modification, and acceptance, and found that younger patients were less likely to use these strategies. In regard to ethnicity, Haozous et al. [38] found that Native American Indians utilised prayer as a primary coping strategy and described this as a characteristic of this ethnic group. Similarly, in a study conducted by Reddick et al. [39] African American participants were more likely to utilise hope and prayer than Caucasian participants. Im et al. [40] explored differences and similarities across four ethnic groups in the USA and found that Caucasian participants focused on selecting pain treatments and the control of pain, whilst ethnic minority participants coped by normalising and minimising pain. It therefore appears that ethnicity is a factor in how people try to control pain, particularly in relation to using prayer. It is also likely that age plays a role in the selection of coping strategies; however, there is a paucity of research in this area. Further research is warranted to examine the effects of both age and ethnicity on the use of coping strategies in cancer pain.

7.2 Clinical Variables

The studies we reviewed explored differences in coping across types of cancer (e.g., head/neck; prostate; lung etc.) and disease stage. Whilst none of the studies included in our review reported differences between cancer types and coping strategies selected, Whale et al. [41] reported that distraction (keeping busy) whilst planning regular rest periods was the most common strategy selected by participants in their study who were being treated for head and neck cancers. Kwekkeboom [42] described the most used strategies for managing cancer pain amongst females with breast and gynaecologic cancers were body positioning, the therapeutic use of heat, relaxation and distraction. Fischer et al. [43] did not find any significant differences in their study looking at coping across people with head/neck, prostate or lung cancers, although they reported the most commonly used strategy was the use of positive coping statements, similar to the findings of Gaston-Johansson et al. [44] who reported on coping in females with breast cancer. Our review found one study describing correlates of coping strategies in relation to disease stage [45]. This research investigated pain, fatigue and sleep dysfunction along with behavioural self-care strategies to manage fatigue and pain, finding that relaxation and physical exercise were the most commonly used non-pharmacological coping strategies in males with metastasised cancers.

7.3 Psychological Variables

Across the reviewed studies, pain coping strategies were described in relation to four psychological variables: self-efficacy, coping style, verbalising, and pain appraisal:

7.3.1 Self-Efficacy

Buck and Morley [46] did not find that self-efficacy correlated with the attentional coping strategies they measured. In contrast Lin [47] found in their study that self-efficacy was correlated with a range of cognitive (e.g., diverting attention) and behavioural (e.g., relaxation) coping strategies.

7.3.2 Coping Style

Prasertsri et al. [48] looked at correlates, predictors and associates of coping strategies in cancer pain and found that participants had significantly higher catastrophising scores using the CSQ [49] in the "high anxious coping style" and "defensive-high anxious style" than the "repressive coping style" group.

7.3.3 Verbalising

In regard to verbalising, Wilkie and Keefe [50] found significant differences in the use of pain coping strategies (including diverting attention, ignoring pain, praying/hoping, increased activity, self-statements and pain behaviours) between people who verbally expressed pain versus people who did not.

7.3.4 Pain Appraisal

Arathuzik [51] described differences in coping strategy depending upon how participants appraised their pain (threat or challenge). Acceptance and/or reappraisal were most likely to be used by those who appraised pain as threat whilst communication, problem solving, relaxation, visualisation, acceptance and/or reappraisal were more likely to be utilised by those who appraised their pain as a challenge.

7.4 Summary of the Systematic Literature Review

Findings from the systematic literature review suggest that the relationship between coping strategies and psychological variables should be an area for further research. Self-efficacy has been known to be a mediator of pain for many years [52] and, clinically, clarifying whether this is linked to the use of particular coping strategies would be useful. To what extent catastrophising, a known maladaptive coping strategy associated with poor outcomes [53], is associated with (lowered) self-efficacy in relation to managing cancer pain would have clear clinical applications. For example, cognitive behavioural therapy and pain management programs which can identify the use of adaptive and maladaptive coping strategies can utilise coping skills training individually tailored to patient characteristics. Individually tailored CBT programmes for cancer pain have been shown to have increased benefit across a range of pain outcomes than standard CBT [54]. We will revisit clinical implications further at the end of this chapter.

8 Older People and Cancer Pain

Cancer affects older people more than any other group but relatively little is known about their subjective experience of managing pain [55]. Older people are also more likely to be undertreated for pain and it is recognised that there are a multitude of biopsychosocial reasons which contribute to this, including patient, social, environmental and physician factors [56]. Cohort effects are known to influence attitudes towards health and illness [57]; in particular, stoicism is high in current cohorts of older people and may lead to patients under-reporting pain symptoms [58].

Older people are also at risk of a multitude of co-morbid age-related chronic and acute painful conditions including arthritis and circulatory disorders [58] and, as such, the identification and meaning of pain may be problematic. For older people and their carers, it can be difficult to isolate the source of pain due to complexity and co-morbid pain conditions, which can pose a threat to seeking appropriate treatments. Previous research has highlighted how older people can attribute the increasing presence of pain as signalling impending death and highlighted concerns about loss of independence, becoming a burden to others and relinquished social roles [59].

The paucity of research in relation to understanding how older people experience the day-to-day management of their cancer pain led us to focus our research in this area and the study described below is the first to investigate how older people in the UK successfully self-manage their cancer pain at home. Using exploratory methods, we investigated how older people self-manage their cancer pain symptoms as outpatients, as this group has the highest prevalence of cancer pain [34].

Self-management encompasses a range of activities that a patient may be required to undertake in collaboration with healthcare professionals in an outpatient setting, such as adhering to treatment regimes, managing symptoms and coping with the functional, interpersonal and emotional impact of illness [60]. Whilst pain management is largely under the close control of healthcare professionals in inpatient settings, managing pain for outpatients is known to be much more difficult [61] with poorer pain coping reported in outpatient groups [62]. Self-management has potential positive financial benefits as inpatient care requires more financial resources from already stretched healthcare systems. Successful self-management would therefore have clear implications in enabling people with cancer to manage their own symptoms and ultimately to stay at home and reduce, as much as possible, inpatient admissions for uncontrolled and unmanageable pain symptoms. Many people wish to be cared for, and die, at home, and supported self-management for people with cancer has been identified as a priority in the UK [63].

9 An Interpretative Phenomenological Analysis: Self-Management of Cancer Pain in Older People

Our research question focused on how older people self-manage their cancer pain at home, as well as exploring their experiences in relation to potential facilitators, personal strengths and coping strategies [64].

Due to the lack of research within this area, we selected an exploratory approach as the most appropriate to understand experience, interpretation and meanings of cancer pain. Whilst there are many and varied qualitative approaches, we selected Interpretative Phenomenological Analysis (IPA) as the most suitable method for our addressing our research question. The history and methodology and epistemology of IPA has been described in detail in Volume 1 of Meanings in Pain [65] and therefore we will describe this method briefly within this chapter.

IPA [66] is a particularly useful qualitative methodology to research health topics, which have traditionally been dominated by biomedical models [67]. IPA is a methodology used to investigate subjective experience (phenomenology) and is suited to the exploration of pain, as a subjective experience. Themes are identified in the data from a "first case" analysis, and an ongoing process of identifying convergence and divergence from first case themes then leads to the coding of meanings and interpretations to create a table of super-ordinate and sub-ordinate themes. These are refined and eventually linked to existing literature.

Eight participants (seven male) took part in the study and were recruited from a specialist oncology hospital in the UK. Participants were interviewed using a semi-structured interview format, and the questions were specifically open-ended to allow exploration of the perceived nature of their pain, impact, meaning, self-management and future hopes. We used existing clinical and conceptual literature which related to positive psychology concepts, and adjustment to illness and pain, to guide our creation of the semi-structured interview questions [21, 68, 69], which we validated with a service-user group prior to its use.

We asked the participants the following questions, based upon their experience of cancer pain:

- Can you tell me about your pain?
- What is the impact of your pain?
- What does having this pain mean to you?
- What is your attitude towards pain?
- How do you manage your pain at home?
- What personal strengths or qualities help you to manage your pain?
- Have you received any advice on managing your pain?
- What do you think would help you manage your pain better?
- What are your hopes for the future in relation to your pain?

Using IPA, two major themes emerged from the study (Table 10.2), highlighting a process of initial loss of control to regaining a sense of control, and identified subthemes are discussed below with illustrative quotations.

Table 10.2 Themes in the experience of older people self-managing cancer pain

Super-ordinate theme	Sub-ordinate theme
Losing control	1. The pain was excruciating 2. The pain restricts life 3. Burden
Taking back control	1. Inner strengths 2. Using the past to shape the present 3. What I do can help with the pain 4. Support networks 5. Establishing one's own system

9.1 Losing Control

This overarching theme was termed "Losing Control" and participants discussed how the pain experience had been excruciating, particularly in the early stages of their cancer or pre-diagnosis. The pain was experienced as highly aversive and was appraised as threatening in the early stages with often debilitating and distressing impact on functioning. Pain was described in many instances by the use of metaphor and imagery and externalised as though generated outside of the body. Difficulty in locating the pain was a common experience:

> Well when it first started it was very, very hard to bear because it was a very stern pain. When my back first broke I couldn't hardly walk . . .and during this time the pain levels were astronomic . . .the slightest jerk would make me almost scream with the pain, because it was so sharp, and it would almost stop me breathing it was that painful, and that really was pain like I've never felt before. (participant 8)

> The pain is atrocious, you've got pain coming from all over, you can't really tell where it's coming from. (participant 3)

Participants described the restrictions the pain had placed upon their lives. Pain had reduced what was now physically possible, negatively impacting on current social lives and future hopes and dreams. Uncontrolled pain has previously been shown to be distressing when family life and social interactions are negatively impacted [70]:

> The pain really restricted what I was doing so I, I didn't want to go out particularly, I didn't want to sing in the choir, y'know, those sort of things, it just, it just knocked me, knocked me sideways really. (participant 1)

> I think it has affected me socially. . .walking. . .I mean we used to walk right down to our friends every Saturday night, and I couldn't walk round there now. . .I'm just trying to think of the things I used to do. Rambling and bowling and all this that and the other. (participant 6)

> I'd have loved to have gone to Russia, but now my life, my world, is an electric recliner. (participant 2)

The subtheme relating to burden encompassed both the loss of independence and the resultant appraisal that physical dependence would mean familial burden.

Burden can result in significant psychological distress for caregivers, particularly for carers of older people with cancer [71]. Participants discussed trying to hide the emotional impact from their family, which may be linked to stoicism (the ability to endure hardship or pain) and whilst this has been suggested to be a maladaptive strategy [72], it appeared that our participants were employing this strategy to reduce burden on their family members:

> I try to hide a lot of the feelings I was getting 'cause I didn't want to be a burden to my family, so I sort of kept it a secret to some extent, and tried to carry on as normal y'know. (participant 3)

> You are dependent on a lot of help, and in my case as I said before, I was lucky to have a lot of family support and help, so I think that was the only thing with me, that I felt as though I was becoming a burden on my family. (participant 3)

> I'm not independent, I can't wash, you know, not properly, well, I'm just starting to be able to, to do that, but my wife had to wash me. That is significant you know, for anybody I think, because even though it's your nearest and dearest, to have to strip off in the kitchen and be washed, like a little child, you know, it can be a bit embarrassing. (participant 8)

9.2 *Taking Back Control*

The second over-arching theme we termed "Taking Back Control," where participants described the positive strategies they used to manage their pain. The subtheme relating to inner strengths was the most prevalent across the transcripts. People described using determination, stoicism, fighting spirit, persistence and bravery in the face of their pain. Control has long been linked to pain management [73] and those who have an internal locus of control have been shown to have reduced intensity of pain and lower distress [74]:

> That's another key factor, being positive y'know. I never look on the negative side, and again I think that's really important, for everyone to adopt a positive attitude. (participant 3)

> I'm a fighter y'know, I'm determined. I don't want to die, but I sometimes think I don't want to live with this, but no, I plod on. I never sit feeling sorry for myself. (participant 4)

> I'm a very determined sort of a person, fortunately, and I had to alter a few views, because I had to realise. . .that I am just human, I'm not invincible you know. (participant 8)

> You've got to have hope haven't you? Y'know, you've got to, you've always got to think well there's an answer around the corner to this problem y'know. I haven't found it yet, well, I must've done because I cope. I don't know what it is, why I cope, I'm not brave, I'm nothing special, but I just try and not let it get me down. (participant 2)

Participants described how they used past experiences to shape the present experience, almost drawing strength from the past. In particular, using the memory of how family members coped to help them cope, even to the extent of 'honouring' the memory of the deceased:

What I fear is that this cancer at some stage is gonna get worse, and I don't know how I shall cope with it then, but I'll just try as I've done in the past. (participant 1)

She went through all these illnesses, including pancreatic cancer, and she never cried, she was so brave, she was such an example, and I've got to be careful I don't let her down. (participant 1)

If the wife can go through what she went through, then I can go through with this. I mean, I don't do pain very well, I'll tell you now, but if she could cope... (participant 2)

Behavioural coping strategies such as distraction and activity were discussed by our participants, focused on "getting on with life" and engaging in physical activity despite the pain. This gave a sense that participants were moving away from perceived helplessness towards recovery, as they found their actions could positively affect their pain experience, contributing to a sense of agency and emerging self-efficacy.

I have to focus on something y'see, all the time, and somebody said this to me about getting better, do the things you normally do...I'm also focusing on picking me brambles and doing the garden. I can't just sit y'see, even though I'm poorly I have to try and do something. (participant 1)

I do watch the tele, and I do read books...while I'm reading them, I don't notice the pain, but then when I put the book down, I suddenly think, ooh, I've got that pain again. (participant 7)

Doing hands-on things around the house, and that, has helped me tremendously. In fact I'd say that's the key y'know, getting on with something what's not going to make you think about your pain. It doesn't make the pain go away, I'm quite aware of that, but you can cope with it more. (participant 3)

The importance of support networks was highlighted by our participants, in the form of family and healthcare support systems, accessed as and when needed. For some, the complexity of required medication regimes was a challenge and family support was seen as a vital component in the ability to adhere to medication regimes. Consistency, certainty and availability of support systems appeared to be vital to ensure confidence in managing home medical treatments. Previous research has found that adequate social support is positively associated with adjustment and cancer disease progression [75].

Bless my wife, she's a very patient and good lady and she would always just stop everything and just sit with me...she has been excellent at making sure I don't lose sight of reality. (participant 8)

I mean, I have more than twenty tablets taken at different times of the day and you just can't remember to take them. Luckily my wife is very good at that kind of thing and she made out a graph, what time of day to take each tablet. (participant 5)

It depends on how severe the pain is, but like I said, you've always got a back-up here to ring, which is comforting, very comforting, because you know, if the worst comes to the worst, you can always come back here. (participant 3)

The main thing is my girlfriend, she's the great salvation, she's the angel...we all need to love somebody. (participant 1)

The final subtheme in our analysis appeared to illustrate how participants viewed the process of establishing a self-management system that worked for them. This included their ability to select and modify treatments along with the knowledge and understanding of their pain and their own bodies, combining to reach an idiosyncratic self-management system. This demonstrated that self-determination and self-efficacy were key components in adjusting and coping with cancer pain.

> I think the first thing that you need to know is what's caused the pain, you need to know that, so you can understand what the pain is, and I think if you understand what the pain is, you can cope with it better. (participant 6)

> I'm assuming that the medical people in my life like the doctor and the district nurse, they know what they are doing with the pills, and they are doing the best they can for me. (participant 7)

> I know what to do and what not to do, but because the doctor says I got to do this I don't necessarily have to do it. . .I know what suits me. (participant 7)

> I've just developed a system where I take my medication at certain times and I don't wait for the pain to start and then take the medication. I take it before it starts, and that's been the secret of it working for me. (participant 3)

We understood the data as representing a temporal and sequential process that started with an initial perception of lack of control over pain (external locus of control) but led, over time, to a sense of how people could regain control (internal locus of control), underpinned with self-efficacy and mastery of pain. This led participants to successfully manage their cancer pain symptoms and this temporal, dynamic process is illustrated in Fig. 10.1.

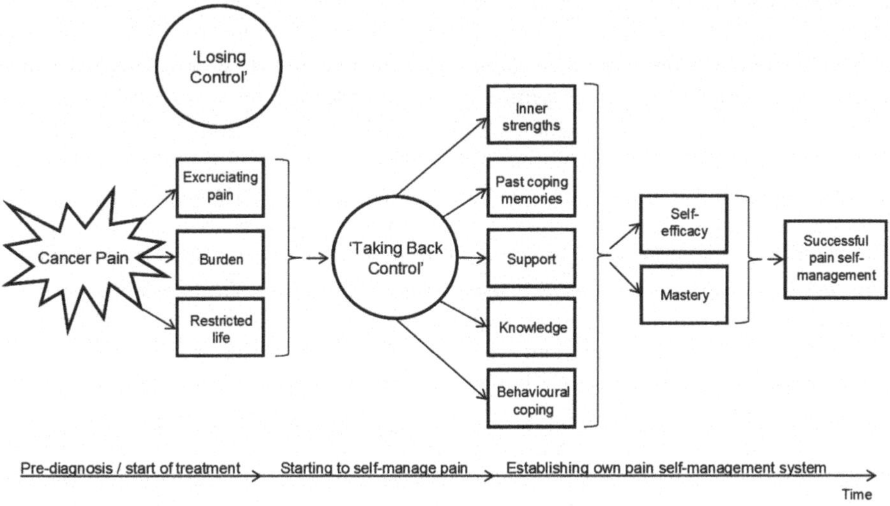

Fig. 10.1 Themes in the process of older adults' successful cancer pain self-management

10 Clinical Implications

Whilst psycho-oncology is a relatively recent discipline, there has been much recent interest in the clinical applications of psycho-social interventions for people diagnosed with cancer. Multiple psychotherapies have been applied within this area, including supportive psychotherapy, cognitive-analytical and third wave therapies such as acceptance and commitment therapy (ACT) and compassion focused therapy (CFT), couples and family therapy, group therapy and narrative approaches [76]. A meta-analysis found that cognitive behaviour therapy (CBT) delivered to patients with breast cancer can positively impact both distress and pain [77].

A review undertaken by Syrjala et al. [11] identified several psychological and behavioural approaches that have strong evidence supporting their efficacy for reducing cancer pain during the diagnosis and treatment phases of the disease. They found that hypnosis, relaxation with imagery, and CBT with relaxation and imagery were particularly helpful approaches. Research would suggest that CBT interventions should target helplessness and focus on increasing positive affect through positive psychological states including fighting spirit and resilience [24].

The founder of the modern hospice movement, Dame Cicely Saunders, described how people with a life-limiting illness experience "total pain," encompassing psychological, social, practical, and spiritual domains [78]. It is clear that in order to address the needs arising from cancer pain, clinicians should ensure all aspects of a person's life and resulting distress are considered. Spirituality can include religious, philosophical, existential, cultural, or personal beliefs, values, and practices. Spiritual care should be included as part of healthcare provision as negative spiritual and religious beliefs have been shown to increase distress and illness burden [79]. To achieve successful pain relief, assessment and treatment of physical pain needs to be done in combination with assessment and treatment of psycho-social and spiritual pain [1]. The palliative care movement has done much to de-medicalise death and dying and ensure that physical symptoms are not prioritised over psycho-social and spiritual needs [78]. Those values should be widely incorporated into healthcare systems worldwide.

Understanding the individual meanings and appraisals of illness warrants careful investigation by the clinician to ensure a good psychological assessment. This in turn can point towards appropriate, person-centred and meaningful interventions. Interventions that focus on helping the search for meaning may help people to derive positive benefits from the cancer experience, and it has been suggested that a successful search for meaning can result in enhanced self-esteem, reduced distress and increased life satisfaction [16]. This may be of particular importance as cancer progresses, with people being increasingly drawn to existential questions such as the meaning of life, sickness, suffering and death [1]. Research suggests that meaning based coping which includes prioritising, down-playing, self-preservation, finding or revising perceived current and/or future meaning, is an important contributor to the coping process. Clinically, it appears that people require support with the continual readjustments and ongoing sense-making of often rapidly fluctuating symptoms to enable them to re-attain positive emotion [27].

If cognitive appraisals and meanings of pain for individuals can be fully understood, then as clinicians we have increased opportunities to markedly improve the pain experience for people experiencing cancer and cancer pain.

11 Conclusions

Our research highlights the importance of recognising the use and function of individual coping strategies. Clinically, there is a clear need for psychologists in healthcare settings to assess for and understand how people experience, appraise and cope with pain. Our own research also points towards the importance of understanding where people are in 'temporal' terms as they actively self-manage pain, i.e., to what extent they are able to draw on personal and social resources in order to (re)gain perceived control over pain. Whilst we have shown that self-efficacy and control are key factors in the successful self-management of cancer pain in our research with older people who manage their cancer pain in community settings, we suggest that caution is exercised: an emphasis on self-management may risk placing too much responsibility for pain management on the individual. For older people, this may contribute to stoical responses such as "I've just got to get on with it," which may have negative impact on help-seeking.

Further research across all areas of cancer pain is warranted, and in particular, how older people experience and cope with pain, due to the paucity of research in this area. The World Health Organisation [6] estimate that over 90% of people with advanced cancer can experience relief from psychological, social, spiritual and physical problems through adequate palliative care. It's a global tragedy that so many, particularly those in low income and developing nations, have little to no access to cancer services, pain relief or comfort through their cancer journey and at the end of life. It is unacceptable with modern developments in pain treatments, both medical and psychological, that so many people continue to experience excruciating cancer pain during the course of their cancer journey. No-one should suffer with poorly controlled or treated pain, cancer related or otherwise, and we call for further research and dialogue in this area to ensure adequate treatment for all.

Note Informed consent was obtained from all individual participants included in the study.

References

1. Cipta AM, Pietras CJ, Weiss TE, Strouse TB. Cancer-related pain management in clinical oncology. J Community Support Oncol. 2015;13(10):347–55.

2. Quaresma M, Coleman MP, Rachet B. 40-year trends in an index of survival for all cancers combined and survival adjusted for age and sex for each cancer in England and Wales, 1971–2011: a population-based study. Lancet. 2015;385:1206–18.
3. Wiffen PJ, Wee B, Derry S, Bell RF, Moore RA. Opioids for cancer pain–an overview of Cochrane reviews. Cochrane Database Syst Rev. 2017;7:CD012592.
4. Van den Beuken-van Everdingen MH, Hochstenbach LM, Joosten EA, Tjan-Heijnen VC, Janssen DJ. Update on prevalence of pain in patients with cancer: systematic review and meta-analysis. J Pain Symptom Manag. 2016;51(6):1070–90.
5. Raphael J, Ahmedzai S, Hester J, Urch C, Barrie J, Williams J, et al. Cancer pain: part 1: pathophysiology; oncological, pharmacological, and psychological treatments: a perspective from the British pain society endorsed by the UK association of palliative medicine and the royal college of general practitioners. Pain Med. 2010;11(5):742–64.
6. World Health Organisation. Cancer [Internet]. Switzerland: World Health Organisation; 2018 [cited 2018 Oct 6]. Available from http://www.who.int/news-room/factsheets/detail/cancer
7. Sudhakar A. History of cancer, ancient and modern treatment methods. J Cancer Sci Ther. 2009;1(2):1–4.
8. Ferlay J, Soerjomataram I, Dikshit R, Eser S, Mathers C, Rebelo M, et al. Cancer incidence and mortality worldwide: sources, methods and major patterns in GLOBOCAN 2012. Int J Cancer. 2015;136(5):359–86.
9. Cancer Research UK. Cancer statistics for the UK [Internet]. London: Cancer Research UK; 2018 [cited 2018 Aug 17]. Available from https://www.cancerresearchuk.org/health-professional/cancer-statistics-for-the-uk
10. Ward SE, Carlson-Dakes K, Hughes S, Kwekkeboom K, Donovan H. The impact on quality of life of patient-related barriers to pain management. Res Nurs Health. 1998;21(5):405–13.
11. Syrjala KL, Jensen MP, Mendoza ME, Yi JC, Fisher HM, Keefe FJ. Psychological and behavioral approaches to cancer pain management. J Clin Oncol. 2014;32(16):1703–11.
12. Katz J, Melzack R. Measurement of pain. Surg Clin North Am. 1999;79(2):231–52.
13. Leventhal H, Meyer D, Nerenz D. The common sense representation of illness danger. In: Rachman S, editor. Contributions to medical psychology, vol. II. New York: Pergamon Press; 1980. p. 7–30.
14. Folkman S, Lazarus RS, Gruen RJ, DeLongis A. Appraisal, coping, health status, and psychological symptoms. J Pers Soc Psychol. 1986;50:571–9.
15. Roesch SC, Rowley AA. Evaluating and developing a multidimensional, dispositional measure of appraisal. J Pers Assess. 2005;85(2):188–96.
16. Lee V, Cohen SR, Edgar L, Laizner AM, Gagnon AJ. Clarifying 'meaning' in the context of cancer research: a systematic literature review. Palliat Support Care. 2004;2:291–303.
17. Park C, Folkman S. Meaning in the context of stress and coping. Rev Gen Psychol. 1997;1 (2):115–44.
18. Chambers SK, Baade P, Youl P, Aitken J, Occhipinti S, Vinod S, et al. Psychological distress and quality of life in lung cancer: the role of health-related stigma, illness appraisals and social constraints. Psychooncology. 2015;24(11):1569–77.
19. Turk DC, Salovey P. Towards an understanding of life with cancer: personal meanings, psychosocial problems, and coping resources. Hosp J. 1985;1(1):73–84.
20. Somerfield MR, McCrae RR. Stress and coping research: methodological challenges, theoretical advances, and clinical applications. Am Psychol. 2000;55(6):620–5.
21. Lazarus RS, Folkman S. Stress, appraisal and coping. New York: Springer; 1984. 456 p.
22. Jensen MP, Turner JA, Romano JM, Karoly P. Coping with chronic pain: a critical review of the literature. Pain. 1991;47(3):249–83.
23. Chirico A, Serpentini S, Merluzzi T, Mallia L, Del Bianco P, Martino R, et al. Self-efficacy for coping moderates the effects of distress on quality of life in palliative cancer care. Anticancer Res. 2017;37(4):1609–15.
24. Greer S. CBT for emotional distress of people with cancer: some personal observations. Psychooncology. 2007;17(2):170–3.

25. Franks HM, Roesch SC. Appraisals and coping in people living with cancer: a meta-analysis. Psychooncology. 2006;15(12):1027–37.
26. Endler NS, Parker JD, Summerfeldt LJ. Coping with health problems: conceptual and methodological issues. Can J Behav Sci. 1993;25(3):384–99.
27. Roberts D, Calman L, Large P, Appleton L, Grande G, Lloyd-Williams M, et al. A revised model for coping with advanced cancer. Mapping concepts from a longitudinal qualitative study of patients and carers coping with advanced cancer onto Folkman and Greer's theoretical model of appraisal and coping. Psychooncology. 2018;27(1):229–35.
28. Dunkel-Schetter C, Feinstein LG, Taylor SE, Falke RL. Patterns of coping with cancer. Health Psychol. 1992;11(2):79–87.
29. Bower JE, Epel E, Moskowitz JT. Biological correlates: how psychological components of benefit finding may lead to physiological benefits. In: Park C, Lechner S, Stanton A, Antoni M, editors. Medical illness and positive life change: can crisis lead to personal transformation? Washington, DC: APA Press; 2009. p. 155–72.
30. Coyne JC, Tennen H. Positive psychology in cancer care: bad science, exaggerated claims, and unproven medicine. Ann Behav Med. 2010;39(1):16–26.
31. Somers TJ, Keefe FJ, Pells JJ, Dixon KE, Waters S, Riordan PA. Pain catastrophizing and pain-related fear in osteoarthritis patients: relationship to pain and disability. J Pain Symptom Manag. 2009;37(5):863–72.
32. Benyon K, Muller S, Hill S, Mallen C. Coping strategies as predictors of pain and disability in older people in primary care: a longitudinal study. BMC Fam Pract. 2013;14(1):67.
33. Appleyard SE, Clarke C. A systematic literature review of self-initiated coping strategies: who chooses what? In: Appleyard SE, Self-management of cancer pain [doctoral thesis] University of Hull, UK; 2014. Available from https://hydra.hull.ac.uk/resources/hull:11347
34. Breivik H, Cherny N, Collett B, de Conno F, Filbet M, Foubert AJ, et al. Cancer-related pain: a pan-European survey of prevalence, treatment, and patient attitudes. Ann Oncol. 2009;20 (8):1420–33.
35. Popay J, Roberts HM, Sowden A, Petticrew M, Arai L, Rodgers M, et al. Guidance on the conduct of narrative synthesis in systematic reviews: a product from the ESRC Methods Programme. London: Institute for Health Research; 2006. 281 p.
36. Bennett MI, Closs SJ, Chatwin J. Cancer pain management at home (I): do older patients experience less effective management than younger patients? Support Care Cancer. 2008;17 (7):787–92.
37. Gagliese L, Jovellanos M, Zimmermann C, Shobbrook C, Warr D, Rodin G. Age related patterns in adaptation to cancer pain: a mixed methods study. Pain Med. 2009;10(6):1050–61.
38. Haozous EA, Knobf MT, Brant JM. Understanding the cancer pain experience in American Indians of the Northern Plains. Psychooncology. 2011;20(4):404–10.
39. Reddick BK, Nanda JP, Campbell L, Ryman DG, Gaston-Johansson F. Examining the influence of coping with pain on depression, anxiety, and fatigue among women with breast cancer. J Psychosoc Oncol. 2005;23(2/3):137–57.
40. Im EO, Lee SH, Liu Y, Lim HJ, Guevara E, Chee W. A national online forum on ethnic differences in cancer pain experience. Nurs Res. 2009;58(2):86–94.
41. Whale Z, Lyne PA, Papanikolaou P. Pain experience following radical treatment for head and neck cancer. Eur J Oncol Nurs. 2001;5(2):112–20.
42. Kwekkeboom KL. Pain management strategies used by patients with breast and gynaecologic cancer with postoperative pain. Cancer Nurs. 2001;24(5):378–86.
43. Fischer DJ, Villines D, Kim YO, Epstein JB, Wilkie DJ. Anxiety, depression, and pain: differences by primary cancer. Support Care Cancer. 2009;18(7):801–10.
44. Gaston-Johansson F, Ohly KV, Fall-Dickson JM, Nanda JP, Kennedy MJ. Pain, psychological distress, health status, and coping in patients with breast cancer scheduled for autotransplantation. Oncol Nurs Forum. 1999;26(8):1337–45.

45. Miaskowski C, Lee KA. Pain, fatigue, and sleep disturbances in oncology outpatients receiving radiation therapy for bone metastasis: a pilot study. J Pain Symptom Manag. 1999;17 (5):320–32.
46. Buck R, Morley S. A daily process design study of attentional pain control strategies in the self-management of cancer pain. Eur J Pain. 2006;10(5):385–98.
47. Lin CC. Comparison of the effects of perceived self-efficacy on coping with chronic cancer pain and coping with chronic low back pain. Clin J Pain. 1998;14(4):303–10.
48. Prasertsri N, Holden J, Keefe FJ, Wilkie DJ. Repressive coping style: relationships with depression, pain and pain coping strategies in lung cancer outpatient. Lung Cancer. 2011;71 (2):235–40.
49. Rosenstiel AK, Keefe FJ. The use of coping strategies in chronic low back pain patients: relationship to patient characteristics and current adjustment. Pain. 1983;17(1):33–44.
50. Wilkie DJ, Keefe FJ. Coping strategies of patients with lung cancer related pain. Clin J Pain. 1991;7(4):292–9.
51. Arathuzik MD. The appraisal of pain and coping in cancer patients. West J Nurs Res. 1991;13 (6):714–31.
52. Bandura A, O'Leary A, Taylor CB, Gauthier J, Gossard D. Perceived self-efficacy and pain control: opioid and nonopioid mechanisms. J Pers Soc Psychol. 1987;53(3):563–71.
53. Leung L. Pain catastrophizing: an updated review. Indian J Psychol Med. 2012;34(3):204–17.
54. Dalton JA, Keefe FJ, Carlson J, Youngblood R. Tailoring cognitive behavioural treatment for cancer pain. Pain Manag Nurs. 2004;5(1):3–18.
55. Chatwin J, Closs J, Bennett M. Pain in older people with cancer: attitudes and self management strategies. Eur J Cancer Care. 2009;18(2):124–30.
56. Deshields TL, Tait RC, Manwaring J, Trinkaus KM, Naughton M, Hawkins J, et al. The cancer pain inventory: preliminary development and validation. Psychooncology. 2010;19(7):684–92.
57. Knight BG, McCallum TJ. Adapting psychotherapeutic practice for older clients: implications of the contextual, cohort-based, maturity, specific challenge model. Prof Psychol Res Pract. 1998;29(1):15–22.
58. Abdulla A, Adams N, Bone M, Elliott AM, Gaffin J, Jones D, et al. Guidance on the management of pain in older people. Age Ageing. 2013;42(1):1–57.
59. McPherson CJ, Hadjistavropoulos T, Lobchuk MM, Kilgour KN. Cancer-related pain in older adults receiving palliative care: patient and family caregiver perspectives on the experience of pain. Pain Res Manag. 2013;18(6):293–300.
60. Foster C, Fenlon D. Recovery and self-management support following primary cancer treatment. Br J Cancer. 2011;105(1):21–8.
61. Ferrell BR, Schneider C. Experience and management of cancer pain at home. Cancer Nurs. 1988;11(2):84-90.
62. Rockett MP, Simpson G, Crossley R, Blowey S. Characteristics of pain in hospitalized medical patients, surgical patients, and outpatients attending a pain management centre. Br J Anaesth. 2013;110(6):1017–23.
63. Department of Health. Improving outcomes: a strategy for cancer. London: Department of Health and Social Care; 2011 [cited 2012 Nov 2]. Available from https://assets.publishing.service.gov.uk/government/uploads/system/uploads/attachment_data/file/213785/dh_123394.pdf
64. Appleyard SE, Clarke C. An interpretative phenomenological analysis of the experiences of older people self-managing cancer pain at home. J Psychosoc Oncol. 2018;36(3):333–49.
65. Snelgrove S. An interpretative phenomenological analysis of non-malignant chronic low back pain. In: van Rysewyk S, editor. Meanings of pain, vol. 1. Cham: Springer; 2016. p. 129–47.
66. Smith JA, Osborn M. Interpretative phenomenological analysis. In: Smith JA, editor. Qualitative psychology: a practical guide to research methods. London: Sage; 2003.
67. Brocki JM, Wearden AJ. A critical evaluation of the use of interpretative phenomenological analysis (IPA) in health psychology. Psychol Health. 2006;21(1):87–108.

68. Peterson C, Seligman MEP. Character strengths and virtues: a handbook and classification, vol. 800. Oxford: Oxford University Press; 2004.
69. Smith JA, Osborn M. Pain as an assault on the self: an interpretative phenomenological analysis of the psychological impact of chronic benign low back pain. Psychol Health. 2007;22 (5):517–34.
70. Chapman CR, Gavrin J. Suffering: the contributions of persistent pain. Lancet. 1999;353 (9171):2233–7.
71. Haley WE. Family caregivers of elderly patients with cancer: understanding and minimising the burden of care. J Support Oncol. 2003;1(2):25–9.
72. Murray G, Judd F, Jackson H, Fraser C, Komiti A, Pattison P, et al. Big boys don't cry: an investigation of stoicism and its mental health outcomes. Pers Individ Dif. 2008;44(6):1369–81.
73. Ferrell BR, Taylor EJ, Grant M, Fowler M, Corbisiero RM. Pain management at home: struggle, comfort, and mission. Cancer Nurs. 1993;16(3):169–78.
74. Crisson JE, Keefe FJ. The relationship of locus of control to pain coping strategies and psychological distress in chronic pain patients. Pain. 1988;35(2):147–54.
75. Usta YY. Importance of social support in cancer patients. Asian Pac J Cancer Prev. 2012;13 (8):3569–72.
76. Watson M, Kissane DW. Handbook of psychotherapy in cancer care. Oxford: Wiley; 2011. 284 p.
77. Tatrow K, Montgomery GH. Cognitive behavioral therapy techniques for distress and pain in breast cancer patients: a meta-analysis. J Behav Med. 2006;29(1):17–27.
78. Ong CK, Forbes D. Embracing cicely Saunders's concept of total pain. BMJ. 2005;331 (7516):576.
79. Puchalski CM, Vitillo R, Hull SK, Reller N. Improving the spiritual dimension of whole person care: reaching national and international consensus. J Palliat Med. 2014;17(6):642–56.

Chapter 11
Common Meanings of Living with Diabetic Peripheral Neuropathic Pain from the Perspective of Patients

Zehra Gok Metin

Abstract Contemporary pain medicine is necessary to explain pain and to help in its treatment; yet, preference for biomedical explanation of pain in the field has meant that attention to the personal experience of pain and to the meanings of pain experience remain a blind spot in knowledge. Thus, the pain literature includes limited information about the common meanings of living with diabetic peripheral neuropathic pain (DPNP) from the perspective of patients. The purpose of this chapter is to describe some of the common meanings of pain in patients with DPNP, as currently reported in the literature, how these meanings interact with other common factors in pain experience, including specific negative emotions or moods (depression, anxiety, anger), or the psychosocial context surrounding pain, and to describe available evidence on the effectiveness of cognitive behavioral therapy (CBT) for patients with DPNP. Further quantitative, qualitative or mixed methods research is needed to more fully understand common experiences of pain in patients with DPNP, and the common meanings ascribed to these experiences.

Clinical Implications: Pain in patients with DPNP involves a range of threatening pain sensations, including sensations with burning, shooting, tingling, or cramping qualities, and additional more cognitive meanings linked to persistent pain, including a sense that pain disrupts daily life in an intrusive way. Pain affects many aspects of daily experience that are meaningful to patients with DPNP; some describe ongoing physical difficulties, others describe work-related problems or challenges in sexual intimacy. Given the heavy personal burden that DPNP imposes on patients and the considerable challenge of managing the condition pharmacologically, clinical use of non-pharmacological therapies such as CBT for painful diabetic neuropathy might be warranted in individual patients.

Keywords Diabetic peripheral neuropathic pain · Interpretative Phenomenological Analysis (IPA) · Lived experiences · Meaning of pain · Pain experience

Z. Gok Metin (✉)
Faculty of Nursing, Medical Nursing Department, Hacettepe University, Altindag, Ankara, Turkey
e-mail: zehra.gok@hacettepe.edu.tr

S. van Rysewyk (ed.), *Meanings of Pain*,
https://doi.org/10.1007/978-3-030-24154-4_11

1 The Visible Side of Diabetic Peripheral Neuropathic Pain

Diabetic peripheral neuropathy (DPN) affects approximately 50% of patients with diabetes and is a major cause of morbidity and increased mortality [1–3]. DPN's clinical manifestations include painful neuropathic symptoms and insensitivity, which increase the risk of foot ulceration, burns, injuries, and amputations [4, 5]. Several recent studies have implicated patients' gender, height, smoking status, alcohol consumption, duration of diabetes, insulin therapy, poor glycemic control, high systolic blood pressure, hyperlipidemia, obesity, peripheral vascular disease, presence of retinopathy and nephropathy, and elevated albumin excretion rates, as risk factors for the development of DPN [1, 3, 5]. Although there is now strong evidence for the importance of nerve microvascular disease in the development of DPN, the risk factors for painful DPN are not well-known. However, emerging evidence regarding the central correlates of painful DPN is now supported by brain imaging [1].

In addition to DPN, painful neuropathy is a common and often distressing complication of diabetes. The International Association for the Study of Pain (IASP) has defined neuropathic pain as "pain initiated or caused by a primary lesion or dysfunction in the nervous system." Neuropathic pain can be characterized as either peripheral or central origin, depending upon the region of the lesion within the nervous system [4]. Neuropathic pain, whether of peripheral or central origin, continues to cause a heavy burden to patients and society by increasing disability and reducing quality of life (QOL) with concomitant increases in use of healthcare resource and costs [3–5].

DPN starts in the toes and gradually moves proximally. Once it is well established in the lower limbs, it affects the upper limbs, with sensory loss following the typical "glove and stocking" pattern of distribution [6]. The patient does not complain of weakness, but when symptoms are present, they tend to be sensory in nature. Symptomatic muscle weakness and motor deficits tend to develop in the later stages of DPN [7].

Painful symptoms such as burning, tingling ("pins and needles," or paresthesia), "shooting" (a sensation similar to receiving an electric shock) or lancing (a "stabbing" sensation) are present in almost a third of patients with DPN and around 20% of all patients with diabetes [1, 7]. These symptoms are generally worse at night and worsen sleep quality [8]. Together with painful symptoms during the day, this often leads to a reduction in an individual's ability to perform daily living activities [8, 9].

Despite numerous conventional treatments, painful DPN causes a significant burden to patients and the healthcare system community. DPN commonly leads to limitations in daily activities and patients tend to have poor satisfaction with treatments [9, 10]. Painful DPN can be extremely distressing to patients and is associated with levels of depression and anxiety [9, 10]. Importantly, symptoms of DPN are not a reliable indicator of the severity of nerve damage. Some patients with severe pain symptoms have little sensory deficit, whereas others with no painful symptoms have completely numb feet, putting them at high risk of foot ulceration.

Insensitivity, or loss of pain, can lead to foot ulcerations, unintentional or serious limb injuries. Patients who lose feeling in their hands cannot sense temperature and often burn themselves while cooking or ironing, and also have difficulty handling small objects. Because of the lower perceived sensation in their feet, patients often sustain puncture or friction wounds or burns that can become infected and/or ulcerated, which can lead to amputation [10].

2 Symptoms, Diagnosis, and Classification of Diabetic Peripheral Neuropathic Pain

Diabetic neuropathies are a group of nerve disorders caused by diabetes that can be classified as peripheral, autonomic, proximal, and focal [11], with peripheral neuropathy being the most prevalent [12]. Diagnosis of DPN is dependent on a variety of factors, including the presence of a characteristic pattern of signs and symptoms. During clinical examination, some people with diabetes have signs of neuropathy, such as reduced nerve conduction velocity, but no symptoms of DPN. If the appropriate clinical tests are not applied and the patient is asymptomatic, diagnosis may be delayed until the patient begins to experience paresthesia, dysesthesias, or neuropathic pain symptoms [13–15]. The diagnosis of DPN begins with a detailed history of sensory and motor symptoms. The quality and severity of neuropathic pain should be assessed using a suitable scale. Clinical examination should consist of inspection of the feet and evaluation of reflexes and sensory responses to vibration, light touch, pinprick and the 10 g monofilament [5].

Peripheral neuropathy may be asymptomatic. Generally, negative and positive sensory symptoms coexist in neuropathic pain. Negative symptoms include deficits of different somatosensory qualities such as tactile hypoesthesia or anesthesia, thermal hypoesthesia, pinprick hypoalgesia, and loss of vibratory sensation. These symptoms are uncomfortable, but not painful. Positive symptoms include stimulus-evoked sensations, such as paresthesia, dysesthesia, hyperalgesia, allodynia, paroxysmal pain, and ongoing superficial pain (see Table 11.1 for definitions of these symptoms or signs) [15].

In mechanical hyperalgesia, normally slightly painful pinprick stimulation is perceived as abnormally painful. Mechanical, dynamic, and mechanical static allodynia is a state in which pain is evoked by light, moving touch or light pressure, respectively. Accordingly, cold and heat allodynia is evoked by normally non-painful cold and warm stimuli. One of the most distressing symptoms that people can experience is neuropathic pain and paresthesia [13]. Pain related to diabetic peripheral neuropathy, which can be DPNP, further impairs quality of life (QOL) in patients who are faced with the treatment challenges and other complications of diabetes [14].

Table 11.1 Symptom or sign descriptions related to pain [6, 16]

Allodynia	Pain due to no noxious stimuli (mechanical, dynamic, or thermal) when applied to the affected area
Anesthesia	Loss of normal sensation in the affected site
Dysesthesia	Spontaneous or evoked unpleasant abnormal sensations
Hyperalgesia	Exaggerated response to a mildly noxious stimulus applied to the affected site
Hyperpathia	Delayed and explosive response to a noxious stimulus applied to the affected site
Hypoesthesia	Reduction of normal sensation
Paresthesia	Non-painful spontaneous abnormal sensations
Phantom pain	Pain from a specific site that no longer exists (e.g., amputated limb) or where there is no current injury
Referred pain	Occurs in a site remote from the source

In addition to spontaneous and stimulus-evoked pain, patients with neuropathic pain complain of numbness, burning, or tingling, or a combination of these; they describe painful electric shock-like, prickly, or "pins and needles" sensations. Patients completing the McGill Pain Questionnaire [12] described their pain using terms such as "punishing-cruel" and "tiring-exhausting." In 1990, Boureau et al. [17] identified six adjectives used substantially more frequently by patients to describe neuropathic pain: "electric shock," "burning," "tingling," "cold," "pricking," and "itching."

3 The Invisible Side of Diabetic Peripheral Neuropathic Pain

Pain is described as an unpleasant, subjective sensory and emotional experience [18]. Pain communicates information about current, impending or perceived tissue damage and is responded to by the individual person as a drive or motivation. More than this, however, pain means a struggle to the person, commonly an ongoing agonizing struggle, of enduring an aversive motivational state without succumbing to it so that it might disable the person and seeking a way to deal with it and remain intact as a person. Pain is not simply a sensation in the individual, but an aversive change that affects the whole person, which is significant or meaningful to the individual [18–20]. Psychophysical and clinical studies show that pain has a common personal meaning of intrusion or threat to self [19, 20]. Thus, pain presents itself to the person in an intrusive or threatening way, disrupting the experienced quality of the person's daily or valued work or social activities, and the drive to achieve present or future goals [21, 22].

A source of major disability and socioeconomic burden [23–26], neuropathic pain substantially decreases health-related quality of life (HRQOL) compared with the general population [26]. Furthermore, there is strong evidence that the presence, severity, and duration of neuropathic pain are associated with impairments in a several important HRQOL domains such as general activity, mood, physical mobility, work, social relations, emotional reactions, sleep, recreational activities, and enjoyment of life [9, 27–29]. Given the nature of neuropathic pain and its potential to affect patient functioning and QOL, the burden of this complication is quite high; it affects health status and QOL as shown in several studies (e.g. [30, 31]).

Diabetic neuropathic pain is characteristically more severe at night and often prevents sleep [32]. Some patients complain about feeling tired during the day because of sleep deprivation [32, 33]. Others are unable to maintain full employment [34, 35]. Severe painful neuropathy can limit exercise capacity, which interferes with daily activities [36]. This is particularly the case when there is an associated disabling, severe postural hypotension due to autonomic involvement [32]. Moreover, the lack of effective therapies and the side effect profile of conventional drugs like tricyclic antidepressants, serotonin noradrenaline reuptake inhibitors, the anticonvulsants gabapentin and pregabalin, and opioids render the treatment unsatisfactory [37]. Not surprisingly therefore, depressive symptoms are common, and some patients experience lower levels of well-being [33, 35, 36, 38].

Currently, there are very few qualitative studies in the pain literature that have studied the lived experiences of patients with DPNP [39, 40]. In [39], conducted in the USA, 4 focus groups and 47 telephone interviews were undertaken to explore patients' meanings and impact of DPNP on personal experience. All respondents ($N = 70$) were over the age of 18 years, read/spoke English, had a diagnosis of DPNP with symptoms for 6 months and a minimum 12-month history of diabetes mellitus, and had a daily pain rating of at least 4 on an 11-point numerical rating scale. Interview transcripts were analyzed thematically based on modified grounded theory principles. In [40], conducted in Turkey, 14 patients with DPNP were interviewed, and Interpretative Phenomenological Analysis (IPA), a method of psychological qualitative research with an idiographic focus, which aims to understand how a specific person, in a specific context, understands a given phenomenon or experience, was used to identify common themes or meanings in patient pain experience. In the next section, I summarize the qualitative findings of these studies, with a focus on personal meaning.

3.1 Common Meanings Related to Walking and Standing in Patients with DPNP

"Yesterday I went shopping with my daughter and visited a few stores. I started getting worse and worse. The pain went down my legs and then into my feet. I also took care to rest while walking. Now, I feel like I have run or played ball. I have so much pain. I'm suffering from pain today because of walking yesterday" (Participant 10, female, 52 years) [40].

"Sometimes my feet will hurt really bad and I can't get up and hardly walk because they're really hurting" [39].

"If I was to sit in a chair all day and put my feet up in the air or, say, soak them in hot water and stuff like this, it eases the pain, but being on your feet is what really kills you"

"While working at home, if I stand for some time, I have unbearable pain. I do my jobs by sitting. The simplest example: while I am cooking or doing other tasks in the kitchen, generally I am trying to complete them while sitting" (Participant 8, female, 58 years) [40].

3.2 Common Meanings Related to Mobility and Exercise in Patients with DPNP

"I can't even get out of bed in the morning unless I have special shoes that are cushioned, because when my feet touch the ground I feel like I am walking on sharp rocks. So, I have that issue, and I guess mine is getting progressively worse because I have been having the knee problems as well lately. So, it seems like mine is worsening, and it's just, you know, more chronic pain, every day" [39].

"It has a big impact on it. [. . .] Where I work, there's a walking trail and usually at 10:00 and at 3:00, I would walk with a coworker. I used to walk with a coworker, but it has limited me to some days I can do it, sometimes I can't. I can't walk as much as I used to or what I would do because of the neuropathy pain" [39].

"The biggest thing: you cannot do any exercise. Simply put, if my son says, 'Dad, let's play ball,' I cannot do that with my son. 'Come on Dad, let's run in the park.' I cannot do that. These are big restrictions for me. For example, I like to ride a bike, but I cannot. I have not been biking for almost 15 years" (Participant 7, male, 50 years) [40].

3.3 Common Meanings Related to Daily-Life Changes in Patients with DPNP

Patients with DPNP reported physical limitations such as trouble standing for very long, difficulty with keeping their balance, bending, climbing stairs and lifting loads, reduced mobility or exercise, and not performing their personal care without getting assistance from someone to help. Most patients with DPNP in [39, 40] reported that their productivity at work or home was negatively impacted by their DPN symptoms. Some patients also specified having serious problems in their sexual life (Sect. 3.3.3). Additionally, many respondents reported challenges with memory and focus (Sect. 3.3.4). Related to this, many respondents had specific limitations in performing routine chores. The limits were often physical, and for others the pain distracted them from achieving daily tasks [39, 40]. Some patients had to reduce their work hours, change jobs to accommodate their DPNP, or resign from their current employed position due to DPNP. Others talked about feeling unwell, being slow, or being habitually late to work due to sleep problems (Sect. 3.3.2). Many of the patients expressed high levels of anxiety, depression, anger, or concern about the future and lower mood due to significant limitations in their physical functioning and the debilitating changes in their cognitive, social, and recreational lives [39, 40].

3.3.1 Common Meanings Related to Energy, Productivity, and Working Life in Patients with DPNP

"I just don't have the energy to get up and do it. I usually make it a point to be able to finish things that I start. I've gotten to a point, sometimes I feel because of the weakness in my leg and it kind of affects my energy, that I'd rather not even start to do it unless it's absolutely mandatory, and I usually try to finish what I do" [39].

"Well I used to work full-time and now I am part time, so, because the numbness, sometimes, the pain, you can't sleep at night, so you get really, really tired, sluggish" [39].

"It feels so painful, but I have to do it. Because I'm working. I do not like being restricted, of course. Either way, I cannot do as much in my workplace as I did before. I used to be very active at work. Now people pity me because of my disease and do not give me responsibility. I am very sad about this situation. I'm going to work as a guest. Sometimes I never want to go to work. I feel useless" (Participant 4, female, 42 years) [40].

3.3.2 Common Meanings Related to Sleep Quality in Patients with DPNP

A majority of respondents reported difficulty with sleep due to their DPNP. Some patients reported waking up in the middle of the night because of pain and not feeling rested in the morning upon awakening [39, 40].

"I have the tingling and numbness, but the way it impacts me the most is sometimes I wake up at night with sharp pain. They're short jabbing pains. They subside and then I'll go back to sleep and then three or four hours later, it'll happen again. So, it sort of wrecks my sleep pattern when that happens. It doesn't happen every night, but once or twice a week" [39].

"After sleeping 7–8 h, even if I wake up fresh, I can fall asleep again after 10 min. The body always wants to sleep because of the constant pain, no matter how much I sleep. My sleep is disturbed due to the numbness in my hands and arms. At that time, I feel like I do not have hands and arms and cannot sleep enough" (Participant 12, male, 44 years) [40].

3.3.3 Common Meanings Related to Sexual Activities in Patients with DPNP

Few participants specified that they experienced problems related to their sexual life, including reluctance, and erectile dysfunction. Consequently, some patients avoided sexual contact due to loss of sensation and reduced sense of intimacy with their partner.

"Well if you're in pain, you're not thinking about trying to make love. You know? You want to, in your mind, but if the pain is hitting—you aren't thinking about it" [39].

"I have a reluctance and erectile dysfunction in sexual life. My wife is very uncomfortable in this situation. She is upset. She's trying to support me. We consulted a doctor together. I started taking medication; then, I took a break. This situation makes me very sad and I am in a bad mood" (Participant 12, male, 44 years) [40].

3.3.4 Common Meanings Related to Focus and Memory in Patients with DPNP

Patients with DPNP expressed difficulties with attention, focus, and memory because of severe pain.

"It's just the uncomfortable feeling, when you are feeling pain inside your body, it just takes your mind off of everything. You can be doing whatever, you can be typing; it just takes your mind off of it" [39].

"My attention is distracted very easily. I have extreme forgetfulness. I forget everything. I forget what I'm talking about and what I am doing. I've never burned a meal until now. Things like that are starting to happen. I forget names, things that I will make. This situation causes a big challenge for my life" (Participant 1, female, 53 years) [40].

3.3.5 Common Meanings Related to Personal Care in Patients with DPNP

Living with chronic pain can cause weakness, fatigue, and low energy levels during the day. These conditions adversely affect many daily living activities, including limitations in cleaning the house or in personal care.

"I had several accidents getting in and out of the bathtub—bathing myself. I just have to slow down now and make sure I'm balanced. I can't move like I used to" [39].

"For me it's shaving. I've cut myself before because my hand went numb" [39].

"I'm tired. For example, when I wake up in the morning, if I have numbness in my hands, I cannot do anything. That day I did not even cut my nails. I could not hold the nail scissors" (Participant 13, female, 49 years) [40].

3.4 *Psychological Aspects of Living With Diabetic Peripheral Neuropathic Pain in Patients with DPNP*

Although the interviews revealed the physical and functional impacts of DPNP on patients, patients also reported common social and psychological experiences. DPNP, like all types of pain, occurs within a physical and psychosocial context that is influenced by several factors within experience. For some patients, DPNP accentuated fears of losing their toes or feet due to amputation at some point in the future. Many patients expressed negative feelings and thoughts about not enjoying life, sudden changes in their sensations, having lower mood, depression, fear of being a burden to their family caregivers, being dependent on others, not going out with their relatives and friends, social isolation, or loneliness [39, 40]. Although the pain sensation can dispose a person with pain to a meaning of threat or intrusion [19], this meaning can be strongly combined with the experienced context of the pain.

3.4.1 Common Meanings Related to Enjoyment and Purpose of Life in Patients with DPNP

"The quality of life, you can't do anything. How can you enjoy your life? I mean, like I said—also, the anxiety with me, I mean, I'm afraid of falling; I'm afraid of cutting my toe, afraid of stepping on something. You know, it's like, how can you do anything and have any kind of [...] enjoyment, you know?" [39].

"I do not enjoy life anymore. I have no wishes. I'm like an 80-year-old woman. It's like the end. I feel like I have not been able to offer much to my husband, children, and friends" (Participant 1, female, 53 years) [40].

3.4.2 Common Meanings Related to Anxiety, Moodiness, and Depression in Patients with DPNP

"And I'll tell you what, the emotional stress that you go through when you hear about people's limbs being cut off, arms, legs, and what have you, I just feel like, is this going to be me one day? Am I going to lose my feet because of it? [...] One thing about this disease, is that it is not just painful. It is painful, and that is the most obvious. But for me, it is the psychological, and emotional turmoil that I go through" [39].

"I feel nervous because you don't know basically what the future holds, if it gets worse, is this pain going to get more intense. The whole numbness and all of this stuff that you read, are you going to get all of the symptoms, or one. So sometimes it is hard, when you start thinking about it, you get some anxiety about it" [39].

"I don't know if I could point out any one thing. There's a lot of depression that goes with it because you don't feel like you want to do anything; you can't do anything. You're just tired all the time, but then that could be due to the lack of sleep; that could be due to the depression. You don't want to actually try anything that would create any physical exertion because it hurts" [39].

3.4.3 Common Meanings Related to Fear and Irritability in Patients with DPNP

"And I am afraid of falling, because sometimes you get a pain or whatever and it just throws me off guard, and a couple times, 'Are you, all right? Are you alright?' 'Yes, I'm fine.' But I am not, really. But I fear falling on the granite floor" [39].

"I get in a lot of trouble with it, especially with my family. And I try and hold my tongue, and it's just like almost impossible" [39].

"Initially, I did not have so much pain, but as it increased, my personality and character changed; my tension and anger reached a top level. I'm becoming too emotional. I start crying even when I speak normally. I feel like I have multiple personalities, both laughing and crying, I experience all things at the same time and it is a very bad situation" (Participant 4, female, 42 years) [40].

3.4.4 Common Meanings Related to Feeling Dependent on Someone or Being a Burden to Family Caregivers in Patients with DPNP

"When I see people, who have lost their hands and feet in the hospital because of the disease, I feel severe anxiety when I have so much pain and fear losing my extremities due to this pain and being a burden on my family" (Participant 6, male, 32 years) [40].

"As long as I have this pain in my hands and feet, I feel that I won't do anything. I do not want to depend on anyone. I'm so afraid of needing support from someone. I am taking my medications. It seems to be cut a bit; then, it starts again. I am always ashamed to say I have pain" (Participant 1, female, 53 years) [40].

3.5 Common Meanings Related to Social and Recreational Activities in Patients with DPNP

DPNP symptoms curtailed some activities, requiring that patients join their friends in a limited way (being slow, for example, due to foot pain and its effects on walking), or preventing them from participating in any activities. Some patients voluntarily distanced themselves from others.

"Well, I pulled back on non-essential activities, going out of the house for non-essential activities. Living here, very close to Manhattan, I really could—I could go out more. I could—recreationally there's a lot more things I could do. So, I really miss social and recreational time" [39].

"Yeah, all of my friends stopped calling me to go play ball on the weekends. Because most of the time I wouldn't. Because, unfortunately, I used to be rather active and it has just kind of settled in, and unfortunately, you've just got to roll with the punches. So yeah, socially it has become a very hindering disease" [39].

"Pain limited my life. I can't see my friends. I'm getting addicted to the house. We used to meet with friends somewhere, eat, drink, and prepare meals. I never want to go out. I go out only for necessary things like shopping or going to the hospital" (Participant 1, female, 53 years) [40].

3.6 Summary of Sect. 3

Participants [39, 40] with pain reported that DPNP caused them considerably intense and persistent pains, which were perceived as unpleasant and distressing, and linked to a common meaning of threat or intrusion to the person. Persistent pain sensations revealed additional meanings of interruption (of life activities), a concern for future consequences, such as concern over amputation of a foot, and a fear of having to endure pain as a burden over time. When they focused on their pain, most patients reported that family relationships often became strained, and they feared being fully dependent on someone else's care. Some patients found it difficult to speak about their illness or explain the intensity of their pain to family members and friends. Impairment in the ability to communicate pain to others, the social life of some participants became correspondingly restricted to the home environment. During periods of severe pain, many participants became angry and worried about the course of their disease. When some participants shifted their focus to themselves as persons, these experiences and meanings led to existential meanings such as, worthlessness, hopelessness, and loneliness. For example, many participants emphasized that significant restriction in their productivity at work or negative feelings about the loss of their jobs because of their pain led to perceptions of being seen by some family members and society as a whole as having less value. All these meanings were closely linked with levels of desire and expectation, leading to a range of negative emotions. For instance, adversity in walking, standing, climbing stairs, exercising or lifting due to pain, led many patients to face difficulty maintaining their daily routines, sleep problems, anxiety, frustration, and fatigue, and an expectation that the 'pain situation' could not improve. These meanings are summarized in Table 11.2.

4 A Model of Diabetic Peripheral Neuropathic Pain Incorporating Common Meanings

Based on the findings described in Sect. 3, Fig. 11.1 (adapted from [19]) outlines the interactions between the basic aspects or dimensions of pain in patients with DPNP, including primary and secondary meanings of pain, and external factors (dashed arrows) that can influence the intensities of these dimensions of experience. Excitatory (+) and inhibitory (−) influences are shown.

5 Cognitive Behavioral Therapy Can Alleviate Diabetic Peripheral Neuropathic Pain

Neuropathic pain is a challenging condition to treat. The variety of causes and affected individuals forms a heterogeneous group. Current management strategies fail to achieve adequate or satisfactory pain management in a high proportion of patients [41–45]. In addition, associated conditions including depression, anxiety and sleeping disorders often increase the adverse effects of DPNP. Previous studies have shown that neuropathic pain interferes with psychological and social well-being and strongly affects daily functioning and overall QOL [44, 46].

Researchers have emphasized that a multidimensional approach is needed to manage the entire spectrum of neuropathic pain, including comorbid conditions. The theoretical basis for such an approach is a biopsychosocial model in which pain results from an interaction of physiologic, psychological and social factors that influence perception of pain and response to pain [47]. As best practice, pain management of DPNP should therefore include psychosocial interventions in addition to the traditional biomedical interventions [44].

Some psychosocial interventions attempt to change the meaning of patients' thoughts, beliefs and behaviors ("re-conceptualization of pain") [48, 49]. Previous research has demonstrated that patients with neuropathic pain differ from those with nociceptive pain in response to pain beliefs about pain, pain-related problems experienced, and factors that increase pain. Patients with neuropathic pain therefore might respond differently to cognitive behavior therapy than patients with nociceptive pain [50]. In this part of the chapter, the effectiveness of cognitive and behavioral interventions for the management of DPNP is evaluated, with a focus on meanings of pain.

Table 11.2 Summary of common meanings in patients with DPNP [39, 40]

Domain	Aspect affected	Illustrative common meanings
Physical	Walking and standing	"Sometimes my feet will hurt really bad and I can't get up and hardly walk because they're really hurting."
	Mobility	"I can't even get out of bed in the morning unless I have special shoes that are cushioned, because when my feet touch the ground I feel like I am walking on sharp rocks."
	Exercise	"The biggest thing: you cannot do any exercise. Simply put, if my son says, 'Dad, let's play ball,' I cannot do that with my son."
	Energy, productivity	"I just don't have the energy to get up and do it. I usually make it a point to be able to finish things that I start."
Daily Life	Work	"Well I used to work full-time and now I am part time, so, because the numbness, sometimes, the pain, you can't sleep at night, so you get really, really tired, sluggish."
	Personal care	"...when I wake up in the morning, if I have numbness in my hands, I cannot do anything. That day I did not even cut my nails. I could not hold the nail scissors."
	Social and recreational activities	"Pain limited my life. I can't see my friends. I'm getting addicted to the house."
		"Well, I pulled back on non-essential activities, going out of the house for non-essential activities."
		"Yeah, all of my friends stopped calling me to go play ball on the weekends. Because most of the time I wouldn't."
	Relationships with others	"When I see people, who have lost their hands and feet in the hospital because of the disease, I feel severe anxiety when I have so much pain and fear losing my extremities due to this pain and being a burden on my family."
	Sexual activities	"I have a reluctance and erectile dysfunction in sexual life. My wife is very uncomfortable in this situation. She is upset. She's trying to support me. We consulted a doctor together. I started taking medication; then, I took a break. This situation makes me very sad and I am in a bad mood."
	Sleep	"The body always wants to sleep because of the constant pain, no matter how much I sleep. My sleep is disturbed due to the numbness in my hands and arms. At that time, I feel like I do not have hands and arms and cannot sleep enough."
Social	Relationships with others	"When I see people, who have lost their hands and feet in the hospital because of the disease, I feel severe anxiety when I have so much pain and fear losing my extremities due to this pain and being a burden on my family."
	Social and recreational activities	"Pain limited my life. I can't see my friends. I'm getting addicted to the house."
		"Well, I pulled back on non-essential activities, going out of the house for non-essential activities."
		"Yeah, all of my friends stopped calling me to go play ball on the weekends. Because most of the time I wouldn't."

(continued)

Table 11.2 (continued)

Domain	Aspect affected	Illustrative common meanings
Psychological/ Cognitive	Focus, memory	"My attention is distracted very easily. I have extreme forgetfulness. I forget everything. I forget what I'm talking about and what I am doing."
	Enjoyment and purpose in life	"The quality of life, you can't do anything. How can you enjoy your life? I mean, like I said—also, the anxiety with me, I mean, I'm afraid of falling; I'm afraid of cutting my toe, afraid of stepping on something."
	Anxiety, moodiness, depression	"I feel nervous because you don't know basically what the future holds, if it gets worse, is this pain going to get more intense."
		"I don't know if I could point out any one thing. There's a lot of depression that goes with it because you don't feel like you want to do anything; you can't do anything. You're just tired all the time, but then that could be due to the lack of sleep; that could be due to the depression."
	Fear, irritability	"And I am afraid of falling, because sometimes you get a pain or whatever and it just throws me off guard. . ."
		"Initially, I did not have so much pain, but as it increased, my personality and character changed; my tension and anger reached a top level. I'm becoming too emotional. I start crying even when I speak normally."

5.1 *Cognitive-Behavioral Therapy*

Cognitive theory applied to pain suggests that patient beliefs and thoughts about their pain or "pain situation" influence their response to pain and the level and type of pain-related distress experienced [51–53]. Cognitive interventions aim to help change unhelpful cognitive and behavioral responses to pain by encouraging people to address and re-evaluate their beliefs and the meanings they hold about pain and their "pain situation" [54–57].

Cognitive-behavioral therapy (CBT) is a short-term and goal-oriented form of psychotherapy that treats problems by modifying the meaning of dysfunctional emotions, behaviors, and thoughts. Pain is influenced by individual cognitions and tissue injury, or perceived tissue injury; the aim of CBT for the treatment of pain is that maladaptive cognitions contribute to the maintenance of emotional distress and behavioral problems.

5.2 *Reducing Pain Related Distress*

People's beliefs and thoughts about themselves and their pain, for example, "What's the point in trying; it always hurts and I never succeed," can contribute to increased distress, reinforce unhelpful behaviors and reduce QOL [58, 59].

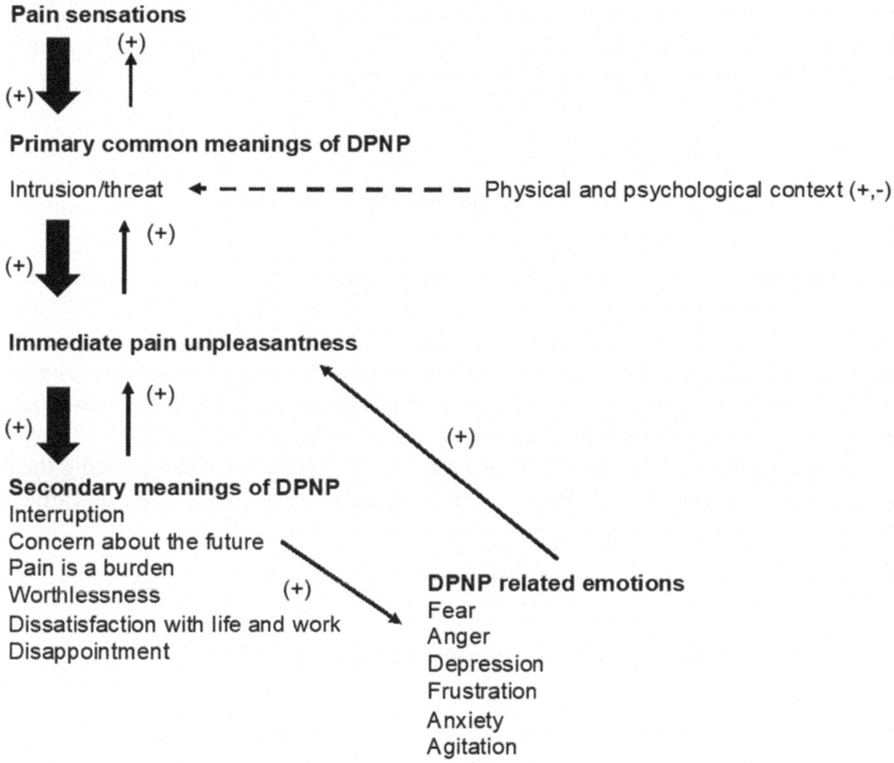

Fig. 11.1 Conceptual framework of DPNP incorporating common meanings of DPNP

5.3 Improving Physical Functioning and Reduce Disability

People's new knowledge about chronic pain and the influence of their beliefs on their behavior is combined with learning about basic anatomy and physiology; in particular, that healing has occurred, and rest does not resolve chronic pain. This helps some people to engage in stretching and exercise activities that will help them to improve their physical function and achieve exercise-related goals.

5.4 Returning to Valued Activities

People with pain often relinquish activities that once provided them with a sense of achievement and/or meaning. Reasons for this include fear of pain and injury resulting in avoidance of activity, pain having often resulted in an inability to reach goals, resulting in disappointment and a sense of failure, and the necessity of having to channel all available energy into activities such as employment

[60]. Personal goal setting of valued activities is an important aspect of cognitive behavioral interventions. Patients are encouraged to set realistic goals and consider the steps they need to achieve if they are to reach their goals.

5.5 *Reviewing and Reducing Medication for Pain*

The majority of people with chronic pain have tried numerous medications in the hope of decreasing the intensity of their pain [61, 62]. Many patients state that medication "only takes the edge off" the pain and some state that the side effects (such as dry mouth, sedation, gastrointestinal problems) outweigh the benefits they derive from their medication. However, some people are reluctant to decrease their medication for fear of the pain increasing and of them not being able to manage. Patients are helped to understand their medication, given advice about taking their medication at a specific time rather than at a pain-contingent fashion, and given support in reducing their medication, should this be a patient goal.

5.6 *Developing Ways to Manage Increases in Pain*

Before coming to pain management programs many people respond to increases in their pain by using medication and rest. Cognitive behavioral pain management helps people to understand the disadvantages of these responses and encourages them to consider, attempt and evaluate alternative strategies during these times to help them break their association between increased pain and decreased activity. It is important to help people elicit, test out, and challenge their cognitions regarding the consequences of an increase in pain level; for example, "I will lose control;" "I cannot cope;" "I am going to be sick;" or, "I will faint in public."

5.7 *Evidence for Cognitive Behavioral Interventions in Chronic Pain*

Morley et al. [51] specified that cognitive behavioral interventions are more effective in restoring function, improving mood and reducing disability and unhelpful pain related behaviors compared with waiting list controls [51]. Additionally, cognitive-behavioral interventions provide significantly greater changes in pain experience (intensity, unpleasantness and sensation), improved cognitive strategies to manage pain and reduced behavioral expression of pain compared with a range of heterogeneous interventions delivered by pain clinics, physiotherapy, occupational therapy and educational packages [51]. McCracken and Turk [63] who, in addition, report an overall decrease in health care costs and an increased chance of returning to work, support these findings [63].

Cognitive behavioral interventions for chronic pain are typically delivered as pain management programs in a group format and by a multidisciplinary team. Although wide variations in the content of these interventions exist across services [51], the main aims of these interventions are to:

- Improve the person's understanding of their persistent pain condition
- Reduce pain related distress
- Improve physical functioning and reduce disability
- Return to valued activities
- Review and reduce medication
- Develop ways to manage increases in pain

The evidence level for the efficacy of psychological interventions for chronic pain is low. Based on Cochrane systematic reviews of psychological interventions for improving pain, affect, and disability in chronic pain for adults, these modalities show moderate-effect sizes of benefit over waiting lists and small or no effects over active comparators for outcomes in pain, disability, and mood [64–66]. The majority of the outcome studies reviewed in the most recent meta-analysis of cognitive behavioral interventions for chronic pain include people with HIV or pain of mixed etiologies [51]. Evans and Fishman [67] developed and investigated the efficacy of "pain management training" for people with HIV-related peripheral neuropathic pain [67]. Their description suggests that the intervention is similar to that currently employed in pain management programs. However, as this was not an empirical study, it is difficult to draw conclusions about the effectiveness of this intervention. A randomized controlled trial compared CBT with supportive psychotherapy for people with HIV-related peripheral neuropathic pain [68]. Both groups reported a significant improvement in pain-related interference. However, the improvement occurred across more domains in the cognitive behavioral group, who also reported a significantly greater improvement in symptoms of depression and several of the subscales of the Brief Pain Inventory [69]. Unfortunately, the authors did not describe the content of the interventions, which again makes it difficult to draw conclusions about the effectiveness of cognitive behavioral interventions for this population.

5.8 *Evidence for Cognitive Behavioral Interventions in DPNP*

Currently, there is only one study that has specifically evaluated CBT in patients with DPNP [70]. Otis et al. [70] conducted a randomized, treatment as usual (TAU), controlled, non-blinded intervention pilot study with a 4-month follow-up in a VA medical center. The study authors hypothesized that participants who received CBT

Table 11.3 Cognitive-behavioral therapy for patients with DPNP [70]

Session 1: Discussing the impact of pain	Reviewing the cycle of pain (i.e., pain, disability, and distress), and establishing goals for treatment.
Sessions 2–4: Educating the participants	The various theories of pain and relaxation techniques including diaphragmatic breathing, progressive muscle relaxation, and visual imagery.
Sessions 7–8: Educating the participants	A time-based pacing technique in which individuals take breaks based on time or amount of work accomplished, rather than taking a break once their pain level has reached an unmanageable level, and regarding the importance of scheduling pleasant activities.
Sessions 9–11: Additional training	In techniques related to anger management, sleep hygiene, and relapse prevention.

would report significant decreases on self-report measures of pain severity, interference, and depressive symptoms from pre-treatment to 4-month follow-up. In that study, participants randomized to the TAU condition did not receive treatment beyond that provided by their primary care provider or other healthcare providers for pain. Participants randomized to the CBT condition participated in 11 sessions of weekly, individually based, cognitive-behavioral pain management therapy. Therapy followed a chronic pain management treatment protocol that included the use of a therapist manual and a corresponding patient workbook containing session content and homework assignments [71]. Each session was 60 min in duration and was delivered by a PhD-level clinical psychologist or a Masters-level therapist. CBT treatment sessions applied in [70] are shown in Table 11.3. Weekly goals were defined at the end of Sessions 1 to 10 and patients were asked to practice the various techniques (e.g., walking 3 times a week, going to the gym, dinner with friends) that were established at the beginning of treatment [70].

Otis et al. [70] examined the changes in pain interference from pre-treatment to 4-month follow-up using the West Haven Yale Multidimensional Pain Inventory (WHYMPI), a 52-item self-report questionnaire consisting of 12 subscales that are applicable across a variety of clinical pain conditions. The Interference subscale (WHYMPI-I) assesses the degree to which pain interferes in vocational, social, family, or marital functioning; the Pain Severity subscale (WHYMPI-PS) assesses level of pain severity or intensity.

The authors found significant decreases in pain interference in patients who received CBT compared to TAU patients. Prior research has identified a common meaning of immediate pain to be intrusion or threat in patients with chronic pain [19–22] and DPNP, as reviewed above [39, 40], and a meaning of interruption focused on disrupted life activities and goal achievement. It is unclear if *interference* in Otis et al. [70] could be viewed as a meaning, similar to *intrusion*, *threat,* or *interruption*, since the authors [70] viewed it in terms of functional change, not experiential meaning. However, it seems reasonable to presume that patients whose daily life is interfered with by pain, experience this interference in terms of an interruption. Otis et al. [70] found no significant differences in the depression scores of patients, irrespective of study group, a result possibly affected by the limited study sample size [70]. Overall, the results of Otis et al. [70] lend modest support to CBT being an effective psychological treatment for reducing pain severity and interference in patients with DPNP, but further study should attempt to identify the extent to which meaning contributes to the effects of CBT in patients with DPNP.

6 Conclusion

Life presents experiences to persons that they endow with meaning. Based on the available scientific literature, this chapter shows that patients with DPNP have experiences that they associate with common meanings, including threat or intrusion, and more cognitive meanings of interruption, concern about the future implications of pain, or worthlessness, which are linked with negative moods, emotions, and expectations. Aspects of life within personal experience that DPNP disrupts include family or marital relationships, sexual intimacy, sleep, memory and the meaning of life.

Continuous evaluation of the patient with DPNP will help to identify how common biomedical, psychosocial, and behavioral factors interact to influence the nature, severity, and persistence of DPNP and its symptoms. Clinical examination of DPNP should focus not only on the location or severity of pain, but also assess mood, physical dysfunction, pain-coping strategies, and social support, and interactions over time between these, the pain sensation and pain-related emotions.

Currently, few qualitative studies have focused on the meanings of DPNP from the patient perspective. Although quantitative studies have evaluated pain severity, depression and QOL in patients with DPNP, this chapter shows that understanding pain in patients with DPNP more fully requires data from both qualitative quantitative measures. Finally, although there is evidence that CBT can reduce the perceived threat or interruption of pain in patients with DPNP, psychological intervention studies that assess DPNP are very limited in number. Mixed method studies testing the efficacy of such interventions, including CBT, to alter the experience, meaning of pain and other common factors in the experience of patients with DPNP are warranted.

This chapter is the first to analyze and synthesize available qualitative data describing some of the common meanings of pain in patients with DPNP. However, since very few such studies have been published this topic, it is impossible to generalize the findings of this chapter to all patients with DPNP.

Acknowledgments I am grateful to Dr. Simon van Rysewyk for comments and edits of an earlier draft of this chapter.

Note Informed consent was obtained from all individual participants included in the study.

References

1. Tesfaye S, Selvarajah D. Advances in the epidemiology, pathogenesis and management of diabetic peripheral neuropathy. Diabetes Metab Res Rev. 2012;28:8–14.
2. Gordois A, Scuffham P, Shearer A, Oglesby A, Tobian JA. The health care costs of diabetic peripheral neuropathy in the US. Diabetes Care. 2003;26(6):1790–5.
3. Davies M, Brophy S, Williams R, Taylor A. The prevalence, severity, and impact of painful diabetic peripheral neuropathy in type 2 diabetes. Diabetes Care. 2006;29(7):1518–22.
4. Tesfaye S, Boulton AJ, Dyck PJ, Freeman R, Horowitz M, Kempler P, et al. Diabetic neuropathies: update on definitions, diagnostic criteria, estimation of severity, and treatments. Diabetes Care. 2010;33(10):2285–93.
5. Tesfaye S. Recent advances in the management of diabetic symmetrical polyneuropathy. J Diabetes Investig. 2010;2:33–42.
6. Merskey H. Classification of chronic pain. Description of chronic pain syndromes and definitions of pain terms. Seattle: IASP Press; 1994. 213 p.
7. McDermott AM, Toelle TR, Rowbotham DJ, Schaefer CP, Dukes EM. The burden of neuropathic pain: results from a cross-sectional survey. Eur J Pain. 2006;10(2):127–35.
8. Quattrini C, Tesfaye S. Understanding the impact of painful diabetic neuropathy. Diabetes Metab Res Rev. 2003;19(S1):S2–8.
9. Gore M, Brandenburg NA, Dukes E, Hoffman DL, Tai K-S, Stacey B. Pain severity in diabetic peripheral neuropathy is associated with patient functioning, symptom levels of anxiety and depression, and sleep. J Pain Symptom Manag. 2005;30(4):374–85.
10. Boulton AJ, Kirsner RS, Vileikyte L. Neuropathic diabetic foot ulcers. N Engl J Med. 2004; 351(1):48–55.
11. Health NIo. National institute of diabetes and digestive and kidney diseases. Diabetes in America. 2nd ed. NIH Publication; 1995(95–1468).
12. Dyck PJB, Norell JE, Dyck PJ. Microvasculitis and ischemia in diabetic lumbosacral radiculoplexus neuropathy. Neurology. 1999;53(9):2113–21.
13. Poncelet AN. Diabetic polyneuropathy. Risk factors, patterns of presentation, diagnosis, and treatment. Geriatrics (Basel, Switzerland) 2003;58(6):16–8, 24–5, 30.
14. Benbow S, Wallymahmed M, MacFarlane I. Diabetic peripheral neuropathy and quality of life. QJM. 1998;91(11):733–7.
15. Melton L III, Dyck P. Diabetic polyneuropathy. In: Dyck PJ, Thomas PK, editors. Diabetic Neuropathy. 2nd ed. Philadelphia: WB Saunders; 1999. p. 255–78.
16. Campbell JN, Meyer RA. Mechanisms of neuropathic pain. Neuron. 2006;52(1):77–92.
17. Boureau F, Doubrere J, Luu M. Study of verbal description in neuropathic pain. Pain. 1990; 42(2):145–52.
18. Thomas PK. Classification of the diabetic neuropathies. In: Gries FA, Cameron NE, Low PA, Ziegler D, editors. Textbook of diabetic neuropathy. Stuttgart: Thieme; 2003. p. 175–7.
19. Price DD, Harkins SW. The affective-motivational dimension of pain A two-stage model. APS J. 1992;1(4):229–39.
20. Cassell EJ. The phenomenon of suffering and its relationship to pain. In: Handbook of phenomenology and medicine. Dordrecht: Springer; 2001. p. 371–90.
21. Crowe M, Whitehead L, Seaton P, Jordan J, Mccall C, Maskill V, et al. Qualitative metasynthesis: the experience of chronic pain across conditions. J Adv Nurs. 2017;73(5):1004–16.
22. Ojala T, Häkkinen A, Karppinen J, Sipilä K, Suutama T, Piirainen A. Chronic pain affects the whole person – a phenomenological study. Disabil Rehabil. 2015;37(4):363–71.

23. Breivik H, Collett B, Ventafridda V, Cohen R, Gallacher D. Survey of chronic pain in Europe: prevalence, impact on daily life, and treatment. Eur J Pain. 2006;10(4):287–333.
24. Elliott AM, Smith BH, Penny KI, Smith WC, Chambers WA. The epidemiology of chronic pain in the community. Lancet. 1999;354(9186):1248–52.
25. Gureje O, Von Korff M, Simon GE, Gater R. Persistent pain and well-being: a world health organization study in primary care. JAMA. 1998;280(2):147–51.
26. Jensen MP, Chodroff MJ, Dworkin RH. The impact of neuropathic pain on health-related quality of life review and implications. Neurology. 2007;68(15):1178–82.
27. Meyer-Rosberg K, Burckhardt CS, Huizar K, Kvarnström A, Nordfors LO, Kristofferson A. A comparison of the SF-36 and Nottingham health profile in patients with chronic neuropathic pain. Eur J Pain. 2001;5(4):391–403.
28. Schmader KE. Epidemiology and impact on quality of life of postherpetic neuralgia and painful diabetic neuropathy. Clin J Pain. 2002;18(6):350–4.
29. Zelman DC, Brandenburg NA, Gore M. Sleep impairment in patients with painful diabetic peripheral neuropathy. Clin J Pain. 2006;22(8):681–5.
30. Haythornthwaite JA, Benrud-Larson LM. Psychological aspects of neuropathic pain. Clin J Pain. 2000;16(2 Suppl):S101–5.
31. Meyer-Rosberg K, Kvarnström A, Kinnman E, Gordh T, Nordfors LO, Kristofferson A. Peripheral neuropathic pain—a multidimensional burden for patients. Eur J Pain. 2001;5 (4): 379–89.
32. Watkins PJ. Pain and diabetic neuropathy. Br Med J (Clin Res Ed). 1984;288(6412):168–9.
33. Novella SP, Inzucchi SE, Goldstein JM. The frequency of undiagnosed diabetes and impaired glucose tolerance in patients with idiopathic sensory neuropathy. Muscle Nerve. 2001;24(9): 1229–31.
34. Watkins P, Edmonds M. Clinical features of diabetic neuropathy. Textbook of diabetes. 1997. p. 1–20.
35. Tesfaye S, Watt J, Benbow SJ, Pang KA, Miles J, MacFarlane IA. Electrical spinal-cord stimulation for painful diabetic peripheral neuropathy. Lancet. 1996;348(9043):1698–701.
36. Robinson N, Yateman N, Protopapa L, Bush L. Employment problems and diabetes. Diabet Med. 1990;7(1):16–22.
37. Finnerup NB, Sindrup SH, Jensen TS. The evidence for pharmacological treatment of neuropathic pain. Pain. 2010;150(3):573–81.
38. Rijken P, Dekker J, Dekker E, Lankhorst G, Bakker K, Dooren J, et al. Clinical and functional correlates of foot pain in diabetic patients. Disabil Rehabil. 1998;20(9):330–6.
39. Brod M, Pohlman B, Blum SI, Ramasamy A, Carson R. Burden of illness of diabetic peripheral neuropathic pain: a qualitative study. Patient. 2015;8(4):339–48.
40. Gok Metin Z, Arslan IE. Diabetic peripheral neuropathic pain from the perspective of turkish patients: a qualitative study. J Transcult Nurs. 2018;29(6):514–22.
41. Smith JA. Qualitative psychology: a practical guide to research methods. London: Sage; 2015. 312 p.
42. Dezutter J, Luyckx K, Wachholtz A. Meaning in life in chronic pain patients over time: associations with pain experience and psychological well-being. J Behav Med. 2015;38(2): 384–96.
43. Vlaeyen JW, Morley S, Crombez G. The experimental analysis of the interruptive, interfering, and identity-distorting effects of chronic pain. Behav Res Ther. 2016;86:23–34.
44. Harden N, Cohen M. Unmet needs in the management of neuropathic pain. J Pain Symptom Manag. 2003;25(5):S12–S7.
45. Jensen TS, Gottrup H, Sindrup SH, Bach FW. The clinical picture of neuropathic pain. Eur J Pharmacol. 2001;429(1–3):1–11.
46. Tölle T, Xu X, Sadosky AB. Painful diabetic neuropathy: a cross-sectional survey of health state impairment and treatment patterns. J Diabetes Complicat. 2006;20(1):26–33.
47. Gatchel RJ. Comorbidity of chronic pain and mental health disorders: the biopsychosocial perspective. Am Psychol. 2004;59(8):795.

48. Turk DC, Swanson KS, Tunks ER. Psychological approaches in the treatment of chronic pain patients—when pills, scalpels, and needles are not enough. Can J Psychiatr. 2008;53(4): 213–23.
49. Scascighini L, Sprott H. Chronic nonmalignant pain: a challenge for patients and clinicians. Nat Rev Rheumatol. 2008;4(2):74–81.
50. Daniel HC, Narewska J, Serpell M, Hoggart B, Johnson R, Rice AS. Comparison of psychological and physical function in neuropathic pain and nociceptive pain: implications for cognitive behavioral pain management programs. Eur J Pain. 2008;12(6):731–41.
51. Morley S, Eccleston C, Williams A. Systematic review and meta-analysis of randomized controlled trials of cognitive behaviour therapy and behaviour therapy for chronic pain in adults, excluding headache. Pain. 1999;80(1–2):1–13.
52. Morley S. Process and change in cognitive behaviour therapy for chronic pain. Pain. 2004; 109(3):205–6.
53. Daniel HC, Van der Merwe JD. Cognitive behavioral approaches and neuropathic pain. Handb Clin Neurol. 2006;81:855–68.
54. Vitiello MV, Rybarczyk B, Von Korff M, Stepanski EJ. Cognitive behavioral therapy for insomnia improves sleep and decreases pain in older adults with co-morbid insomnia and osteoarthritis. J Clin Sleep Med. 2009;5(04):355–62.
55. Glombiewski JA, Hartwich-Tersek J, Rief W. Two psychological interventions are effective in severely disabled, chronic back pain patients: a randomised controlled trial. Int J Behav Med. 2010;17(2):97–107.
56. Holroyd KA, O'donnell FJ, Stensland M, Lipchik GL, Cordingley GE, Carlson BW. Management of chronic tension-type headache with tricyclic antidepressant medication, stress management therapy, and their combination: a randomized controlled trial. JAMA. 2001;285(17): 2208–15.
57. Hoffman BM, Papas RK, Chatkoff DK, Kerns RD. Meta-analysis of psychological interventions for chronic low back pain. Health Psychol. 2007;26(1):1–9.
58. Crombez G, Vlaeyen JW, Heuts PH, Lysens R. Pain-related fear is more disabling than pain itself: evidence on the role of pain-related fear in chronic back pain disability. Pain. 1999; 80(1–2):329–39.
59. Lamé IE, Peters ML, Vlaeyen JW, Kleef M, Patijn J. Quality of life in chronic pain is more associated with beliefs about pain, than with pain intensity. Eur J Pain. 2005;9(1):15–24.
60. de Jong JR, Vlaeyen JW, Onghena P, Goossens ME, Geilen M, Mulder H. Fear of movement/(re) injury in chronic low back pain: education or exposure in vivo as mediator to fear reduction? Clin J Pain. 2005;21(1):9–17.
61. O'Connor AB, Dworkin RH. Treatment of neuropathic pain: an overview of recent guidelines. Am J Med. 2009;122(10):S22–32.
62. Quilici S, Chancellor J, Löthgren M, Simon D, Said G, Le TK, et al. Meta-analysis of duloxetine vs. pregabalin and gabapentin in the treatment of diabetic peripheral neuropathic pain. BMC Neurol. 2009;9(1):6.
63. McCracken LM, Turk DC. Behavioral and cognitive-behavioral treatment for chronic pain: outcome, predictors of outcome, and treatment process. Spine. 2002;27(22):2564–73.
64. Eccleston C, Crombez G. Advancing psychological therapies for chronic pain. F1000Res. 2017;6:461.
65. Vlaeyen JW, De Jong JR, Onghena P, Kerckhoffs-Hanssen M, Kole-Snijders AM. Can pain-related fear be reduced? The application of cognitive-behavioral exposure in vivo. Pain Res Manag. 2002;7(3):144–53.
66. Butler AC, Chapman JE, Forman EM, Beck AT. The empirical status of cognitive-behavioral therapy: a review of meta-analyses. Clin Psychol Rev. 2006;26(1):17–31.
67. Evans S, Fishman B. Pain management training: a cognitive-behavioral approach to treating HIV-related peripheral neuropathy. J Cogn Psychother. 1997;11(4):251–61.

68. Evans S, Fishman B, Spielman L, Haley A. Randomized trial of cognitive behavior therapy versus supportive psychotherapy for HIV-related peripheral neuropathic pain. Psychosomatics. 2003;44(1):44–50.
69. Daut RL, Cleeland CS, Flanery RC. Development of the Wisconsin brief pain questionnaire to assess pain in cancer and other diseases. Pain. 1983;17(2):197–210.
70. Otis JD, Sanderson K, Hardway C, Pincus M, Tun C, Soumekh S. A randomized controlled pilot study of a cognitive-behavioral therapy approach for painful diabetic peripheral neuropathy. J Pain. 2013;14(5):475–82.
71. Otis J. Managing chronic pain: a cognitive-behavioral therapy approach. New York: Oxford university press; 2007. 83 p.

Chapter 12
Connotations of Pain in a Socio-Psycho-Biological Framework

John Quintner, Melanie Galbraith, and Milton Cohen

Abstract Introduction From the dawn of civilization pain has received special attention as part of "dis-ease." In recent years, pain has emerged as a clinical problem worthy of study in its own right. Changing concepts on the phenomenology of pain have flowed as a continuous process from the most ancient writings on medicine to the present day. Because of its multidimensional properties, pain is now viewed in a socio-psycho-biological framework.

Methods The major milestones in the history of pain theory from a Greco-Judeo-Christian perspective have been reviewed in order to provide a background to the different connotations or meanings of pain that have emerged in Western society.

Results Many contemporary connotations of pain mirror those found in the language used by ancient Greek philosophers and others. However, some have evolved to reflect negatively upon pain sufferers.

Conclusion As these negative connotations can give rise to stigmatisation of pain sufferers, it is time to remove them from the pain discourse and replace them with options that have sound foundations in biology.

Clinical Implications When the many connotations of pain are viewed in a socio-psycho-biological framework, it is apparent that the fluidity in the language of pain available to the ancient Greek philosophers can also be found in the English language. The consequent evolution of negative connotations of pain in Western society demand their replacement by scientifically valid connotations that are rooted in biology.

Keywords Chronic pain · Connotations · Etymology · Socio-psycho-biological

J. Quintner (✉)
Arthritis & Osteoporosis WA, Shenton Park, WA, Australia

M. Galbraith
Biosymm Physiotherapy, Rivervale, WA, Australia

M. Cohen
St Vincent's Clinic & UNSW, Darlinghurst, NSW, Australia

S. van Rysewyk (ed.), *Meanings of Pain*,
https://doi.org/10.1007/978-3-030-24154-4_12

1 Introduction

All pain is one malady with different names (Antiphanes, ca. 400 BC)

In Greek mythology POINE (*poena* in Latin) was the personified spirit of retribution, vengeance, punishment and penalty for the crimes of murder and manslaughter [1]. Considering this etymology of the English word "pain," it is understandable that it might convey connotations[1] of punishment and moral failings.

However, the "pain" language used by the ancient Greeks conveyed other important meanings. They distinguished between the feeling of pain, called *ponos*, and pathos or wretchedness (called *lupê*). But *ponos* could also be used to mean illness, poverty and low social status, while *lupê* could mean distress and sometimes grief [2]. Similarly, the Latin word *dolor* could mean both physical pain and grief while *aegritudo*[2] could mean illness, sickness, grief or sorrow. Such fluidity of meaning in this context is considered to be one of language's benefits:

> The more varied the expressions of pain in a given language are, the more opportunities there are for an individual to talk about pain and communicate it to his or her doctor [3, pS133].

These etymologies might explain why there are so many connotations of the word "pain" in the English language. In this chapter, we will attempt to identify and place them in a socio-psycho-biological perspective [4]. Although these connotations can be traced back to ancient meanings of the word "pain," some have evolved to be associated with negative implications for pain sufferers, which can include personal rejection, social exclusion and stigmatisation. It is notable that there appear to be no contrasting *positive* connotations, let alone any that have been beneficial to patients with pain.

2 The Socio-Psycho-Biological Framework of Pain

Chronic pain is increasingly being appreciated through a socio-psycho-biological framework in which pain is reframed as "an interpersonal, inherently social process" [4]. However, a satisfactorily concise diagnostic formulation to embrace these three dimensions has not as yet been devised. Nevertheless, connotations of pain influence and at least in part determine the approaches to those experiencing pain by those in the health care professions, as well as by those responsible for disability certification, providing income support, treating mental health issues, and resolving family

[1]Connotation refers to the emotional or psychological associations a word carries with it. These associations are expressed in the feelings, thoughts and images the word suggests or provokes. They can be positive, negative, or neutral.

[2]Cited as an example of the richness of that vocabulary. Unfortunately, there is no derivative of *aegritudo* in English.

conflicts. In the following sections we situate the respective connotations in the most appropriate of the three dimensions.

2.1 *Social Connotations of Pain*

As Hardy et al. [5] point out: "...the culture in which a man finds himself becomes the conditioning influence in the formation of the individual reaction patterns to pain...A knowledge of group attitudes toward pain is extremely important to an understanding of the individual reaction" [5, p. 23].

2.1.1 Pain as a Curse

In the Western (Judaeo-Christian) heritage, theologians have been most influential in shaping our modern attitudes towards pain, through connoting it as a moral issue or concern.

From the Biblical writings of the Old Testament, there is no doubt that even if pain is seen as an instrument of divine punishment, it is also a curse [6]. Though both pain and suffering are a part of the fate of mankind, as determined by God, and could result from sin, there is no necessity to seek out pain to achieve atonement. Conversely, as there is no virtue in bodily anguish, it is appropriate to seek means of combating pain [7].

2.1.2 Stoicism

In ancient Stoic philosophy, the wise man is able to apply reason to remain free of passions and perturbations of the mind [8]. In the eyes of some philosophers and poets, he represents the ideal human being [9].

Some of the early Stoics believed it was possible to eliminate pain from one's life through the exercise of correct judgment. However, the later Stoics made it clear that even the most sagacious person may have to endure pain with forbearance [2], as is obviously the case today.

Contemporary versions of Stoic philosophy include ideals of taciturnity and self-sufficiency, principles that can be employed by pain sufferers [10]. They closely align with personal ideologies, values, and behaviours that are commonplace across many industrial nations [11].

2.1.3 Pain as Enlightenment

The Greek philosopher Antisthenes (445–360 BCE), a disciple of Socrates and a founder of the Cynic school of philosophy,[3] encouraged his students to accept willingly the burden of pain that occurs in their soul's search for virtue. They were encouraged to reject all conventional desires for health, wealth, power, fame and to divest themselves of all possessions and property [12].

The philosophy of asceticism, or suffering for the benefit of humanity, is a part of some early Christian perspectives [13] that viewed pain as offering a means of enlightenment. Suffering was then not only a sign of grace but also an attribute eagerly to be sought [6]. In fact, according to Bronkhorst [14] "the ideal martyr is a victor who, in spite of being subjected to the most atrocious tortures, remains courageous and does not flinch."

Ascetism persists in modern society but in forms that do not require adherence to systems of religious belief. Individuals seeking spiritual growth and enlightenment may do so through food abstinence, which may lead to the development of anorexia [15], as well as through sexual abstinence, seclusion from society, and renunciation of possessions [12]. Other forms of ascetism include the production of pain through undertaking exhausting exercises, performing self-laceration, which in extreme circumstances can include self-flagellation and even castration [12].

2.1.4 Pain as Punishment

A return to the concept of pain as punishment can be traced to the teachings of the mediaeval Christian theologian Aurelius Augustinus (354–430 CE), known as St. Augustine, recognised as the founder of the Western spirit [16]. He attributed the evils of mankind to the Biblical Fall in the Garden of Eden, the "original sin":

> Banished (from Paradise) after his sin, Adam bound his offspring also with the penalty of death and damnation, that offspring which by sinning he had corrupted in himself, as in a root; so that whatever progeny was born. . .would drag through the ages the burden of Original Sin. . .wallowing in evil, it was falling headlong from one wickedness to another; and joined to the faction of the angels who had sinned, it was paying the most righteous penalty of its impious treason [16, p. 149].

He imposed a difficult heritage upon Christianity, that of a basic character flaw in man's nature [16]:

> A religion which teaches men and women to regard their humanity as chronically flawed and can alienate them from themselves. Nowhere is this alienation more evident in his denigration of sexuality in general and women in particular [16, p. 149].

[3] By the beginning of the third century BCE, six main philosophical schools of thought had emerged: Platonism, Aristotelianism, Scepticism, Cynicism, Epicureanism and Stoicism.

Augustine's opinion reinforced that of the Christian theologian Tertullian (160–220 CE), who had castigated women as "evil temptresses, an eternal danger to mankind" [16]:

> Do you not know that you are each an Eve? The sentence of God on this sex of yours lives in this age: the guilt must of necessity live too. You are the devil's gateway, you are the unsealer of that forbidden tree, you are the first deserter of the divine law. . .You so carelessly destroyed man, God's image. On account of your desert, even the son of God had to die [16, p. 150].

No doubt influenced by his own early life experiences, Augustine believed that "man is so weak and corruptible that redemption is possible only through Divine Grace, and he maintained this view with uncompromising severity" [17, p. 55]. His influence upon Christian theology has been profound. He set out to confront humans with the "sorry truth of their inner selves" [18, p. 155]. His central conclusions—that human beings are driven by inaccessible instinctual drives, especially sexuality and aggression, which exist in tension with morality and conscious thought—have been so internalized that they are often taken to be common sense [18, p. 155].

The internalisation and repression of these powerful emotional forces, which can be traced back to the theme of original sin, helps to explain "the aching and never satisfied self" and the phenomenon of "social guilt," which are so commonly experienced in our contemporary Western society [18].

2.1.5 Pain as Guilt

Augustine's lasting legacy also extended to the feeling of religious guilt stemming from man disobeying God's command in the Garden of Eden [19]. Generally, the feeling of guilt can be understood as a reaction for having done wrong or having wanted to do wrong to someone else and it provokes expectations of punishment (*ponos*). In psychoanalytic parlance, the guilty person thus looks for punishment; furthermore, their guilt inhibits further aggressiveness and has a paralysing effect [20].

From their study of patients with chronic low back pain, Serbic and Pincus [21] found that guilt was a common experience in this context and there is no reason to believe that it is confined to those in this diagnostic category. Using their "pain-related guilt scale," they found that patients were feeling guilty not only about behaviours that result from the pain experience and that limit their roles and activities but also about failing to recover, and indeed, about experiencing the pain itself. These responses reflect deep-rooted cultural expectations that may have caused these patients to feel guilty for being unable to meet moral values and standards that include being able to obtain an explanation for their pain, being left without a socially acceptable medical diagnosis, having to accept a changed identity, inability to meet the physical and emotional demands of family and work, inability to counter unfair judgements made by colleagues, and failure to be restored to full health following medical treatment [22].

2.1.6 Malingering

Claudius Galen (c130–c210), a towering figure in the history of Western medicine, attributed all pain to the "interruption of continuity," thus implying that pain arises from noxious stimulation causing tissue damage [23]. He easily distinguished patients with "real" pain from those who were faking pain (i.e., malingering) on the grounds that the latter group made every effort to avoid treatment that was perceived to be in any way unpleasant and/or risky to them [24].

The influential Arabic physician Avicenna (980–1037 CE) was aware that pain might persist long after the original noxious stimulus was no longer present. He classed such pain as "not true pain" and cautioned doctors not to treat it because the cause no longer existed [25]. The distinction between "true" and "false" pain thereby assumed great importance in determining the way mediaeval physicians might respond to these patients.

Malingering was an issue of concern for doctors during the nineteenth and early twentieth centuries [26, 27]. Many authorities took a moralistic approach to malingerers in the military services and considered them "reprehensible, corrupt and detestable" due to their bad influence upon their fellow soldiers. Methods of detection included administration of anaesthetic agents, strong galvanic currents, cautery and blistering [27]. Sir John Collie [26], an English medico-legal examiner in the 1920s opined:

> Many deserving cases do not receive proper consideration because unworthy persons have had expended upon the sympathy and help which are the legitimate rights of the true sufferer. . .This work deals with the dark side of the human nature. . . [26, p. v].

Mendelson and Mendelson [28] concluded that, at least in the medico-legal context, there are no valid methods of detecting malingering presenting as pain despite the numerous attempts to prove otherwise.

2.2 *Psychological Connotations of Pain*

2.2.1 Mental Pain

The construct of mental pain can be traced back to the Greek term *lupê* (meaning distress or grief). It is no less real than somatic pain but has not attracted the same amount of attention from researchers [29]. Other terms for the same phenomenon include psychic pain, psychological pain, emptiness, psychalgia, and "psychache." [29] Such pain can be expressed in metaphors borrowed from physical pain, such as "heartache and "broken heart." In his famous book, Man's Search for Meaning published in 1963, Austrian neurologist and psychiatrist Victor Frankl [30] identified "mental pain" as a metaphor for existential frustration, resulting from the sufferers' inability to see meaning in their life. He did not consider them as suffering from a mental disease. Biro [31] pointed out that these people as identified by Frankl [30] are not mentally ill, deceitful or weak and that neuroscientifically somatic pain and psychological pain are indistinguishable [32].

2.2.2 Hysterical Pain

Hippocrates (460–379 BCE) was the first physician to use the term hysteria [33]. He recognised the difference between the compulsive movements of epilepsy, caused by a disorder of the brain, and those of hysteria attributed—as it occurred only in women—to "abnormal movements of the uterus in the body" [33, p. 111]. It was then believed that this affliction derived from poisoning of the uterus by "venous humours due to a lack of orgasms" and the development of "uterine melancholia" [33], or, in other words, a form of *lupê*.

Thomas Sydenham (1624–1689), the great seventeenth century English physician, also described the symptoms of then-called hysteria and recognised that symptoms of hysteria, including pain, may simulate "almost all forms of organic diseases" [17, p. 95]. A contemporary example of this connotation is the proposed connection between fibromyalgia (chronic widespread pain), as a lifelong condition commencing with childhood suffering, and hysteria [34]. However, Merskey [35] considered hysteria to be an unreliable term when applied as a pain diagnosis.

2.2.3 Weak Mind

Sydenham characterised his hysterical patients as wanting in "the firmness of the spirits" and "those to whom nature has given a soft or weak governing of mind, or it has been so rendered by a long series or continuance of disease" [36]. Steered by psychologist Wilfred Fordyce [37], an IASP Task Force in 1995 reconceptualised low back pain that persisted beyond 6 weeks as "activity intolerance" and recommended discontinuance of reimbursement and wage replacement on the grounds that without monetary reward such behaviour would then cease. Although not explicit, it could be inferred that "activity intolerance" was being used as a euphemism for "lack of will power" or a "weakness of the mind." Following her extensive review of the evidence, Gamsa [38] concluded that any psychological disturbance observed in pain patients is more likely to be a consequence than a cause of their experience of chronic pain. She warned against the danger of ascribing psychological causation to those with intractable pain.

2.2.4 "Pain-prone" Personality

Engel [39] proposed the psychoanalytic construct of "pain-prone" individuals who exhibited, in varying degrees, guilt for which pain served as an atonement, a masochistic character structure as evidenced by "the large number of painful injuries, operations, and treatments they had undergone," repressed aggression, conflict over their sexual impulses indicating a "sado-masochistic type of sexual development," and unconscious identification with an actual or fanciful love object. These patients also could fulfil criteria for a number of psychiatric diagnoses, such as "conversion hysteria, depression, hypochondriasis and paranoid schizophrenia, or mixtures of these" [39].

When Szasz [9] outlined his concept of the "painful person," he referred to one whose "humanity is intimately related to, or is wholly dependent upon, his being in pain and suffering." He gave two examples in the Western world: Jesus Christ, who it is said through his suffering rescued mankind from sin and the Virgin Mary who epitomised the suffering, soulful, or painful mother (*Mater Dolorosa*).

After reviewing the evidence claiming that such a personality exists, Naylor et al. [40] rejected this and other psychodynamic formulations of the construct on the grounds that they lacked scientific objectivity and could be confounded by some of the somatic items included in the measuring instrument (i.e., the Minnesota Multiphasic Personality Inventory).

2.2.5 Pain "Somatiser"

The ancient Greeks considered somatic and psychic illnesses as constituting a unity [41]. Mental and psychiatric conditions had no separate status of causality, physiology or regimen of treatment [42]. All were considered authentic illnesses and received validity in the nosology [42]. However, in the Platonic notion, the mind is said to rule over the body, which is but the executer of desires and ideas. Although he was not a physician, Marcus Tullius Cicero (106–43 BCE), a Roman philosopher, boldly stated that bodily ailments could be a result of emotional factors [17].

The Viennese neurologist and psychoanalyst Wilhelm Stekel (1886–1940) adopted this idea and coined the term "somatization" to describe a bodily disorder considered to originate from a deep-seated "neurotic" cause. In this psychodynamic sense, somatisation connotes "an immature defence mechanism" whereby psychic derivatives are "converted" into bodily symptoms [43].

The psychiatrist Lipowski [44] acknowledged the psychoanalytic origins of the concept but viewed somatization also as a maladaptive mode of coping with life stress and expanded the concept by suggesting that the patient incorrectly interprets the symptom (e.g., pain) as signifying serious physical illness. Barsky [45], another psychiatrist, described the "somatizing personality," as characterized by long-standing absorption with one's body and disease, which remains unassuaged by reassurance, negative laboratory results, and conventional symptomatic treatment.

Pain is said to be the commonest presenting complaint of somatising individuals [46]. In this view, pain is seen as the somatic expression of intra-psychic conflict resulting in the tensing of the muscles in a hyper-aroused body [46]. But this is a hypothesis of causation, not a description, which if valid, would allow all those with chronic pain to be so labelled. It would be equally invalid to propose that "somatising" is the commonest feature of individuals with pain [47].

The casualty of this set of connotations is the pain sufferer, who must wear the inherent victim-blaming consequence of the label of "somatisation," which in fact only reflects the medical observer's "psychologisation" of their clinical problem.

2.2.6 Pain "Catastrophiser"

The word "catastrophe" is derived from the Latin *catastrophe* and the Greek *katastrophe*, both of which can signify "a reversal or overturning of what is expected." In modern usage, the word has come to connote a "disaster." The earliest written example of catastrophic thinking, which appears in the cuneiform writing on a clay tablet from Nippur describing the prayer of a king's daughter in Babylon: "Pain has seized my body. May God tear this pain out" [13, p. 8].

Sullivan et al. [48] conceptualized catastrophising as a maladaptive cognitive style and used the term in reference to "an exaggerated negative orientation towards noxious stimuli." Quartana et al. [49] were more cautious and suggested that "pain catastrophizing" may not be a unique psychological construct "above and beyond negative affectivity in general, or with regard to other negative pain-related cognitive process variables, such as fear of pain."

2.3 Somatic Connotations of Pain (the "Biological" in "Socio-Psycho-Biological")

2.3.1 From Object Intrusion to Nociception

In many ancient cultures, when no evidence of bodily injury could be detected, the cause of pain was commonly attributed to a bodily intrusion by demons or divine spirits [23]. For example, the biblical Job attributed his pain to the "arrows of the almighty within me."(Job 6.4) Democritus (460–362 BCE) postulated the existence of "sensitive soul atoms" throughout the body, thereby foreshadowing the theory of nociception [23, p. 20]. He also viewed pain as an intrusion object, "the sharp hooked irregularity of which stimulates the body by invading its vessels" or *poroi* (i.e., its openings) [23, p. 19].

Our current understanding of the free nerve endings that form a plexus under the skin is largely due to the work of Charles Sherrington [50], the famous neurological scientist, in the early years of the twentieth century. He concluded from his animal experimental research: "There is considerable evidence that the skin is provided with a set of nerve-endings whose specific office it is to be amenable to stimuli that do the skin injury, stimuli that in continuing to act would injure it still further" [50, p. 40]. Because they responded to noxious stimuli, he called these endings "nocicipient." (he later called them "nociceptors").

Despite the insights from the study of nociception, the ancient belief in forms of "object intrusion" as responsible for pain can be seen today in some techniques of alternative medicine, where "noxious materials" are said to be expelled from the body through mechanical means, such as cupping [51].

2.3.2 Humoral Imbalance

According to the Hippocratic corpus (c 390 BCE), pain is felt when one of the four humoral elements (i.e., blood, phlegm, yellow bile and black bile) is either deficient or present in excess. Pain was then connoted as a lack of harmony that required a restoration of the humoral balance [3]. The body reacted to this alteration and attempted to correct the imbalance by mixing (physically and chemically blending) the constituents in order to recreate a perfect fusion of all the humours without any excesses remaining [52]. For 1600 years, physicians and physiologists remained bound in the shackles of this incomprehensible theory of the four cardinal humours, which had never been shown to exist [53]. But the idea lived on and is found in the concept of homeostasis, a term coined by Walter Cannon [54] for processes that living organisms use to actively maintain fairly stable conditions necessary for survival.

2.3.3 From Passions to Emotions

The word "emotion" first became established as a category of mental states during the nineteenth century [8]. Prior to this, numerous unsuccessful attempts had been made to distinguish "troubling desires and passions," both of which had biblical connotations, from the morally neutral "milder affections and sentiments" [8]. The Platonic philosophical tradition regarded "emotion" as the enemy of reason, and hence of judgement, truth and morality [55]. However, reflecting shifts in intellectual authority across cultures, the words "passions" and "emotions" have been assigned different roles in the production of mental experiences and psychological theories [8]. In the early-nineteenth century, Thomas Brown, the Scottish philosopher, subsumed the "appetites," "passions," and "affections" under a single category: the "emotions" [8]. Another Scottish philosopher, Charles Bell argued that an emotion was a movement of the mind producing certain changes such as grief, joy or astonishment, which could become visible through "outward signs" on the face or body. In other words, emotion was a "thing" with agentive properties [56]. By the middle of the nineteenth century, Bain [57] offered this definition: "Emotion is the name here used to comprehend all that is understood by feelings, states of feeling, pleasures, pains, passions, sentiments, affections" [57, pp. 4–5].

In line with Bell's theory, William James [58] defined the various emotions as vivid mental feelings of visceral changes brought about directly by the perception of some object in the world. In other words, he proposed that emotions are based upon an evaluation of the state of the body in response to a pertinent stimulus [59]. However, critics raised a number of issues, including failure to differentiate between the different emotions and to distinguish emotions from non-emotions, and denial of the role played by cognitive factors in generating emotions. Finally, in response to criticism by Irons [60], who argued that emotions were irreducible attitudes of the whole person, James was forced to virtually retract his own theory [8]. In any event, James' theory was necessarily limited by the absence of any concept of what constitutes "feelings" at a cellular level, so that in effect his proposal that "emotions are based upon an evaluation of the state of the body" was—and is—untestable.

Notwithstanding that problem, Craig [59, 61] proposed a fusion of James's speculative concept of emotion as a consequence of bodily change (of the entire body and not just the viscera) and the concept of homeostasis. He conceptualised emotions as mechanisms through which the integrity of the body (self) can be maintained in the face of conditions that threaten to disturb its homeostasis. He considered pain to be one of many possible such emotions, including temperature, itch, visceral distension, muscle ache, hunger, thirst, "air hunger" and sensual touch.

Craig [59] cited neurobiological findings suggesting that pain is not part of the exteroceptive somatosensory system but rather "is represented in an unforeseen, novel pathway in humans that is part of a hierarchical system of interoception subserving homeostasis, the sense of the physiological condition of the body (interoception) and the subjective awareness of feelings and emotion." In other words, "pain is a specific emotion that reflects homeostatic behavioral drive, similar to temperature, itch, hunger and thirst" [61]. The specificity of this disturbance of homeostasis is attributed to signalling through the summated activation of nociceptors; that is, by tissue damage.

Craig's formulation can be seen as an attempt to unify (old) specificity theory with (new) convergence theory, favouring the former [62]. However, it is not made clear what distinguishes pain from the other "homeostatic" emotions. Furthermore, this proposed "fusion" needs to be seen as both teleological—towards a desirable outcome—and open-ended, as it does not define what is successful "homeostatic behavioral drive." In summary, emotions are seen here to be interoceptive responses to bodily need and reflect patterns of activity in the brain that, in some unexplained way, are said to correspond with social, cognitive, or fictive (i.e., imaginary) circumstances [63].

2.3.4 From Affect to "Affective Primes"

By contrast, Panksepp [64] did not attempt to answer the perennial question "What is an emotion?" Instead, his studies concerned the affective (i.e., subjective feelings or *qualia*) component of emotion in animals, which he saw as being grounded on "motivational-behavioural" neural networks (in subcortical regions) that evolved long before humans. Using a metaphor borrowed from Information Technology, he conceptualised affects as "evolutionarily conserved core executive routines guiding the living mind" [65].

Panksepp described seven "affective primes" perhaps functioning as nodes in a foundational distributed integrative system. They are held to be responsible for a "rich interweaving of affective signals into every aspect of conscious life—from what we find interesting, to what we avoid, what we crave and seek out, and from what we learn forever to what we forget next week—indeed, virtually every aspect of the mind" [64]. In this context, pain is considered to reflect a basic aversion to tissue damage and is a component of the FEAR (*sic*) response, a forward-looking extension of homeostasis and organism preservation.

According to Panksepp's connotation, pain is but one component of a primary biological process and becomes an irreducible experience, as was postulated by Irons [60]. In his formulation, which is teleological,[4] [66] secondary processes are said to include the learning and memory mechanisms (circuitries) of the brain (situated in the basal ganglia), and the tertiary processes to include higher order (i.e., neocortical) cognitive functions that allow many thought-related symbolic functions [67]. However, linking a brain area to a specific behaviour is not, in and of itself, evidence that the area produces a phenomenological experience.

3 Conclusion

When the many contemporary connotations of pain are viewed in a socio-psycho-biological framework, it can be seen that the fluidity in meaning enjoyed by the ancient Greek philosophers has been perpetuated and expanded. They are evidence of the figurative language that people can seize upon in order to overcome some of the obstacles they face when attempting to communicate their pain and other unpleasant sensations [68]. Through the influential opinions of mediaeval theologians, moral judgements became unavoidable intrusions into the clinical engagement between clinicians and pain sufferers [69]. Since that time, negative connotations of pain and of pain sufferers have continued to exert a needless influence on the attitudes of both parties. By contrast, positive connotations that might assist both those experiencing pain and those who seek to minister to them seem to be absent. Exposure of these connotations as negative is the first step towards removing them from the socio-cultural landscape and thereby reducing the burden of stigma borne by many pain sufferers. Plausible biologically-based connotations are now emerging as options with which to replace them, but their scientific validity remains to be determined.

This analysis is necessarily neither complete nor comprehensive. It is focused upon literature that is relevant to Western culture through a Graeco-Judaeo-Christian lens. It has shown that when connotations of pain are dissected an array of distinctive but different states is revealed, all invested with negativity. If nothing else, this dissertation emphasises the importance of developing a universally applicable definition of pain that is simultaneously phenomenologically accurate, biologically sound, morally neutral and universally applicable.

[4]Teleology, originating with Aristotle [66], is the study of the ends or purposes that things serve.

Acknowledgement We gratefully acknowledge the advice and many contributions from Professor Brian Griffiths.

References

1. Gunzburg R. Poena, the goddess of divine retribution. Eur Spine J. 2015;24:415–6.
2. Garrett JE. Is the sage free from pain? Volga J Philos Soc Sci. 1999;6:1–12.
3. Santoro D, Bellinghieri G, Savica V. Development of the concept of pain in history. J Nephrol. 2011;24(S17):S133–6.
4. Carr DB, Bradshaw YS. Time to flip the pain curriculum. Anesthesiology. 2014;120(1):12–4.
5. Hardy JD, Wolff HG, Goodell H. Pain Sensations and Reactions. Baltimore: Williams and Wilkins Company, 1952. 435 p.
6. Jakobovits I. Jewish Medical Ethics. New York: Bloch Publishing; 1975. 465 p.
7. Low JF. Religious orientation and pain management. Am J Occup Ther. 1997;51(3):215–9.
8. Dixon T. Emotion: the history of a keyword in crisis. Emot Rev. 2012;4(4):338–44.
9. Szasz T. Pain and pleasure: a study of bodily feelings. 2nd ed. Syracuse, NY: Syracuse University Press; 1988. 303 p.
10. Edwards CH. The suffering body: philosophy and pain in Seneca's letters. In: Porter JI, editor. Constructions of the body in classical antiquity. Ann Arbor: Michigan UP; 1999. p. 252–68.
11. Pathak EB, Wieten S, Wheldon CW. Stoic beliefs and health: development and preliminary validation of the Pathak-Wieten stoicism ideology scale. BMJ Open. 2017;7:e015137.
12. Deezia BS. Ascetism: a match towards the absolute. IAFOR J Ethics Relig Philos. 2017;3(2):85–98.
13. Fülöp-Miller R. Triumph over pain (trans. Eden and Cedar Paul). London: Hamish Hamilton, 1938. 438 p.
14. Bronkhorst J. Ascetism, religion and biological evolution. Method Theory Stud Relig. 2001;13:374–418.
15. Chadbourne BC. Ascetism in the modern world: the religion of self-deprivation. Inquiries J. 2014;6(3).
16. Armstrong K. A history of god. From Abraham to the present: the 4000-year quest for god. London: Vintage Books; 1999. 460 p.
17. Alexander FG, Selesnick ST. The history of psychiatry: an evaluation of psychiatric thought and practice from prehistoric times to the present. London: George Allen and Unwin; 1967. 471 p.
18. Boyce J. Original sin and the making of the western world. Melbourne: Black Inc; 2014. 208 p.
19. Pratt JB. The ethics of St Augustine. Ethics. 1903;13(2):222–35.
20. Alexander F. Remarks about the relation of inferiority feelings to guilt feelings. Int J Psychoanal. 1938;19:41–9.
21. Serbic D, Pincus T. The relationship between pain, disability, guilt and acceptance in low back pain: a mediation analysis. J Behav Med. 2017;40:651–8.
22. Toye F, Seers K, Allcock N, Briggs M, Carr E, Andrews J, et al. A meta-ethnography of patients' experience of chronic non-malignant musculoskeletal pain. Health Serv Deliv Res. 2013;1(12).
23. Keele KD. Anatomies of pain. Springfield: Charles C Thomas, Publisher; 1957. p. 16–40.
24. Finger S. Origins of neuroscience: a history of explorations of brain function. Oxford: Oxford University Press; 2001. p. 148–64.
25. Tashani OA, Johnson MI. Avicenna's concept of pain. Libyan J Med. 2010;5:1–4.
26. Collie J. Fraud in medico-legal practice. London: Edward Arnold; 1932. 276 p.

27. Connor H. The use of anesthesia to diagnose malingering in the 19th century. J R Soc Med. 2001;99(9):444–7.
28. Mendelson G, Mendelson D. Malingering pain in the medicolegal context. Clin J Pain. 2004;20 (6):423–34.
29. Tossani E. The concept of mental pain. Psychother Psychosom. 2013;82:67–73.
30. Frankl VE. Man's search for meaning. New York: First Washington Square Press; 1963. 160 p.
31. Biro D. Is there such a thing as psychological pain? And why it matters. Cult Med Psychiatry. 2010;34:658–67.
32. Eisenberger NI. The neural basis of social pain: evidence for shared representations with physical pain. Psychosom Med. 2012;74(2):126–35.
33. Tasca C, Rapetti M, Carta MG, Fadda B. Women and hysteria in the history of mental health. Clin Pract Epidemiol Ment Health. 2012;8:110–9.
34. Villalpando MIB, Sotres JFC, Manning HG, Gonzalez AA. Fibromyalgia: a functional somatic symptom or a new way to conceptualize hysteria? Salud Mental. 2005;28(6):41–450.
35. Merskey H. Psychiatry and pain: causes, effects, and complications. In: Merskey H, Loeser JD, editors. The paths of pain 1975–2005. Seattle: IASP Press; 2005. p. 421–31.
36. Pearce JMS. Sydenham on hysteria. Eur Neurol. 2016;76:175–81.
37. Fordyce WE. Back pain in the workplace: management of disability in non-specific conditions. Task force on pain in the workplace. Seattle, Washington: IASP Press; 1995. 75 p.
38. Gamsa A. Is emotional disturbance a precipitator or a consequence of chronic pain? Pain. 1990;42(2):183–95.
39. Engel GL. Psychogenic pain and the pain-prone patient. Am J Med. 1959;26(6):899–918.
40. Naylor B, Boag S, Gustin SM. New evidence for a pain personality? A critical review of the last 120 years of pain and personality. Scand J Pain. 2017;17:58–67.
41. Fornaro M, Clementi N, Fornaro P. Medicine in psychiatry in Western culture: ancient Greek myths and modern practices. Ann General Psychiatry. 2009;8(21):1–8.
42. Fabrega H. The concept of somatization as a cultural and historical product of Western medicine. Psychosom Med. 1990;52:653–72.
43. Mai F. Somatization disorder: a practical review. Can J Psychiatr. 2004;49(10):652–62.
44. Lipowski ZJ. Somatization: the experience and communication of psychological distress as somatic symptoms. Psychother Psychosom. 1987;47:160–7.
45. Barsky AJ. Amplification, somatization, and the somatoform disorders. Psychosomatics. 1992;33:28–34.
46. Merskey H. Pain disorder, hysteria or somatization? Pain Res Manag. 2004;9(2):67–71.
47. Quintner JL, Cohen ML. Response to Gerald Aronoff ("Myofascial pain syndrome and fibromyalgia: a critical assessment and alternate view"). Clin J Pain. 1999;15:155–7.
48. Sullivan MJ, Bishop SR, Pivik J. The pain catastrophizing scale: development and validation. Psychol Assess. 1995;7(4):524–32.
49. Quartana PJ, Campbell CM, Edwards RR. Pain catastrophizing: a critical review. Expert Rev Neurother. 2009;9:745–58.
50. Sherrington CS. Qualitative difference of spinal reflex corresponding with qualitative difference of cutaneous stimulus. J Physiol. 1903;30(1):39–46.
51. Kim J-I, Lee MS, Lee DH, Boddy K, Ernst E. Cupping for treating pain: a systematic review. Evid Based Complement Alternat Med. 2011;2011:467014.
52. Hart G. Descriptions of blood and blood disorders before the advent of laboratory studies. Br J Haematol. 2001;115:719–28.
53. Richet C. An address on ancient humorism and modern humorism. Br Med J. 1910;2 (2596):921–6.
54. Cannon W. The wisdom of the body. New York: WW Norton; 1932. 693 p.
55. Gray JA, Porter R. Emotion. In: Bullock A, Trombley S, editors. The new fontana dictionary of modern Thought. 3rd ed. London: Harper Collins Publishers; 2000. p. 266–8.
56. Bell C. Essays on the anatomy and philosophy of expression. London: John Murray; 1824. 280 p.

57. Bain A. The emotions and the will. London: John W Parker & Son; 1859. 605 p.
58. James W. What is an emotion? Mind. 1884;9:188–205.
59. Craig AD. How do you feel? Interoception: the sense of the physiological condition of the body. Nat Rev Neurosci. 2002;3(8):655–66.
60. Irons D. The psychology of ethics. Edinburgh: W Blackwood & Sons; 1903. 176 p.
61. Craig AD. A new view of pain as a homeostatic emotion. Trends Neurosci. 2003;26(6):303–7.
62. Quintner JL, Buchanan D, Cohen ML, Katz J, Williamson O. Pain medicine and its models: helping or hindering. Pain Med. 2008;9:824–34.
63. Strigo IA, Craig AD. Interoception, homeostatic emotions and sympathovagal balance. Philos Trans R Soc B. 2016;371:20160010.
64. Panksepp J. The basic emotional circuits of mammalian brains: do animals have affective lives? Neurosci Biobehav Rev. 2011;35:1791–804.
65. Watt DF. Reflections on the neuroscientific legacy of Jaak Panksepp (1943–2017). Neuropsychoanalysis. 2017;19(2):183–98.
66. Johnson MR. Aristotle on teleology. Oxford: Clarendon Press; 2005. 352 p.
67. Alcaro A, Carta S, Panksepp J. The affective core of the self: a neuro-archetypical perspective on the foundations of human (and animal) subjectivity. Front Psychol. 2017;8:1424.
68. Bourke J. Pain: metaphor, body, and culture in Anglo-American societies between the eighteenth and twentieth centuries. Rethink Hist. 2014;18(4):475–98.
69. Scheurich N. Moral attitudes and mental disorders. Hast Cent Rep. 2002;32(2):14–21.

Chapter 13
Is "Chronic Pain" a Meaningful Diagnosis?

Milton Cohen and John Quintner

Abstract Introduction "Chronic pain" and "chronic primary pain" in particular are both about to appear in the international taxonomy, ICD-11. While this is likely to have great utility at administrative and possibly clinical levels, the question arises as to whether people experiencing chronic pain also will benefit.
Methods The definitions (denotations) of "chronic pain" as symptom, disease and diagnosis are refracted through the constraints of language, especially figures of speech, in order to ascertain if "meaning" is possible.
Results "Chronic (primary) pain" as a taxonomic entity remains tied to the definitional link of "pain" itself to actual or apprehended tissue damage. This situation does not assist the plight of those experiencers of pain in whom tissue damage cannot be demonstrated.
Conclusion While the definition of pain itself is constrained by the link to tissue damage, there can be no satisfactory denotation of "chronic pain" for those who live with that experience. As such, meaning remains elusive.
Clinical Implications "Chronic pain" and "chronic primary pain" in particular are about to appear in the international taxonomy, ICD-11. While this is a welcome development towards the recognition of chronic pain as a societal and clinical problem, with positive diagnostic, therapeutic and administrative implications, the question also arises, whether this "entity" carries meaning for those who experience it. This article argues that, while pain itself is defined as being linked to tissue damage, there can be no satisfactory denotation of "chronic pain" for those who live with that experience. As such, meaning remains elusive.

Keywords Chronic pain · Taxonomy · Meaning

M. Cohen (✉)
St Vincent's Clinic & UNSW, Darlinghurst, NSW, Australia
e-mail: M.Cohen@unsw.edu.au

J. Quintner
Arthritis and Osteoporosis WA, Shenton Park, WA, Australia

S. van Rysewyk (ed.), *Meanings of Pain*,
https://doi.org/10.1007/978-3-030-24154-4_13

1 Introduction

There can be no doubt that a human condition called "chronic pain" exists. Yet, its fundamental nature continues to resist description, dissection and even definition. However, what is meant by a "condition" and how might that be different from a "disorder," a "disease" or a "diagnosis"? This vagueness has perplexed both the experiencer and the observer of "chronic pain," leading to the stigmatisation of one, the frustration of the other and the unhappiness of both. For the purpose of this chapter, a "medical condition" will be defined simply as an existing state of ill health or unwellness.

In the upcoming revision of the International Classification of Diseases, 11th edition (ICD-11), chronic pain, defined as pain that has persisted for 3 months, is accorded an entry as a taxonomic entity and therefore as a *de facto* diagnostic label in its own right. In developing this classification of chronic pain as an identifiable condition, the joint International Association of Pain (IASP)/World Health Organisation (WHO) Task Force allocated the many and diverse examples of that condition into seven subcategories [1].

In accordance with WHO criteria, the task force attempted to give first priority to aetiology of pain, followed by underlying pathophysiological mechanism, and finally the body site in which pain is experienced. Also employed was the WHO principle of "multiple parenting," which allows any one example to be included under more than one category. This was particularly apt as clinically, even when clearly associated with another diagnosis, chronic pain itself can be as important a source of suffering and a therapeutic challenge as any underlying disease process.

The ICD-11 classification identified six clusters of "secondary" chronic pain:

- chronic cancer-related pain [categorised by aetiology]
- chronic postsurgical and posttraumatic pain [categorised by aetiology]
- chronic neuropathic pain [categorised by mechanism]
- chronic secondary headache or orofacial pain [categorised by body site]
- chronic secondary visceral pain [categorised by body site]
- chronic secondary musculoskeletal pain [categorised by body site]

This process exposed the challenge of categorising chronic pain that cannot be associated with or attributed to another condition, thereby logically attracting the label "chronic primary pain," which was defined as:

> ...chronic pain in one or more anatomic regions that persists or recurs for longer than 3 months and is associated with significant emotional distress (e.g., anxiety, anger, frustration, or depressed mood) and/or significant functional disability (interference in activities of daily life and participation in social roles) and that cannot be better accounted for by another chronic pain condition [2].

In stark contrast, however, to the entities subsumed under "chronic secondary pain," in which the label "chronic pain" is comorbid with another diagnosis, the label "chronic primary pain" raises the pain itself to the level of a diagnostic entity. In other words, the symptom—chronic pain—becomes both the condition and the diagnostic entity. Given the blurring between symptom, condition and diagnosis that attends "chronic pain," the question arises whether the term conveys meaning. This issue turns on the definition of pain itself.

2 The Denotation[1] of "Chronic Pain"

"Pain" was defined in 1979 by the IASP as "a sensory and emotional experience associated with tissue damage or described in terms of such damage." The significant operational concepts here are firstly pain defined as an "experience," to indicate its intrinsically subjective nature, which itself entails a challenge to adequate description in language [4], and secondly the link to "tissue damage."

Although this definition is said to have captured the essence of the pain experience [5], it has come under recent challenge, notably from Williams and Craig [6], who consider that other dimensions of pain should be made explicit and prefer:

> A distressing experience associated with actual or potential tissue damage with sensory, emotional, cognitive and social components;

and from Cohen, Quintner and van Rysewyk [4], who among other things consider that the link to tissue damage is problematic and offer:

> A mutually recognisable somatic experience that reflects a person's apprehension of threat to their bodily or existential integrity.

Historically, chronic pain has been variously conceptualised as pain lasting or recurring for more than 3–6 months [7] or as that which persists past normal healing time [8]. An implicit presumption of the latter conception is that an "injury"—or episode of acute tissue damage—is a necessary prerequisite for chronic pain, as described in the fundamental IASP definition. Any temporal qualifier of "pain," such as "chronic," cannot do other than perpetuate that fundamental definition including the link with tissue damage. It follows that the denotation of "chronic primary pain" in fact comprises four elements:

1. pain: the experience is actually or is considered to be associated with tissue damage;
2. chronic: pain that has persisted for 3 months;
3. primary: "cannot be better accounted for by another chronic pain condition," which can be taken also to mean "tissue damage has eluded detection to date" or "continues to be associated with tissue damage;"
4. "associated with significant emotional distress and/or significant functional disability."

---

[1] John Stuart Mill [3] defined a non-connotative term (i.e., a denotation) as one which signifies a subject only, or an attribute, whereas a connotative term is one which denotes a subject and implies an attribute.

As element (4) also occurs in instances of "chronic secondary pain," the pathognomonic feature of "chronic primary pain" is element (3), which can be seen also to be fundamentally constrained by the denotational link of "pain" to "tissue damage." The question arises, does this definitional link of "chronic pain" to "tissue damage" convey meaning?

3 The Meaning of Chronic Pain

3.1 The "Problem" of Language

Language is the medium of conferring meaning. This speaks to the importance of communication between patient and clinician, where difficulties can arise due to the discordance between the personal experience of illness and the scientific understanding of it, and to the impossibility of defining a limit as to what should be communicated [9].

Medical language plays a strong role in the description of illness: a simple change in terminology can result in different interpretations of a known condition in terms of its seriousness and potential implications [10], while vagueness in terminology in situations of uncertainty can amplify patient distress [11].

Underlying this conundrum is the issue of "meaning," a word that challenges itself. Two broad types of meaning have been identified: denotative, which refers to how a word, a sign or sentence is to be decoded; and existential, which refers to meanings of life or for living [12]. Even in this chapter, which is necessarily limited to the denotative meaning of the word "pain," some symbols have different meanings depending on the context, and "pain" is one such symbol. To quote Baumeister and Landau [12]:

> . . .symbols point to other symbols in networks, indeed ones that require *a community of brains that can create shared understandings*. The connections that constitute those networks are social conventions, laws and the like. Hence a symbol's meaning inheres in its position within a network rather than its resemblance to physical things [12, p. 2].

Furthermore, to paraphrase Baumeister and Landau's next sentence, the meaning of "pain" within a network of symbols is independent of any specific physical facts (such as nociception, neural pathways or brain activation).

In the current discussion, the main "symbol" is language. It has long been appreciated that the meaning of language is not to be found in dictionaries, which consist of a list of substituted symbols: "This can be substituted for that in such and such circumstances" [13, p. 207]. However, formal attempts to define the term "meaning" have resulted in confusion, as evidenced by a "widely divergent set of opinions in the writings of the best philosophers" [13, p. 248].

Although it has been suggested that the human mind is motivated to share thoughts and ideas with others, and that people use meaning through the vehicle of language to connect socially with other people [14], the question of what linguistic symbols actually *mean* needs to be further explored. This has been referred to as the "aboutness" of language [15, p. 89].

Philosopher Ludwig Wittgenstein [14] took a bold step by defining the meaning of a word as the way it is used in the language (§43). In his earlier work, he had conceived of language as mediating between thought and the reality that it is about. Employing a novel analogy of the function of language in communication, he wrote, "Uttering a word is like striking a note on the keyboard of our imagination" [14, §6]. However, in his later work, Wittgenstein moved away from seeing meanings as mental states—pictures, descriptions, or images of thoughts, objects or actions "in the mind"—but rather as being enshrined in the words themselves that are being used, that are always linked to personal experiences as actions and outcomes involved in social practices:

> When I think in language, there aren't 'meanings' going through my mind in addition to the verbal expressions; the language is itself the vehicle of thought [14, §329].

That is to say, Wittgenstein demoted imagination or mental projection of relevant scenarios (German: *Vorstellung*) to only an auxiliary role in generating meaning as a fundamental ingredient of language.

Contemporary philosopher Horst Ruthrof [15] argues that Wittgenstein's highly influential definition of linguistic meaning as "use" can only be a partial account of what actually occurs. Ruthrof deprecates Wittgenstein's demotion in importance of mental states such as imagination (German: *Vorstellung*) [15], recalling his reference to "image-mongery" (*Vorstellerei*), which "is of no interest to us" [14, §390]. According to Ruthrof [15], Wittgenstein seems to be torn between foregrounding *Vorstellung* as a methodological device and dismissing it as a legitimate component of meaning, since "nothing is more wrong-headed than calling meaning a mental activity!" [14, §693].

By contrast, as Ruthrof points out, "at the very heart of his [Wittgenstein's] philosophising, readers are invited to conjure up and participate in variations of possible scenarios" [15, p. 72] of word uses that can guide understanding what a word means. In support of his argument, Ruthrof [15] points to the utility of phrases such as "We can imagine...," "It is easy to imagine...," "Think how...," or "Suppose...." Luis-Flores [16], another philosopher, also does not regard as trivial the role of *Vorstellung* in Wittgenstein's *Philosophical Investigations* [14]. From his viewpoint, "it has to be understood as a practice (*Praxis*), as a behaviour (*Verhalten*), and as an activity (*Handlung*)" [14, p. 218]. Ruthrof attempts to meld Wittgenstein's insistence on "let the use teach you the meaning" [17, p. 181] with a promotion of *Vorstellung* back to be an important ingredient of language:

> If you are able to imagine what I am talking about and the way I am saying it, then there is meaning; if not, there is not. And vice versa, if I am able to imagine what others are talking about and the way they do so, an event of linguistic meaning has occurred: if not, it has not [17, p. 1].

It follows that in order to use language meaningfully, "members of a speech community must be able to share, to a high degree, the way in which language and nonverbal readings are to be associated with one another" [17, p. vi]. Nonverbal readings include the activation of linguistic expressions by various stimuli—visual, aural, tactile, olfactory, gustatory—such as eye contact, tone of voice, facial expression, gestures, and body posture, collectively known as "body language," itself a set of symbols that also reflect social conventions.

When concluding his detailed exegesis of Wittgenstein's approach, Ruthrof suggests "…the search for meaning in interpretive use of language turns out to be a search for what language is *about*, or its extensional connection. What is at stake in defining meaning as "use" is the process between language and its aboutness" [15, p. 89]. In this view language, especially if novel, acts as a set of instructions for exploring pathways towards different possible meanings, and is therefore an important tool (instrument) to negotiate meanings in everyday discourse [18].

From this analysis, which recalls the importance of context and "a community of brains" as quoted above, it might be concluded that determining meaning in a clinical context is also ultimately subject to negotiation between patient and clinician. This is especially so with a pain experience, which can be communicated only through metaphor and simile, as well as by "body language." Provided both parties are familiar with the same community sanctioned figures of speech or symbols, they may agree upon a common meaning.

Ruthrof [19] introduced the term "sufficient semiosis" when referring to the situation when there is community agreement on sufficient common ground for communication to take place. It explains how ordinary communication is possible without the necessity of invoking truth conditions:

> Sufficient semiosis is the principle of a chain of mental representations and reasoning in support of synthetic judgments. The point at which such a chain is regarded as sufficient is determined not logically but only pragmatically, when a community is satisfied that the judgment is well grounded in shared social experience [19, p. 49].

However, in order to successfully negotiate linguistic meaning, Ruthrof [20] also sees the necessity for a set of rules to determine what language is *about*. These include "rules governing well-formed linguistic expressions, rules for a well-formed domain that can function as the goal of linguistic *aboutness* and rules for combining the right clusters from both domains."

The following sections explore insights gained from medical epistemology that may facilitate patient-clinician negotiation of certain word meanings. In this context, the set of rules are those recognised by biomedicine with its focus on making a diagnosis; that is, establishing the presence of a disease or disease process. The rules are about arriving at a diagnosis through the pattern of sequential data acquisition, creation of the mental abstraction or "problem representation," generation of hypotheses, and the search for and selection of illness scripts [21].

3.2 *Can "Chronic Pain" Be Denoted as a Symptom?*

The term "symptom" is derived from the Greek expression *sumptoma*, conceptualised as an occurrence or happening, but also as an implied prediction or portent as to what might befall that person [22]. Symptoms are experiential reports presented to the clinician by the patient in the course of the medical consultation [23, 24]. As undeniable "natural facts," symptoms constitute core components of the diagnostic process [25] and can be regarded as cultural and communicative acts. In this case, they can be seen as an attempt by the patient to enable the clinician to experience something akin to their own experience [26], which is the essence of empathy [27].

However, as symptoms may be impossible to convey by literal language, they are frequently expressed as metaphors or similes [28]. Metaphors for pain include such descriptors as "cutting," "burning," "stabbing," and "piercing." Common similes that are used include chest pain described as like "a band or tightness around the chest," headache as "a pounding within my skull," and abdominal pain as "something is gripping me." The clinician is then charged with the responsibility of trying to understand the meaning of the symptom when thus conveyed [28]. In a multicultural context, the potential for misunderstanding complicates this responsibility [29].

Traditionally, symptoms have been interpreted as indicators of disease [30]. As has been argued above, "chronic secondary pain," although now elevated in taxonomy to a clinical problem in its own right, can still be seen as a symptom of an underlying disease process. However, that interpretation is challenged when there is no evidence of underlying disease, as in "chronic primary pain" (element (3) above). Clearly, "chronic primary pain" cannot be a symptom of itself, so from the point of view of the experiencer, there is no fixed framework of reference—or denotation—that might facilitate communication.

3.3 *Can Chronic Pain Be Denoted as a Disease?*

Chronic pain has been framed as a distinct medical condition. Bonica [31] regarded intractable pain as "a destructive force that often imposes severe emotional, physical, economic, and sociological stresses on the patient as well as his family as well as on society." Other conceptualisations of chronic pain have included "pain without apparent biological value," which contains a teleological if not also moral judgment, and "persistent pain that is as a rule not amenable to treatments based upon specific remedies" [7], which begs the question of "specific remedies" and almost justifies therapeutic nihilism.

These arguments signaled a major conceptual shift in pain medicine, from viewing chronic pain as a presenting symptom of disease to an entity-in-itself or a physical "thing" possessing agency; that is, an instrument able to exert causal power or elicit other effects. There is a tension therefore between pain defined as an experience and pain conceptualised in clinical medicine as a disease, which is clearly a "thing" [32]. Dorland's Medical Dictionary, 31st edition [33], offers this definition of disease: "any deviation from or interruption of the normal structure or function of any body part, organ, or system that is manifested by a characteristic set of symptoms and signs and whose etiology, pathology, and prognosis may be known or unknown."

Despite the insistence on the presence of "a characteristic set of symptoms and signs," this and similar dictionary denotations fail to convey that the concept of "any deviation from or interruption of the normal structure or function of any body part, organ, or system" is in fact a high-level abstraction or stipulation that is largely consensus-driven *within* the medical community [34, 35]. In this light, those who argue chronic pain is a disease in its own right seek to superimpose a well-trodden map of biomedicine over a conceptually complex landscape, which is experienced, lived and felt by the person with pain [36]. A contemporary example of imposing a "biomedical consensus" is the condition labelled as "myofascial pain," which is used as a surrogate for an as yet unknown but presumed disease process [37].

The proposition by Siddall and Cousins [38] of pain-as-a-disease arose out of their dissatisfaction with pain being classed only as a nonspecific symptom of disease, taken together with an undue deference being paid to the biomedical imperative to identify and treat any underlying disease. Having reviewed the evidence to date on the physiology of nociception and how that might be perturbed, they concluded: "Persistent pain has its own pathology, symptoms and signs and is therefore a disease entity" [38]. However, as argued by Cohen et al. [36], this assertion raises the question, which "characteristic" clinical features apart from the complaint of pain constitute the "disease" of pain? These authors argue that, for chronic pain to be considered a disease, "it is not sufficient to invoke a syndrome or a grouping of symptoms of which pain might be a component, especially when the diversity of conditions associated with pain is considered." They conclude that on neither pathological nor clinical criteria can pain-as-a-disease be justified.

3.4 Can Chronic Pain Be Denoted as a Diagnosis?

If chronic pain is not a disease, it follows that "chronic primary pain" is also not a disease. The question arises whether it can nonetheless be denoted as a diagnosis. Ever since the latter's emergence as a scientific discipline in Ancient Greece, diagnosis has played a pivotal role in Medicine [39]. In modern times, diagnosis has become inextricably related to specific disease formulations that serve simultaneously to legitimise both the patient's experience and the physician's status in society as a healer [40]. Diagnosis, from the Greek *dia* denoting "through" and *gnosis*, denoting "knowledge" [36], originally referred to the process of identifying in the body underlying causal associations considered responsible for the patient's complaints and then conferring an "appropriate" label [41]. In its simplest form, medical diagnosis consists of linking symptoms reported by the patient and a lesion observed in the tissue (*the clinicopathologic inference*) [41].

In addition to identifying the steps to be taken for such identification, diagnosis has become connoted as implying a deeper inherent knowledge, if not also certainty [31]. Despite themselves lacking definite physical properties (mass, velocity, chemical composition, etc.) and therefore being detached from the physical world of things, medical diagnoses through their meaning to the patient can exert real effects on that world [12].

Medical diagnoses tend to be concrete concepts, simultaneously implying understanding of a condition and the therapeutic pathway to be followed. Diagnostic labels thus not only become facts (or "things") in the real world but also form the basis for ways of structuring reality through particular interpretations, conjectures and hypotheses. The knowledge implied by a diagnostic label is utilised in two ways. Firstly, it provides a potential source of meaning for the patient and thus the basis for narratives and stories that will have meaning to others whose voices also need to be heard and acknowledged, such as family members, society, ethicists, insurance companies and others [42]. As expressed by Jutel: [43]

> From the patient's perspective, the diagnosis and the metaphors employed to explain that diagnosis should clarify the patient's story of symptoms and help to render his or her experiential world comprehensible [38, p. 942].

Secondly, by structuring the approach of the physician to the problem, the knowledge implicit in the diagnostic label "underlies the socially indispensable tasks of diagnosis and prognosis and the rationalization of therapeutic practice" [35]. But do these principles apply to chronic pain? Does any denotation of the term "clarify the patient's story" or, for that matter, "help to render. . .comprehensible" the world of the patient or of the physician? And does that term introduce rational therapy? Following the rules or procedures for making a diagnosis, as laid out by Bowen [21], chronic pain would form part of the process of data collection, but illness scripts would default to those associated with tissue damage, according to the IASP definition of pain. So, it may be concluded that a paradox has been reached, in which chronic pain itself is conceptualised as an experience but not a physical "thing," while the use of that same term as a diagnostic label, which is a "thing" that has agentive properties, identifies only one interpretation, which does not capture that experience.

This reflects an aspect of pain that has been rarely discussed, that pain is an example of *aporia*. From the Greek meaning "without a path, a passage or a way," an aporia in philosophy is a puzzle or an apparently insoluble impasse in a discussion. It may also refer to a state of being perplexed when confronting such an impasse. As Quintner et al. [32] wrote:

> Yet, as clinicians, it is necessary to engage the aporia of pain because of our ethical obligation to the person (as "*the other*") in pain. We do know some things about pain: it is not an illness, not a disease, not an emotion, not a memory, but rather all these things although not susceptible to any one such reduction. By contrast, the only persons who expertly know pain are the persons "in pain"; yet, their efforts to communicate such expertise is limited by the lack of language to express their pain [32, p. 830].

4 Conclusion: Is "Chronic Pain" a Meaningful Diagnosis?

Chronic pain is now appreciated through a socio-psycho-biological framework [44]. However, a satisfactory and concise diagnostic formulation to embrace these three dimensions has not yet been devised. Most medical conditions tend to be classified in terms of disturbed anatomy and/or physiology and/or biochemistry, to the exclusion of psychological, social and cultural factors. Because pain is experienced in the body, its classification has tended to be expressed in somatic terminologies in accordance with the clinicopathological method [45]. The ICD-11 formulation of chronic (primary) pain can be seen as an attempt to include common social and psychological dimensions of pain, but it cannot escape the primacy of the somatic or biomedical dimension of pain. Perhaps this is inevitable while the definition of pain itself is linked to "tissue damage," both formally and in figures of speech. That "chronic pain" and "chronic primary pain" are to be established in ICD-11 is undoubtedly a boon for all "observers" of that experience, including clinicians, researchers, funders, administrators and policy-makers. But do these taxonomic terms have meaning for the "experiencers" of pain?

To invoke Ruthrof's theory of how language can achieve meaning through use and imaginability (i.e., sufficient semiosis), do these terms imply agreement within a language community sufficient for ordinary communication to take place without the necessity of endless explication invoking formal arguments where true premises guarantee true conclusions? This may well be the case for the discourse between "observers" of pain but not for the "experiencers." For them, promoting chronic pain as a disease or a diagnosis fails to capture denotational meaning let alone existential meaning which, as suggested by Baumeister and Landau [12], involves purpose, value, mattering, continuity, and coherence. As has been argued in this chapter, each conceptualisation of chronic pain is traditionally premised upon the necessary existence of somatic pathology, which is determined independently of context and/or how the experience is for the patient. This construal implies that to be diagnosed with "chronic pain" may in fact not be meaningful to the person with that experience.

In order to overcome this impasse, pain itself must be redefined to free that experience from any specific physical fact [12] on the one hand and, on the other, to recognise the social context in which it occurs, a context that is mediated in verbal beings primarily through non-physical symbols endowed with meaning. Contemporary redefinitions of pain, as proposed by Williams and Craig [6], and Cohen, Quintner and van Rysewyk [4], represent two options to stimulate discussion of this important issue.

This analysis of whether "chronic pain" has meaning has in fact exposed the conjunction of two imponderables: the meaning of "meaning" and the "aporia" of pain. In seeking to limit "meaning" to its denotative (or definitional) usage, any argument is constrained by the definitional linkage of "pain" to tissue damage, which presents the problem of finding the words to describe the experience when tissue damage is unknown or unknowable. Another constraint is the differing perspectives of the "experiencer" and "observer" of pain, especially when both parties reside in "a community of brains that can create shared understandings."

The strength of this analysis lies in the attempt to refract different clinical conceptualisations of "pain"—as symptom, disease or diagnosis—through theorising on the meanings of medical language. This chapter can be seen as an exploratory attempt by clinicians to help other clinicians navigate the tortuous path that they and their patients with pain confront daily. The limitation of the analysis lies in attempting to be concise when discussing these two imponderables at a theoretical level, although, to paraphrase Wittgenstein, in pondering the imponderable, we endlessly trace an outline around the frame through which we view things.

References

1. Treede RD, Rief W, Barke A, Aziz Q, Bennett MI, Benoliel R, et al. A classification of chronic pain for ICD-11. Pain. 2015;156(6):1003–7.
2. Nicholas M, Vlaeyen JWS, Rief W, Barke A, Aziz Q, Benoliel R, et al. The IASP taskforce for the classification of chronic pain. The IASP classification of chronic pain for ICD-11: chronic primary. Pain. 2019;160(1):28–37.
3. Mill JS. A system of logic, ratiocinative and inductive. In: Robson JM, editor. The collected works of John Stuart Mill, vol. 2. London: University of Toronto Press; 1843. p. 5.
4. Cohen M, Quintner J, van Rysewyk S. Reconsidering the IASP definition of pain. Pain Rep. 2018;3(2):e634.
5. Derbyshire SWG. The IASP definition captures the essence of the pain experience. Pain Forum. 1999;8(2):106–9.
6. Williams ACC, Craig KD. Updating the definition of pain. Pain. 2016;157:2420–3.
7. Merskey H, Bogduk N, editors. Classification of chronic pain: descriptions of chronic pain syndromes and definition of pain terms. 2nd ed. Seattle: IASP Press; 1994. 222 p.
8. Bonica JJ. The management of pain. Philadelphia: Lea and Febirger; 1953. 592 p.
9. Tallis R. Hippocratic oaths: medicine and its discontents. London: Atlantic Books; 2004. p. 43–73.
10. Young ME, Brookes L, Norman GR. Found in translation: the impact of familiar symptom descriptions on diagnosis in novices. Med Educ. 2007;14:1146–51.
11. Peters A, McEwan BS, Friston K. Uncertainty and stress: why it causes diseases and how it is mastered by the brain. Prog Neurobiol. 2017;156:164–88. https://doi.org/10.1016/j.pneurobio.2017.05.004.
12. Baumeister RE, Landau MJ. Finding the meaning of meaning: emerging insights from four grand questions. Rev Gen Psychol. 2018;22(1):1–10.
13. Ogden CK, Richards IA, editors. The meaning of meaning. 8th ed. New York: Harcourt, Brace and World, Inc.; 1946. 363 p.
14. Wittgenstein L. Philosophical investigations. Anscombe GEM, Hacker PMS, Schulte J, translators. Oxford: Blackwell; 2009. 246 p.
15. Ruthrof H. Language, Vorstellung, and meaning as use. Linguist Philos Invest. 2011;10:60–92.
16. Luis-Flores H. The imagination's piano in Wittgenstein's Philosophische Untersuchungen. In: Wittgenstein and the future of philosophy. A reassessment after 30 years. Papers of the 24th international Wittgenstein-symposium. Kirchberg am Wechsel: Austrian Ludwig Wittgenstein Society; 2001. p. 218–23.
17. Jackendoff R. A user's guide to thought and meaning. Oxford: Oxford University Press; 2012. 274 p.
18. Ruthrof H. The body in language, vol. 7. London: Cassell; 2000. 193 p.

19. Ruthrof R. Semantics and the body: meaning from Frege to the postmodern. Toronto: University of Toronto Press; 1997. 321 p.
20. Ruthrof H. Language and imaginability. Newcastle Upon Tyne: Cambridge Scholar Publishing; 2014. 270 p.
21. Bowen JL. Educational strategies to promote clinical diagnostic reasoning. NEJM. 2006;355: 2217–25.
22. Temkin O. The Double Face of Janus. Baltimore: The Johns Hopkins University Press, 1977. p. 419–40.
23. King LS. Medical thinking: a historical preface. Princeton: Princeton University Press; 1982. p. 91–104.
24. Nessa J. About signs and symptoms: can semiotics expand the view of clinical medicine? Theor Med Bioethics. 1996;17:363–77.
25. Eriksen TE, Risør MB. What is called symptom? Med Health Care Phil. 2014;17(1):89–102.
26. Semino E. Descriptions of pain, metaphor and embodied simulation. Metaphor Symbol. 2010; 25(4):205–26.
27. Stein E. On the problem of empathy. Stein W, transl. Washington: ICS Publications; 1989. 135 p.
28. Periyakoil VS. Using metaphors in medicine. J Palliat Care. 2008;11(6):842–4.
29. Masukume G, Zumla A. Analogies and metaphors in clinical medicine. Clin Med. 2012;12: 55–6.
30. Miller FG, Colloca L. Semiotics and the placebo effect. Perspect Biol Med. 2010;53:509–16.
31. Bonica JJ. Introduction to the first world congress on pain. In: Bonica JJ, Albe-Fessard DG, editors. Advances in pain research and therapy, vol. 1. New York: Raven Press; 1976. p. xxvii–xix.
32. Quintner JL, Buchanan D, Cohen ML, Katz J, Williamson O. Pain medicine and its models: helping or hindering. Pain Med. 2008;9:824–34.
33. Dorland W. Dorland's illustrated medical dictionary. 31st ed. Philadelphia: WB Saunders Company; 2007. 535 p.
34. Rosenberg CE. The tyranny of diagnosis: specific entities and individual experience. Milbank Q. 2002;80:237–60.
35. Body R, Foex B. On the philosophy of diagnosis: is doing more good than harm better than "primum non nocere"? Emerg Med J. 2009;26:238–40.
36. Cohen ML, Quintner JL, Buchanan DA. Is chronic pain a disease? Pain Med. 2013;14:1284–8.
37. Quintner J, Bove G, Cohen M. A critical evaluation of the "trigger point" phenomenon. Rheumatology. 2015;54:392–9.
38. Siddall PJ, Cousins MJ. Persistent pain as a disease entity: implications for clinical management. Anesth Analg. 2004;99:510–20.
39. Kleisiaris CF, Sfakianakis C, Papathanasious IV. Health care practices in ancient Greece: the hippocratic ideal. J Med Ethics Hist Med. 2014;7:6.
40. Crookshank FG. The importance of a theory of signs and a critique of language in the study of Medicine. In: Ogden CK, Richards IA, editors. The meaning of meaning. 8th ed. New York: Harcourt, Brace and World; 1946. p. 337–55.
41. Bichat X. Anatomie générale appliquée à la physiologie et à medicine, vol. 3. Paris: Broson, Gabon et Cie; 1825.
42. Hanne M. Diagnosis and metaphor. Perspect Biol Med. 2015;58:35–52.
43. Jutel A. Sociology of diagnosis: a preliminary review. Sociol Health Ill. 2009;31:278–99.
44. Carr DB, Bradshaw YS. Time to flip the pain curriculum? Anesthesiol. 2014;120:12–4.
45. Sullivan M. The problem of pain in the clinicopathological method. Clin J Pain. 1998;14(3): 197–201.

Chapter 14
The Meaning of Pain Expressions and Pain Communication

Emma Borg, Nathaniel Hansen, and Tim Salomons

Abstract Both patients and clinicians frequently report problems around communicating and assessing pain. Patients express dissatisfaction with their doctors and doctors often find exchanges with chronic pain patients difficult and frustrating. This chapter thus asks how we could improve pain communication and thereby enhance outcomes for chronic pain patients. We argue that improving matters will require a better appreciation of the complex meaning of pain terms and of the variability and flexibility in how individuals think about pain.

We start by examining the various accounts of the meaning of pain terms that have been suggested within philosophy and suggest that, while each of the accounts captures something important about our use of pain terms, none is completely satisfactory. We propose that pain terms should be viewed as communicating complex meanings, which may change across different communicative contexts, and this in turn suggests that we should view our ordinary thought about pain as similarly complex. We then sketch what a view taking seriously this variability in meaning and thought might look like, which we call the "polyeidic" view. According to this view, individuals tacitly occupy divergent stances across a range of different dimensions of pain, with one agent, for instance, thinking of pain in a much more "body-centric" kind of way, while another thinks of pain in a much more "mind-centric" way. The polyeidic view attempts to expand the multidimensionality recognised in, e.g., biopsychosocial models in two directions: first, it holds that the standard triumvirate—dividing sensory/cognitive/affective factors—needs to be enriched in order to capture important distinctions *within* the social and psychological dimensions. Second, the polyeidic view attempts to explain (at least in part) *why* modulation of experience by these social and psychological factors is possible in the

E. Borg (✉) · N. Hansen
Reading Centre for Cognition Research, Department of Philosophy, University of Reading, Reading, UK
e-mail: e.g.n.borg@reading.ac.uk

T. Salomons
Department of Psychology, Queen's University, Kingston, ON, Canada

S. van Rysewyk (ed.), *Meanings of Pain*,
https://doi.org/10.1007/978-3-030-24154-4_14

first place. It does so by arguing that because the folk concept of pain is complex, different weightings of the different parts of the concept can modulate pain experience in a variety of ways. Finally, we argue that adopting a polyeidic approach to the meaning of pain would have a range of measurable clinical outcomes.

Clinical Implications: First, by making a subject's tacit beliefs about pain explicit it will be possible to create a more open, shared space for pain communication (particularly between clinicians and patients) and support a move away from purely quantitative measures of pain towards more discursive pain narratives. Secondly, the polyeidic view might provide a mechanism for predicting who will do well or badly from cognitive interventions for pain management, allowing more efficient use of healthcare resources. Finally, the polyeidic approach might also contribute to the creation of more nuanced cognitive interventions by elucidating the pre-conscious beliefs that influence a subject's experience of pain.

Keywords Polyeidic theory · Chronic pain · Communication of pain · Meaning of pain terms

1 Introduction

In her 1926 essay "On being ill" Virginia Woolf laments the poverty of our language for pain—"let a sufferer try to describe a pain in his head to a doctor and language at once runs dry," she notes [1]. Looking at studies of patient-doctor exchanges about pain, particularly those dealing with chronic pain, it seems that Woolf's worry is born out.[1] Both patients and clinicians frequently report problems around communicating and assessing pain, with patients expressing dissatisfaction with their doctors and doctors often finding exchanges with chronic pain patients difficult and frustrating [2–5]. Yet we know that positive patient-clinician interaction matters to both parties and that a patient's sense that they are being listened to can increase their overall sense of well-being, as well as promoting adherence to lifestyle changes and medical interventions that lead to reduced levels of experienced pain [6–8]. So, how could we go about improving pain communication, and thereby enhance quality of life, particularly for chronic pain patients? This chapter explores that question by reflecting on what might be learned from philosophical accounts of the meaning of pain terms, seeing how these views impact on practical issues around pain communication and shed light on a newer model of how to think about pain communication (one that we hope might deliver concrete clinical improvements).

Joanna Bourke, in a 2014 New York Times article, describes being in hospital and telling a friend that her pain is "beyond words," only to be reminded by her

[1]Scarry E. *The Body in Pain: the making and unmaking of the world.* [New York: Oxford University Press. 1985. 400 p.] provides perhaps the most well-known contemporary defence of the view that pain cannot be captured in language.

friend that she has been talking about her suffering for the past hour [9]. Perhaps, her friend empathetically notes, the problem isn't that people can't speak about pain but rather that witnesses refuse to hear. We want to propose a mid-ground between Bourke's remark and her friend's perceptive response: the problem with pain communication and assessment, we suggest, is not that pain is entirely beyond words, nor is it that hearers simply refuse to listen. Rather it is that speakers and hearers need to be aligned in how they are thinking about the multiple different dimensions of pain in order to really hear what one another says.[2]

2 Philosophical Accounts of the Meaning of Pain Expressions

In most areas of discourse, it seems that for communication to take place a hearer must understand what a speaker means by her terms. For instance, if I say "I am going to the bank" intending to mean that I am going to the riverbank, but you take me to have said that I am going to a financial institution, it seems clear that communication has failed. Thus, in this section, we consider the question of what it would mean for two people to mean the same thing by their pain terms. Philosophers of language have suggested a number of different possible criteria for judging that two expressions have the same meaning and we survey three common proposals that may be made for pain terms: sameness of reference, sameness of associated descriptive or cognitive content (what philosophers often call "sense"), and sameness of affective or experiential effects (what philosophers sometimes call "expressivist" content).

2.1 Sameness of Reference

Some terms in a language (those we might label "descriptive phrases") seem to serve to pick out objects via the properties those objects possess. So, for instance, a description like "the current president" picks out a person just in case they have the property of being the current president and the expression seems able to pick out different people at different times or places. (e.g., selecting Barack Obama if uttered

[2]A word of clarification on the discussion that follows: although we will often phrase things in terms of "the meaning of pain terms," what we are really interested in in this chapter is pain communication. For those familiar with philosophical distinctions, our interest is in pragmatic content (the complete, context-dependent message a speaker conveys by her utterance) rather than purely semantic content (the literal meaning of words and phrases). Thus, we don't intend to take a stand here on whether the distinctions we draw are ones that are ultimately best modelled as part of the semantics of pain expressions or are rather part of the pragmatics of pain communication. Although this is a crucial question, we don't have space to pursue it here, so simply set it to one side.

in America in 2015, or Olusegun Obasanjo if uttered in Nigeria in 2000) On the other hand, some expressions, like proper names or what philosophers call indexical or demonstrative expressions (terms like "I," "she" or "that girl"), seem (at least on some of their uses) to pick out things in the world independent of those objects' descriptive profiles. So, the name "Barack Obama" picks out a particular individual whether or not he is now, or indeed ever became, US President, and it hangs on to that individual regardless of any other changes in his properties (such as where he lives or who he works for, etc.). John Stuart Mill (better known for his creation of the ethical theory "Utilitarianism") argued that the meaning of one of these latter types of expressions—what we will call a "referring term"—is exhausted by the object to which it attaches. Since, on the Millian view, there is supposed to be nothing more to the meaning of a referring term than the object to which it refers, two referring expressions which refer to the same object *must* have the same meaning.

So, one option for pain terms would be to claim that they are simple referential expressions—they label the pains to which they refer—and thus two pain terms mean the same just in case they both refer to one and the same pain.[3] A patient and a clinician could thus be assured at least the basic starting point of successful communication as long as the pain terms they use in their conversation do in fact coincide in reference, i.e., so long as they both pick out the same pain object. However, this simple view faces objections from two different directions: first, it is unclear that pain expressions really can be treated as genuine referring terms, and second, the Millian view of proper names itself faces some serious challenges. In what follows, we consider only the first of these worries.[4]

Thinking about the nature of pain first, if we are to treat pain terms as referring expressions this imposes on us a view about what pains are—they must be things, capable of being referred to from both the first-person perspective and the third person perspective (i.e., the patient must be able to refer to a discrete pain that she feels, while the clinician must be able to refer to one and the same pain, even though she does not feel it). This way of thinking about pains is certainly codified in many of the linguistic forms used to talk about pain; so, for instance, I might say that "I have a headache" or that "My backache is really bad today." These possessive forms of linguistic expression treat pains as objects (compare "I have a potato")—ordinary language does, it seems, treat headaches and backaches as things on a par with ordinary, physical objects, things which individuals stand in possession relations to.

However, despite its intuitive appeal, reflection shows that this simple referential view of pain also faces some potential problems. First, looking at the grammar of our language, it is unclear whether we should treat pain terms as count nouns (like

[3]Bourke, J. *The Story of Pain.* [Oxford: Oxford University Press. 2014. 416 p.] rejects this reifying model, where pain is conceptualised as an entity that can be referred to. Instead she argues for an adverbial approach where pain expressions qualify verbs; as she writes (2014: 7) "pain is not an intrinsic quality of raw sensation; it is a way of perceiving an experience". See also Tye M. Pain and the Adverbial theory. Am Phil Quart 1984;4:319–327.

[4]For an overview of the problems with the Millian view per se, see the entry on "Names" in the Stanford Encyclopaedia of Philosophy (https://plato.stanford.edu/entries/names/).

"dog") or mass terms (like "water"), for both uses seem possible. A count noun is a term where we can give a numerical answer to the question "How many F's are there?"—so we can ask "How many dogs are there in the park?" but "How many waters are there in the glass?" seems ill-formed. For pain, then, we can speak of "A pain in the foot" or say, "I have one pain in my foot and another in my thigh," where "pain" clearly operates as a count noun. However, we also say things like "There is pain around here" or "I've been having pain all night" where "pain" seems to operate as a mass term [10 p. 12]. More problematic still, some expressions treat pain as a process rather than as an object. For instance, instead of "I have a pain in my ankle," I could say "My ankle hurts" or "I have a sore ankle," yet these linguistic forms make "ankle" the thing being referred to, with "hurting" a process this object is undergoing, or soreness a property this object has. In English, speakers seem quite happy to switch between these two alternative modes of expression, even in the process of describing a single episode of pain [10]. While in some other languages it seems that process constructions are favoured over object uses ([11] Chap. 6). So, looking just to our language, it seems that there is evidence for different stances on the kind of referential expressions pain terms might be: count nouns, mass terms or labels for processes.

Second, even if we treat pain terms as referring to discrete objects, we still need to know what kind of objects those might be. Given the common assumption that to have a pain is to have a certain kind of sensation, we might intuitively think that a given utterance of a pain term refers to a particular episode of that sensation, however this then raises the question of how clinicians, when talking with patients in pain, manage to refer to sensations which the patient, but not they, are currently having. This raises the fundamental dichotomy of pain talk: the contrast between the private (an internal experiential state) and the public (third-person discussions of pain), and this contrast formed the basis of a prolonged attack on the referential view of pain terms from the philosopher Ludwig Wittgenstein. In his seminal book *Philosophical Investigations*, Wittgenstein sketched a number of arguments that purport to show that we cannot in fact treat pain terms as referring to internal, private sensations [12].

For instance, Wittgenstein asks us to consider a man who keeps a diary in order to record his sensations. On day one the man has a sensation of pain which he decides to label "S," and he thus writes "S" down in his diary to record this occurrence. A few days later, the man has another sensation and he is now faced with the question of whether he should label this sensation "S" as well or introduce a new name "S∗" for it. What the man needs to decide is whether his current sensation is the same as, or different from, the one that he experienced earlier, but Wittgenstein argues that there is simply no criterion by which the man can be said to be going right or wrong here. Imagine that the man decides the current sensation does feel the same as the last one and so he labels it "S" again. There is no possibility of overruling the man here, we must judge that he acts correctly. On the other hand, imagine that the man decides the sensation feels different from the previous one, and so he labels it "S∗." Here again it seems we have no option but to accept what he does as correct—the subject is the sole arbiter of whether the two sensations feel the same to him. Yet this seems to entail that there is no genuine criterion of correctness available in this case—whatever the man decides to do is right. Compare this with the man who learns that the vehicle he is looking at is called "a bus." If he later sees a train and decides it is the same kind of thing as the vehicle he saw earlier and so calls it "a bus," we can judge that he has made a mistake. There is a public consensus as to what counts as going on correctly with a term like "bus" and this doesn't include applying it to trains. If the subject is the only person who can decide how he should go on, this means that there is simply no objective criterion of right or wrong when it comes to labelling internal experiential states. Yet in the absence of such a criterion Wittgenstein suggests that the whole practice of naming breaks down. Naming, he suggests, depends on a shared practice, where it is possible to criticise someone for getting the practice right or wrong, and if this is absent then the very act of naming itself is meaningless. As he writes [12, 92e, 257]:

> When one says "He gave a name to his sensation" one forgets that a great deal of stage setting in the language is presupposed if the mere act of naming is to make sense. And when we speak of someone's having given a name to pain, what is presupposed is the existence of the grammar of the word "pain"; it shows the post where the new word is stationed.

According to Wittgenstein, if pain terms named private sensations they would constitute what he calls a "Private Language"—a language made up of words where only the speaker has access to what those words mean—but, he argues, such a language is not in fact possible.[5]

If we accept Wittgenstein's worries, one option would be to preserve the simple referential view of pain terms but argue that they refer not to some hidden mental state but to something more observable, such as public pain behaviour. This would be to adopt a "behaviourist" model for the meaning of pain terms, whereby saying that someone is in pain means simply that there has been a certain kind of environmental stimuli (e.g., an event causing tissue damage) and that the subject is now displaying certain kinds of behaviour (e.g., they rub the affected limb, say that they are in pain, etc.). The behaviourist picture does seem to capture something intuitively correct about the meaning of pain and pain terms, for ordinary folk do often take appropriate pain behaviour as criterial in attributing pain to someone else (for instance, expressing preference for nonverbal behaviour over verbal behaviour when judging/interpreting the credibility of pain displays [13]. Furthermore, as some authors have recently noted, a public or social aspect to pain (recognition of which lies behind a move towards this kind of behaviourist model) may indeed be fundamental to pain [14, 15]. However, although such an approach would avoid Wittgenstein's worries about privacy (and indeed some people have read Wittgenstein as favouring a behaviourist model, although he himself seems clearly to reject

[5]Echoes of Wittgenstein's worry can also be found in Elaine Scarry's rejection of the referential model for pain terms. As she writes in *The Body in Pain: the making and unmaking of the world* [New York: Oxford University Press. 1985, p. 162]: "[P]ain is not 'of' or 'for' anything—it is itself alone. This objectless-ness, the complete absence of referential content, almost prevents it from being rendered in the language."

such an attribution), still the behaviourist model seems far from satisfactory. For a start, we are willing to allow that someone can be in pain without displaying the appropriate kinds of behaviour; for instance, someone may be stoically not crying out, etc., or they could be paralysed in such a way that displaying pain behaviour is not possible for them. Yet—as the Note accompanying the IASP definition of pain makes clear—the lack of pain behaviour in these cases doesn't force us to conclude that the subjects are not experiencing pain. On the other hand, we also allow that someone can engage in appropriate pain behaviour without actually being in pain: a footballer might roll around on the ground in apparent agony simply in order to get his opponent booked. So, while appropriate behaviour is a good guide for attributions of pain, it seems wrong to think that the meaning of pain expressions can be given simply by reference to this behaviour.[6]

As noted above, there seem to be two common elements to paradigm cases of pain: pain sensations and bodily damage. Thus, a final candidate for the referential objects of pain expressions might be bodily injuries themselves. Pain is most commonly experienced in the context of an injury, or potential injury, to the body making tissue damage or nociception the most intuitive "ground truth" in terms of making an attribution about pain.[7] However the problem with treating injury as the referential object of pain expressions is what Melzack and Wall termed the "variable link" between pain and injury [16]. In cases such as episodic analgesia or congenital insensitivity to pain, severe injury can occur without any pain. Conversely, phantom limb pain occurs without injury to the site at which the pain is experienced. Furthermore, many chronic pain disorders occur without any apparent injury or clear pathology. While we cannot, in these cases, rule out the possibility that peripheral pathology exists undetected, such cases demonstrate the difficulty of using injury as the referential object for reported pain.[8]

To summarise, the referential model of pain terms does indeed seem to capture an intuitive truth about our use of pain expressions: pain terms are often used in a paradigm referential manner and subjects do seem naturally to conceptualise pain in this referential way. However, evidence from natural language also supports treating

---

[6]Perhaps a better option then would be to reject the view of sensation states upon which Wittgenstein's objections are premised, whereby they are essentially private, hidden states (i.e., that they are what philosophers would term "Cartesian mental states," from Descartes theory of mind). We won't explore this option in what follows but note that this may be the kind of move Wittgenstein himself favoured.

[7]As the IASP definition of pain notes (https://www.iasp-pain.org/Education/Content.aspx?ItemNumber=1698#Pain), pain is commonly thought of as an experience "associated with actual or potential tissue damage, or described in terms of such damage". Though, as Aydede M. Defending the IASP definition of pain [Monist 2017; 4: 439–464] points out, this latter point is probably better phrased as being *describable* in terms of such damage, rather than requiring actual description in these terms.

[8]Furthermore, recent work shows that, in certain kinds of hypothetical scenarios, people are willing to use pain terms to describe a target individual even when it is stipulated that the target has not undergone any relevant injury; see Borg E, Harrison R, Stazicker J, Salomons T. Is the folk concept of pain polyeidic? Mind Lang. 2019. Online first: https://doi.org/10.1111/mila.12227.

pain terms in a variety of different referential ways (as count nouns, as mass terms, and as process expressions) and as having a variety of different potential referents (pain sensations, pain behaviour, and bodily injury). It seems that we might expect a truly satisfactory account of pain terms to capture this potential variety of meaning.

2.2 *Sameness of Sense*

Whereas Mill thought that the meaning of a name was wholly given by the object to which it referred, Gottlob Frege, the nineteenth century philosopher and logician, held that the meaning of a name is given (at least in part) by its descriptive content. So, for instance, take the names "Cary Grant" and "Archie Leach": although these two names coincide in reference (they pick out one and the same man), they seem to differ in meaning as they pick out that person in different ways or via different properties. (e.g., "Cary Grant" picks out a famous film-star, but "Archie Leach" picks out someone non-famous, perhaps thought of simply as the brother of John Leach) Frege's insight was to realise that someone who believed that "Cary Grant was a famous film star" might perfectly rationally not believe that "Archie Leach was a famous film star" (Frege's example concerned Hesperus and Phosphorous), yet this seems to show that there must be more to the names' meaning than just the objects to which they refer, since if reference was all there was to the names, it would be hard to see how someone could understand both names and yet fail to realise that they refer to one and the same object. Names, then, Frege suggested come with or abbreviate some kind of descriptive content. Frege labelled this the name's "sense." On this model, what is required for sameness of meaning is sameness of sense: it is not (or at least, not just) that two token terms must pick out the very same thing in the world, the way that they pick out that thing must match. With regard to pain terms, this would mean that two people must be thinking of a state in the same way—under the same description—in order for them to converse successfully about pain.

This view apparently commits us to the idea that there is a unique quality which all pains have in common and which we can use to describe them (so that pains have a unique *painful* way of feeling). While intuitively this claim seems extremely plausible, we should note that it, too, is not without problems. First, one could query whether there really is an underlying phenomenological identity between, say, a stabbing pain in the stomach and a dull ache in the head (this recognition—of the vast range of sensations we are willing to class as pains—is sometimes known as the *heterogeneity problem*).[9] Although it is undeniable that we do classify quite distinct experiences under the shared label "pain," one might wonder whether this is due to

[9]As Clark (p.184) puts it: "what is the sensory resemblance between the intense freezing pain of an almost frozen foot and the diffuse hot pain of a sunburned back?" [Clark, A. Painfulness is not a quale. In: Aydede M, editor. Pain: New Essays on Its Nature and the Methodology of Its Study. Cambridge, MA: MIT Press; 2005; pp. 177–198.]

all the experiences having a common, shared phenomenology, instead of some other (potentially higher-level) form of commonality. More problematic, however, is the worry that positing a unique phenomenal quality shared by all and only pains seems to lead to something of an explanatory cul-de-sac, for there is little more which could be said to explain or make clear this special experiential sense. This is exemplified by the use of the term "algosity" by Field to label pain's characteristic phenomenology [17]. Despite the clear benefits of identifying the experiential source of our judgements about what is and isn't painful, the term never went into wide use. The most likely explanation for this lack of traction is that the term didn't describe what the characteristic quality "was," only what it "was not": essentially algosity was defined as the negative sensory and affective quality that separates pain from other sensory experiences that are experienced as unpleasant (e.g., itch and dysthesia). The term essentially functioned as a conceptual placeholder for Fields' argument that sensory and affective qualities of pain are inseparable and that there is an affective quality inherent in the (so-called) "sensory-discriminative" dimension of pain that is, in fact, pain's defining feature. This pedagogical utility, however, did not translate into scientific or clinical utility, as it did not provide any guideline for identifying or isolating this phenomenological quality in a way that would allow us to compare or contrast the algosity present in, for example, two different pain experiences. We are thus left with the view that two token pain terms mean the same because they both express the same ineffable sense, and we might feel that this hasn't taken us very far down the road of explaining the meaning of pain terms. Again, then, although the view that, for pain terms, sameness of meaning means sameness of sense captures something intuitively compelling about our use of pain terms, it is unclear that this can give us the full or complete story.

2.3 *Expressivist Model*

Finally, perhaps the notion of sameness of meaning is special for pain terms because they perform a special role in our language, neither referring to objects nor describing them, but expressing our response to a sensation. In this way, saying "It hurts" or "I'm in pain" would be more like a cry or a groan, or saying "ow!" In this way two speakers might mean the same by their pain terms if they both expressed the same kind of basic affective response to a stimulus. This sort of expressivist view was canvassed (though rejected) by Wittgenstein [12, §244]:

> [H]ow does a human being learn the meaning of the names of sensations? Here is one possibility: words are connected with the primitive, the natural, expressions of the sensation and used in their place. A child has hurt himself and he cries; and then adults talk to him and teach him exclamations and, later, sentences. They teach the child new pain-behaviour.

However, this expressivist view seems problematic for a number of reasons. First, saying "I am in pain" is an articulate use of language, the expressions making up the utterance belong to the general domain of English and can be combined with other words of English to make different, meaningful grammatical structures. For instance, "I am in pain" can be a premise in a valid logical inference (which requires it to express a truth-evaluable claim), but exclamations of pain cannot be used in this way (compare "I am in pain therefore someone is in pain" with "Ow! Therefore someone is in pain"). Furthermore, saying "I'm in pain," or using richer language to describe one's experiential states (saying "This is a searing pain" or "There is a persistent ache at this point in my lower back") is a voluntary linguistic action (a speech act), unlike more primitive interjections. Finally, the model of pain terms as simple expressions of an affective response also seems to fit badly with Melzack's recognition that more intense pains are described with more words—clearly to make sense of this phenomena it seems that the words must be contributing semantic content (or meaning) to the utterances in which they occur [18, p. 201].

As noted above then, it seems that each of these philosophical ideas captures something correct about how we use pain terms. Perhaps unsurprisingly, each account gains its credibility from answering to some intuitively correct dimension of our use of pain expressions. Thus, an account of pain communication which ignores any of these accounts does so at the risk of missing some genuine feature of our use of pain terms. It follows, though, that treating any one of them as giving us the complete, privileged account of the meaning of pain terms will also lead to serious problems. Treating all pain terms as simple referring expressions, which pick out discrete, countable objects, means ignoring the very many uses of pain terms which don't fit this model (e.g., where pain terms refer to mass qualities, or to processes). Treating all pain terms as having their meaning given by some ineffable private sensation requires us ignoring the social dimension of pain and leads to an explanatory dead-end. Treating pain terms as akin to grunts or cries requires ignoring the genuine semantic contribution that can be made by pain terms. What we want to suggest, then, is that to improve matters here we should adopt a model which permits all the kinds of variability in meaning sketched above, recognising that each of the proposals answer to just one element amongst a multitude of important dimensions people recognise in their thought and talk about pain. What we need is a model of meaning for pain terms and pain communication which is capable of recognising that pain talk is multidimensional and that what matters for successful communication is recognising which of the many dimensions of pain are to the fore (and in what way or to what degree). In the next section, we want to provide an initial sketch of what one such a model might look like, before turning in §§3–4 to explore some of the potential clinical advantages of adopting this kind of approach.

3 The Polyeidic Approach to Pain Communication

Part of the problem with the philosophical views canvassed above, we claim, is that they assume too one-dimensional a view of pain communication, seeing speakers as *either* referring to a hidden internal state *or* referring to some process bodily parts can undergo *or* as describing a unique phenomenal character *or* doing something akin to grunting and crying, etc. We would like to suggest that a better model would be one that allows that people allude to different dimensions of pain in their communicative acts and that communication goes better when both parties recognise the dimensions in play and agree on their relative importance. Thus, we want to explore what we term a "polyeidic" (or "many-ideas") view.

The polyeidic account draws inspiration from three places: first, some work in philosophy which recognises a degree of complexity to bodily sensations in general; second, the common recognition in psychology that pain is a multidimensional phenomenon; third, the recognition that pain terms are scalar (that is to say, pain terms allow us to rank pains with respect to one another; for instance, we can say "My black eye is more painful than my broken nose," or "This hurts more than that"). Taking these points in order: some philosophers have recognised that sensations in general may have different dimensions or aspects. Thus, it seems that a token experience of a sensation has a sensory or experiential component (how it feels), and an affective or emotional component (how that phenomenological component is perceived by the subject), it may have a motivational force and it will have a duration and cause, and be influenced by context and by individual beliefs or background, etc. So, for instance, Brogaard holds that "bodily sensations are partial descriptions of emotions and tactile experiences or other events of the body" [19 p. 1]; see also Merlau-Ponty's 1958 criticism, of the idea that experience or bodily sensation could be treated as simple, instead of shot-through with meaning [20].[10] On this kind of view, then, pain, as an archetypal sensation, should also be expected to have some kind of internal complexity. Furthermore, within psychology, the specific idea that pain is a multidimensional phenomenon, involving sensory and emotional experience, modulated by psychological, social and contextual factors, is well-rehearsed (e.g., Melzack and Casey [21]). Thus, some authors explicitly recommend a "biopsychosocial model of pain," where the biological underpinnings of pain are recognised as just one determinate of a subject's pain experience, with biological factors modulating, and in turn being modulated by, an individual's psychological make-up and social context [5]. The polyeidic view attempts to expand the multidimensionality recognised in the biopsychosocial model in two directions: first, it holds that the standard triumvirate—dividing sensory/cognitive/affective factors—is insufficiently nuanced. For, as we suggest below, there are important distinctions to be drawn *within* the social and psychological dimensions. Second, the polyeidic view attempts to explain (at least in part) *why* modulation of experience by these social and psychological factors is possible in the first place, by pointing to the nature of the folk concept of pain as inherently complex (i.e., as containing parts which can be operated on). The possibility of modulation emerges, the polyeidic view claims, *because* the ordinary, folk concept of pain is itself multifaceted.

To begin to see how this might be possible, the polyeidic view turns to the third point above—the recognition that pain terms are scalar in nature. Recently in

[10] We don't want to commit here to a view about whether experiential and affective elements can in fact be held apart or must comprise a single dimension.

philosophy of language and linguistics there has been a great deal of work on scalar terms in general and it has become apparent that scalar terms come in different varieties: some (such as "rich") rank objects on just a single scale, while others (such as "intelligent," "healthy," or "red") are capable of ranking objects across a range of different dimensions. So, for instance, it is possible for one object to be "bluer" than another if it is closer to a paradigm blue, or if it has a greater colour intensity, or if more of its surface is coloured blue; or again, if we rank one food item as "more healthy" than another, it makes sense to ask "healthy in what respect?" as there are different ways in which something can be healthy or non-healthy [22–24]. In a similar fashion, then, the polyeidic view suggests that pain terms incorporate a range of different dimensions along which pain can be ranked. The polyeidic view thus claims that:

(a) the concept of pain is a concept that amalgamates a number of distinct dimensions
(b) these ideas or dimensions can conflict with one another[11]
(c) people hold latent positions on these rankings
(d) rankings are subject to predictable contextual influence
(e) positions on these dimensions have direct behavioural effects.

The polyeidic view takes seriously Melzack and Torgerson's insight that "The word 'pain'... refers not to a specific sensation which can vary only in intensity, but to an endless variety of qualities that are categorized under a single linguistic label" [25, p. 50].

Adopting a polyeidic model for the concept of pain makes it clear that there are two pressing questions that we need to settle if we are to clarify how we think and talk about pain[12]:

1. what exactly are the dimensions of pain?
2. how should we model these dimensions?

Providing an adequate answer to either of these questions unfortunately goes far beyond the scope of the current chapter, but we would like to provide an initial sketch of some possible answers here.

---

[11] So, for instance, the idea that pain is, on the one hand, a mental state, but also, on the other, a state which has a (non-brain) bodily location when instantiated, seems to involve a prima facie conflict; see Borg E, Harrison R, Stazicker J, Salomons T. Is the folk concept of pain polyeidic? [Mind Lang. 2019. Online first: https://doi.org/10.1111/mila.12227. 2019] §4, for further discussion of this idea.

[12] We might perhaps hope that answering questions (1) and (2) could help to provide an account of pain with a rather greater degree of normative force than that currently deployed in nursing, according to McCaffery and Beebe 1989: 7, whereby "Pain is whatever the experiencing person says it is, existing whenever the experiencing person says it does." [McCaffery M, Beebe A. Pain: clinical manual for nursing practice. St. Louis Missouri: CV Mosby Company. 1989. 353 p.]

3.1 The Dimensions of Pain

On the first question: there has been a relatively significant amount of work in both philosophy and psychology that directly or indirectly addresses the question of the dimensions of pain. For instance, a seminal work like the McGill Pain Questionnaire recognises a number of distinct sensory dimensions of pain, including descriptors like *burning*, *throbbing* and *stabbing* as ways of classifying different pains, while the IASP *Classification of Chronic Pain* 1986/1994 offers an extremely detailed taxonomy of chronic pain conditions, each with its own set of classifying features.[13] Furthermore, it is possible to see different philosophical theories of pain as focusing on different dimensions of a more generalised notion of pain. So, for instance, the imperative account of pain (see, e.g., Klein 2015) stresses the motivational aspect of pains (and thus is well-suited to classic acute pain, such as that arising from broken bones, burns, etc., but sits less comfortably with chronic pains and atypical cases like pain asymbolia) [26], while an evaluative approach, such as that proposed by Bain stresses the affective aspects of pain—the characteristic unpleasantness of pains—and thus offers a good explanation of their motivational force and their characteristic phenomenological profile, but has less to say about the bodily dimension of pains and their locative properties [27, 28]. What we suggest is that further work enumerating exactly what the dimensions of pain are and the extent to which any of the associated aspects are necessary or sufficient for pain will be useful (as this will help informed decision making in currently controversial cases, such as the debate over whether social pain is really a form of pain at all, or something more like distress). So, a truly adequate account of pain might need to distinguish some or all the following dimensions as relevant to our thinking about pain:

- mental/bodily dimension: accommodating the fact that folk often think of pains as paradigm mental, experiential states yet also as worldly states located in non-brain-based body parts [29].
- public/private dimension: accommodating the fact that folk often do think of pains as essentially private and yet they also often think of pain as the most publicly accessible of mental states (connecting to Wittgenstein's claim in the *Philosophical Investigations* that, in the right circumstances, there is just no room for doubting whether another is in pain when they seem to be in pain), with pain potentially having an ineliminable social aspect [14, 15].
- conscious/unconscious dimension: accommodating the fact that folk often think of pain as necessarily felt (initially rejecting the idea that someone could have a pain they are unaware of), yet, on prompting, will also usually allow the possibility of pain a person is not aware of (e.g., allowing that a subject might have a

[13]See also the 3-way definition of pain in Sternbach R. Pain: a psychophysiological analysis. [New York: Academic Press. 1968. 204 p.], the account of pain as "sensation plus affect," in Szasz T. Pain and Pleasure: a study of bodily feelings [New York: Basic Books. 1975. 372 p.], and Leder's account of what he calls "the experiential paradoxes of pain", Leder D. The experiential paradoxes of pain. [J Med Phil. 2016;41:444–60.]

pain they are momentarily unaware of when distracted, or that a person might be woken up by a pain, which seems to show folk allow that pains can exist before a person is aware of them).[14]

- aversive/non-aversive dimension: accommodating the fact that folk typically think of pains as unpleasant and thus aversive yet will also allow the possibility that pains are not always aversive (as in the case of eating spicy food or other forms of masochism).
- purposeful/non-purposeful dimension: recognising that folk think of pains as typically useful (i.e., as designed in order to indicate bodily damage) but also as potentially useless (e.g., in chronic pain conditions).
- motivational/non-motivational: accommodating the idea that pains are held to be motivational (they impel us to action to protect, favour or treat the injured part of the body) [26], yet folk allow that pain can sometimes fail to be motivational (e.g., when considering the case of pain asymbolia).
- various sensory or affective dimensions (including the degree of intensity, duration, etc.): recognising that pains can have different more fine-grained phenomenal properties which allow them to be ranked against one another along these dimensions (e.g., the characteristics used in the McGill Pain Questionnaire).

Once we recognise the rich complexity inherent to thinking about pain, the possibility opens up that at least some of the problems of poor pain communication have their roots in a failure to properly appreciate differences between individuals concerning how they stand on these multiple dimensions of pain. In the final two sections of this paper we thus consider the potential clinical implications of adopting a polyeidic view of thought and talk about pain.

4 Assessing Pain

The subjectivity of pain and the difficulties involved in effectively communicating pain pose an ongoing problem for clinicians who want to correctly diagnose and effectively treat their patients' pain. Developing a reliable and clinically useful method of measuring and communicating pain has been a major goal for pain researchers and has led to the creation of a family of pain rating scales [30].

The simplest of these scales are single-dimensional, and employ either ordinal, interval, or ratio scales that are intended to measure the *intensity* of a patient's pain. For example, the Verbal Rating Scale used by the University Hospital of Wales NHS Trust "is a four-point scale of no pain = 0, mild = 1, moderate = 2 or severe = 3."[15]

[14]The much-debated issue of "fish pain" is relevant here, as a cornerstone of the argument that fish don't feel pain is that they lack the requisite neural machinery necessary for consciousness and therefore can't feel pain. An acceptance of this argument by folk would seem to suggest that they hold that consciousness is necessary for pain.

[15]http://www.paincommunitycentre.org/article/pain-assessment-tools

Verbal Rating Scales are ordinal scales; there may be differences in magnitude between the intervals 0 and 1, and 1 and 2, etc., which are not captured by rating pains on the scale [31, 32].

Numeric rating scales anchor the endpoints of a numerical scale with "no pain" and "worst pain imaginable," and ask patients to rate their pain with a number from 0 to 10. Such scales might appear to be interval scales, which could represent something the ordinal verbal rating scale cannot, namely that equal differences between points on the scale represent equal differences in intensity of pain. For example, if numerical pain rating scales are genuine interval scales, the difference between "no pain" (a rating of 0 on the scale) and a rating of 1 should be the same as the difference between "worst pain imaginable" (a rating of 10 on the scale) and a rating of 9. But it is implausible that patients interpret numerical rating scales as involving equal distances between points, or even that in ideal conditions experimenters could establish such equal intervals between degrees of pain. Without such equal intervals, numerical rating scales should be treated simply as ordinal scales with more values than standard verbal rating scales.

Visual analogue scales employ either a vertical or horizontal line, typically 10 cm long, anchored at either end with descriptions like those used on the numerical pain scales: "no pain" and "worst pain imaginable" [30, p. S240].[16] Patients rate their pain by marking a point along the scale line, which clinicians can then measure with a ruler to determine the patient's pain rating from 0 to 100. Katz and Melzack claim that visual analogue scales are superior to both verbal analogue scales and numerical scales [33, p. 5]. They write:

> A major advantage of the visual analogue scale is its ratio scale properties. In contrast to many other pain measurement tools, equality of ratios is implied, making it appropriate to speak meaningfully about percentage differences between VAS [visual analogue scale] measurements obtained at multiple points in time or from independent samples of subjects.

A ratio scale requires, in addition to equality of intervals, an equality of ratios. That is, if a patient rates a migraine at 8 cm on the scale, then a broken rib that was rated at 4 cm would be 50% as painful. It is this property that Katz and Melzack give as an advantage over the verbal analogue and numerical rating scales of pain, and they cite the study conducted in Price et al. [34] in support of treating visual analogue scales for rating pain as ratio scales [34]. Price et al. [34] found that when patients were asked to indicate ratios of the intensity of pain sensations generated by heat pulses applied to the skin by a contact thermode, their responses indicated they were using the visual analogue scale as a ratio scale. For example, they consistently were able to identify a heat stimulus as producing pain that was twice as intense as a lower

[16]Kenny DT, Trevorrow T, Heard R, Faunce G. Communicating pain: Do people share an understanding of the meaning of pain descriptors? [Austral Psychol. 2006 Nov 1;41 (3):213–8] found that the pain descriptor that participants rated the highest was "unimaginable", which problematizes the use of "worst pain imaginable" as the anchor for the upper end of pain scales and raises the possibility that pain is a dimension (like height or cost) that does not have an upper bound.

level stimulus. But even if participants are able to consistently assess pain ratios, there is still the question of how they are understanding the meaning of the pain that they are measuring. For example, Williams et al. [35] document both inter- and intrapersonal variation in participants' understanding of what the upper endpoint of the visual analogue scale means [35, p. 461]:

One subject gave each of the following three descriptions [of the upper endpoint of the visual analog scale] at different points in the interview: "the worst pain I have yet experienced;" "the most severe pain I can imagine;" "the worst pain you yourself have experienced...probably the worst you could experience." The meaning assigned to the upper endpoint of the scale will affect the meaning of ratios of that upper value: 50% of "the worst pain I have yet experienced" is a very different value than 50% of "the most severe pain I can imagine," or "the worst pain you could experience."

4.1 A Multi-Dimensional Pain Rating Scale

One objection to the use of single-dimensional pain scales is that different types of pain may be *incommensurable*. The blinding, incapacitating experience of a migraine is very different than the mostly aching, but occasionally piercing pain of a broken rib, and "the pain of a toothache is obviously different from that of a pin-prick" [18 p. 278]. It therefore may not make sense to ask patients to rate the pain from migraines and broken ribs on the same scale.[17] Chronic pain also poses problems for comparing different types of pain on the same measurement scale, as Ballantyne and Sullivan observe: "When pain is chronic, its intensity isn't a simple measure of something that can be easily fixed" [36 p. 2099]. Indeed, although the overwhelming emphasis in the field is on quantitative assessment methods, in light of chronic pain the limitations of these approaches are clear. For although such measures are essential to understanding or targeting mechanisms or benchmarking pain management, they are not designed to describe personal experiences of pain, or essential personal attributes of pain, such as the burden or meaning of pain. Consequently, patients with chronic pain often do not feel understood by their health care providers [37].

The McGill Pain Questionnaire (MPQ) improves on single-dimension scales as it was designed to measure the variety of different dimensions of pain experienced by patients. The developers of the MPQ endorse a multi-dimensional conception of the meaning of "pain." As noted above, Melzack and Torgenson hold that [25, p. 50]: "The word 'pain'...refers not to a specific sensation which can vary only in intensity, but to an endless variety of qualities that are categorized under a single linguistic label." The MPQ was generated by asking participants to evaluate

[17]Price et al. [34], however, had no problem asking patients to compare the intensity of the pain generated by a contact thermode with the intensity of chronic back pain.

102 pain-related words and group together those that were qualitatively similar, yielding 20 categories, which are in turn organized into three classes: *sensory*, *affective*, and *evaluative* expressions. Participants were then asked to rate the words in each category in terms of "how much pain each word represents" [25, p. 52]. There was agreement among different groups of participants regarding the relative intensity conveyed by certain pain-associated words. For example, "crushing" was consistently rated as representing more pain than "pinching," and "pounding" was consistently rated as representing more pain than "flickering." On the assumption that intensity is a dimension along which the sensory, affective, and evaluative dimensions of pains can be assessed, the MPQ allows for four dimensions within which pains can be rated.

There is no principled reason for not expanding the dimensions of pain measurement permitted by the MPQ to include additional factors discussed above (perhaps expanding assessment from quantitative measures to more qualitative pain narratives) [38, 39]. For example, while the MPQ tracks qualitative aspects of pain and arranges them in terms of intensity, two pain sufferers or those treating pain may differ in terms of whether they classify pain as mental or as bodily, or in terms of some continuum between the two. They might also differ in the extent to which they think their sensations are meaningful signs of something wrong with the body (i.e., believing that pain always constitutes a warning about bodily damage) or the extent to which they view their pain as motivational. These kinds of differences might affect what kinds of treatments a patient or clinician might think is appropriate for alleviating pain.

5 Clinical Implications of the Polyeidic View

Among the assumptions of the polyeidic view are that individuals hold latent positions on these hypothesized dimensions and that these positions have direct effects on their pain-related behaviours, including their communication about pain. Nowhere are these modulatory effects more relevant than when an individual seeks medical care for their pain. In what follows, we will briefly discuss some ways that a polyeidic perspective and measurement of polyeidic dimensions might be clinically relevant.

Chronic pain patients are intense users of medical services. Communications between patients and their medical team (particularly at the primary care level) are critical for access to specialty care, adequate therapeutic interventions (pharmacological and non-pharmacological), compensation claims, and social support. In the absence of patient-specific information, physicians rely on schemas to make decisions, which leads to less effective treatment, particularly for women and visible minorities [40]. It is clear that effective communication is critical to appropriate care, but unfortunately, these interactions are frequently frustrating for patient and medical staff alike [1–4]. Many clinicians find it hard to accept pain without evidence of pathology and many patients feel stigmatised by clinicians who dismiss their reports or narratives of pain [41]. This situation can endure when clinicians apply current mechanism-based paradigms to pain reports/narratives that cannot be sufficiently explained [42].

One source of poor clinical communication is basic differences between patients and clinicians in beliefs about pain, which have the potential to change behaviours and to lead to fundamental differences in how a patient's symptoms (and their behavioural responses to those symptoms) might be viewed. To illustrate how such differences might affect a clinical interaction, we present hypothetical differences along three of the putative dimensions suggested above:

Mental/Bodily Dimension A physician has been trained to view pain as a symptom that will direct clinical investigation to some peripheral pathology which, once found and resolved, should result in the eradication of the pain. When testing indicates there is no evidence for peripheral pathology of imminent concern, the physician adopts a sceptical indifference. He tells the patient that the pain is "all in your head," with the implication that the patient need not worry about it. This advice makes no sense to the patient, whose concerns are triggered by the pain and anguish they feel, regardless of whether tissue damage is apparent.

Public/Private Dimension A patient's wound is being examined by a nurse practitioner. She asks how it has been healing, to which he responds "fine, no problems." The nurse practitioner, believing that the patient would communicate a problem if there were one, sends him home without any further treatment. Afterwards the patient's wife asks why he didn't mention the nearly constant pain he has been experiencing at the wound site. He responds that "nobody wants to sit and listen to my bellyaching; besides, she's not going to understand."

Purposeful/Non-purposeful Dimension A doctor prescribes a medication known to be effective for alleviating the chronic pain symptoms a patient has been experiencing. When the patient returns a month later, she asks how the medication worked. He tells her that he tried it once and it worked, but that he never took it after that. She asks why, he replies "I didn't want to mask the pain with drugs."

In each of these hypothetical examples, the clinical encounter has been ineffective based simply on differences in pain related beliefs that neither party may even be aware they hold. In the first, the physician has a far more body-centric view of pain than the patient. His concern about the pain is a function of its connection to injury, while hers is related to the feeling of suffering. In the second case, the patient views pain as something that one should keep to themselves and is sceptical that such a private experience even can be communicated, while the nurse practitioner not only presumes that pain can be effectively communicated, but that anyone who is able to do so, will do so. In the third case, the patient views pain as purposeful: It is trying to tell him something important and if he "masks" the pain with drugs, that message might not be delivered, and he might suffer further damage as a result, a position at odds with the doctor's belief that the pain is non-purposeful, and not conveying any message that justifies continued suffering.

In each of these cases, it is likely that clinician and patient leave the clinical encounter not quite understanding why it was unsuccessful, perhaps thinking the other party irrational or lacking in empathy. In fact, these behaviours are rational if viewed as reflections of these individuals' latent beliefs about the essential nature of pain. Were a measure available that could outline these beliefs, the outcomes of these encounters would be less mysterious, and the parties would have the opportunity to reflect on their own beliefs and whether the resulting behaviours are adaptive or not.

An intriguing possibility that arises is that beliefs that lead to ineffective communications and/or maladaptive behaviours could be targeted for intervention. In fact, changing a patient's emotions and behaviours by altering maladaptive beliefs is the goal of cognitive behavioural therapy (CBT) and other cognitive therapies which have been demonstrated to be effective for pain [43–45]. Within the CBT framework, it is not necessary that the beliefs be accessible to the holder. A key principle is that there are reflexive or automatic patterns of thought that influence how pain makes us act and feel. For example, an individual might become anxiety ridden about a minor pain in their ankle. Within a cognitive behavioural conceptualisation, this response might be mediated by automatic thoughts that the therapist must teach the patient to become aware of (for example, in this instance, the individual might feel that pain is always a sign of serious damage and, as such, will likely render them unable to maintain their beloved jogging schedule). Because these automatic thoughts are reflexive or pre-conscious, they remain largely opaque and difficult to recognize for patient and therapist alike. The polyeidic approach outlines a novel and clinically tractable new framework for elucidating these pre-conscious beliefs, first by defining measurable dimensions along which these beliefs might differ between individuals, and second by widening the scope of beliefs to include beliefs about pain's essential nature.

The fact that these beliefs might be reflexive and not readily accessible to the patient suggests some guidelines for what a measure of these latent dimensions might look like. Given that these beliefs might reflect latent positions along dimensions that both patient and clinician might not have consciously considered, it should not be presumed that a patient's superficial agreement with statements made by the physician (such as those that might be outlined in an "our approach" style mission statement) necessarily suggest agreement. Nor should it be presumed that such statements necessarily reflect the clinician's latent position in a way that is determinate of their behaviours (in the same way that, for example, someone might acknowledge that a person with conflicting political views might be good dinner company, while consistently avoiding such situations in reality). This latter point also raises another desired characteristic of a measure of these latent positions, namely that they have predictive validity. If such a measure mapped out dimensional positions but did not explain or predict pain related behaviours, it would have little practical or clinical utility, and might simply reflect abstract truisms that patient or clinician have been taught. Finally, in terms of utility, it is important that measurement of these latent positions be reliable and not dependent on the clinical dynamics that such measurements are intended to improve. As such, measurement with a standardized scale is preferable to simply obtaining the information qualitatively as part of routine clinical interview interviewing. The latter would be subject to clinician biases, or patients' desire to please their clinician, and would extend qualitative assessment in clinical settings which are already frequently time constrained.

6 Conclusion

Many extant philosophical views on the meaning of pain terms do capture important aspects of the content these expressions may communicate. However, we have suggested that it is important not to adopt too univocal a stance on the nature of this content, since pain terms can and do perform a range of different roles in natural language. A better approach would be to recognise the variety of dimensions of meaning which ordinary speakers associate with pain terms and which, we suggest, reflect underlying differences in the way that people conceptualise pain.

We have briefly outlined what a view taking seriously these different dimensions might look like (the "polyeidic" view) and have sketched how a view like this could have clinical utility. We suggest that, as part of a comprehensive pain assessment approach (which includes talking, listening, and observing patients, and having them complete standardised questionnaires, tasks, and physiological measures), the polyeidic approach could improve patient-clinician communication about pain, building stronger, more trusting patient-clinician relationships. This in turn can improve a patient's openness to trying cognitive-based pain management techniques and enhance the potential efficacy of such treatments. Given the current well-rehearsed problems with surgical and pharmaceutical treatments for chronic pain, this would be a valuable result.[18] However, it is also clear that further interdisciplinary research on the polyeidic approach is needed, first to establish properly the operative dimensions of our ordinary concept of pain, second to evaluate how the stances an individual adopts on these dimensions can act to promote or to constrain the pain sufferer's ability to engage in top-down modulation of their experience, and third to explore what kinds of psychoeducational interventions are most effective in mitigating harmful beliefs and enhancing those that are helpful.

[18]For the outcomes associated with surgical interventions see e.g., Taylor RS, Taylor RJ. The economic impact of failed back surgery syndrome. Brit J Pain. 2012;6:174–181. See also Brox J, Nygaard O, Holm I, Keller A, Ingebrigsten T, Reikeras O. Four-year follow-up of surgical versus non-surgical therapy for chronic low back pain. Ann Rheum Dis. 2010;69:1643–1648. For outcomes associated with long-term opioid medication, see e.g., Chou R, Deyo R, Devine B, Hansen R, Sullivan S, Jarvik JD, et al. The effectiveness and risks of long-term opioid treatment of chronic pain. Evid Rep Tech Assess (Full Rep). 2014;218:1–219. https://www.ncbi.nlm.nih.gov/books/NBK258809/. Also, Dowell M, Haegerich T, Chou R. CDC Guideline for Prescribing Opioids for Chronic Pain—United States. JAMA. 2016;315:1624–1645.

References

1. Woolf V. On being ill. The Criterion. 1926;4(1):32–45.
2. Levinson W, Stiles W, Inui T, Engle R. Physician frustration in communicating with patients. Med Care. 1993;31:285–95.
3. Matthias M, Parpart A, Nyland K, Huffman M, Stubbs D, Sargent C, et al. The patient-provider relationship in chronic pain case: providers' perspectives. Pain Med. 2010;11:688–97.
4. Upshur C, Bacigalupe G, Luckmann R. "They don't want anything to do with you": patient views of primary care management of chronic pain. Pain Med. 2010;11:1791–8.
5. Hadjistavropoulos T, Craig K, Duck S, Cano A, Goubert L, Jackson P, et al. A biopsychosocial formulation of pain communication. Psychol Bull. 2011;137:910–39.
6. Dorflinger L, Kerns R, Auerbach S. Providers' roles in enhancing patients' adherence to pain self management. Transl Behav Med. 2013;3:39–46.
7. Butow P, Sharpe L. The impact of communication on adherence in pain management. Pain. 2013 Dec;154:S101–7.
8. Mistiaen P, van Osch M, van Vliet L, Howick J, Bishop FL, Di Blasi Z, et al. The effect of patient–practitioner communication on pain: a systematic review. Eur J Pain. 2016;20 (5):675–88.
9. Bourke J. How to talk about pain. New York Times. 2014 July 12.
10. Halliday M. On the grammar of pain. Functions Lang. 1998;5:1–32.
11. Wierzbicka A, Goddard C. Words and meanings. Lexical semantics across domains, languages, and cultures, vol. 336. Oxford: Oxford University Press; 2013. 336 p.
12. Wittgenstein L. Philosophical investigations. 1st edn. Anscombe E, translator. London: Macmillan. 1953. 256 p.
13. Poole GD, Craig KD. Judgments of genuine, suppressed and faked facial expressions of pain. J Pers Soc Psychol. 1992;63:797–805.
14. Williams ACC, Craig K. Updating the definition of pain. Pain. 2016;157:2420–3.
15. Cohen M, Quintner J, van Rysewyk S. Reconsidering the International Association for the study of pain definition of pain. Pain Rep. 2018;3:e634.
16. Melzack R, Wall PD. The challenge of pain, vol. 368. London: Penguin; 1983. 368 p.
17. Field H. Pain: an unpleasant topic. Pain Supple. 1999;6:S61–9.
18. Melzack R. The McGill Pain Questionnaire: from description to measurement. Anesthesiology. 2005 July;103:199–202.
19. Brogaard B. What do we say when we say how or what we feel? Philosophers' Imprint. 2012;12:1–22.
20. Merleau-Ponty M. The phenomenology of perception, vol. 696. London: Routledge; 1958. 696 p.
21. Melzack R, Casey K. Sensory, motivational and central control determinants of pain. In: Kenshalo D, editor. The skin senses. Springfield, IL: Charles Thomas; 1968. p. 223–39.
22. Hansen N, Chemla E. Color adjectives, standards, and thresholds: an experimental investigation. Ling Phil. 2017;40:239–78.
23. Sassoon G. A typology of mutlidimensional adjectives. J Sem. 2013;30:335–80.
24. Solt S. Multidimensionality, Subjectivity, and Scales: Experimental Evidence. In: Castroviejo E, McNally L, Sassoon GW, editors. The Semantics of Gradability, Vagueness, and Scale Structure. Dordrecht: Springer. 2018. 293 p.
25. Melzack R, Torgerson W. On the language of pain. Anesthesiology. 1971;34:50–9.
26. Klein C. What the Body Commands. Cambridge, MA: MIT Press. 2015. 224 p.
27. Bain D. What makes pains unpleasant? Philos Stud. 2013;166:69–89.
28. Bain D. Why take painkillers? Nous. 2017;53:462–90.
29. Borg E, Harrison R, Stazicker J, Salomons T. Is the folk concept of pain polyeidic? Mind Lang. 2019. https://doi.org/10.1111/mila.12227.
30. Hawker G, Mian S, Kendzerska T, French M. Measures of adult pain. Athritis Care Res. 2011;63(S11):S240–2.

31. Jensen T, Karoly P. Self-report scales and procedures for assessing pain in adults. In: Turk DC, Melzack R, editors. The Handbook of Pain Assessment. New York, NY: Guildford Press. 1992; p. 135–51.
32. Stevens S. On the theory of scales of measurement. Science. 1946;103:677–80.
33. Katz J, Melzack R. Measurement of pain. Surgical Clinics. 1999;79:231–52.
34. Price D, McGrath P, Rafii A, Buckingham B. The validation of visual analogue scales as ratio scale measures for chronic and experimental pain. Pain. 1983;17:45–56.
35. Williams ACC, Davies HT, Chadury Y. Simple pain rating scales hide complex idiosyncratic meanings. Pain. 2000;85(3):457–63.
36. Ballantyne JC, Sullivan MD. Intensity of chronic pain — the wrong metric? New Eng J Med. 2015;373:2098–9.
37. Werner A, Malterud K. It is hard work behaving as a credible patient: encounters between women with chronic pain and their doctors. Soc Sci Med. 2003;57:1409–19.
38. Morse JM. Using qualitative methods to access the pain experience. Br J Pain. 2015;9:26–31.
39. Schiavenato M, Craig KD. Pain assessment as a social transaction: beyond the "gold standard". Clin J Pain. 2010;26:667–76.
40. Tait R, Chibnall J, Kalauokalani D. Provider judgments of patients in pain: seeking symptom certainty. Pain Med. 2009;10:11–34.
41. De Ruddere CKD. Understanding stigma and chronic pain: a state-of-the-art review. Pain. 2016;157:1607–10.
42. Woolf CJ, Bennett GJ, Doherty M, Dubner R, Kidd B, Koltzenburg M, et al. Towards a mechanism-based classification of pain? Pain. 1998;77:227–9.
43. Turk DC, Meichenbaum D, Genest, M. Pain and behavioral medicine: A cognitive-behavioral perspective (Vol. 1). Guilford Press. 1983. 452 p.
44. Eccleston C, Williams AC, Morley S. Psychological therapies for the management of chronic pain (excluding headache) in adults. Cochrane Database Syst Rev. 2009;2:CD007407. https://doi.org/10.1002/14651858.CD007407.pub2.
45. Thorn B. Cognitive therapy for chronic pain: a step-by-step guide. New York: Guilford. 2017. 355 p.

Chapter 15
On Saying It Hurts: Performativity and Politics of Pain

Grant Duncan

Abstract Pain and pleasure affect us all. Knowing this with empathy, and acting upon it, civilises us. Without such empathy, pain can become a means of domination and injustice. Moreover, pain is expressed and responded to in all social contexts, and the word "pain" has diverse meanings, depending on the associated activities. To observe various ways in which we say that it hurts, and the many meanings of *pain*, I follow ordinary-language philosophy, particularly Ludwig Wittgenstein and John L Austin, and I consider a range of social and historical contexts, from the closest intimacy, everyday chatter, the clinic, and beyond, to the domain of public policy and human rights. This addresses our verbal expressions of pain, their lived contexts and effects, within relationships and among social groups, altering mutual obligations, eliciting actions and reactions, and thence creating moral, legal and political norms. My aim, then, is to consider the social and political implications of ordinary performative pain-talk, in particular regarding the relationship between pain and justice, public policy, human rights and law.

Clinical Implications: Clinicians' roles in disability assessment and pain management are often affected by extrinsic concerns such as financial incentives and the risks of substance abuse. This chapter addresses those complex issues, illustrating how they have arisen in particular historical circumstances, and with political and economic causes and consequences. While unable to provide ready-to-hand clinical solutions, a pragmatic ordinary-language approach to conversations and disagreements about pain helps to chart the grounds on which these matters are debated.

Keywords Pain · Language · Speech-acts · Public policy · Human rights · Compensation · Political theory · Trust

G. Duncan (✉)
College of Humanities and Social Sciences, Massey University, Auckland, New Zealand
e-mail: L.G.Duncan@massey.ac.nz

S. van Rysewyk (ed.), *Meanings of Pain*,
https://doi.org/10.1007/978-3-030-24154-4_15

1 Pain, Certainty and Doubt

Pain and pleasure affect us all. Knowing this with empathy, and acting upon it, civilises us. Without such empathy, pain can become a means of domination and injustice. But we can only recognise and empathise with the pain of others in as much as they express it openly. This chapter looks more closely at the *linguistic* expression of pain and the associated political problems of justice and equality.

René Descartes (1596–1650) is normally cited for advancing a dualist and mechanistic theory of pain, which is reflected in some traditional biomedical approaches, but has since been debunked by contemporary biopsychosocial theory [1, 2]. In his *Meditations*, Descartes referred initially to (what we now call) "phantom-limb pain" in order to raise doubt about the existence of his body. If I can feel pain in a limb that has been amputated, then perhaps all pain sensations are only deceiving me, and so the body itself may not exist at all. However, later he used pain as evidence in his effort to prove the opposite: that his soul was intimately "intermingled" with a body. He reasoned that I cannot doubt that I think or that I feel pain. Pain signifies to the mind that there is something wrong with the body, and so the two substances, mind and body, must be united—somehow.

No one has since resolved the basic problem that Descartes posed: How do events in the nervous system relate to our *consciousness* of pain? But his dual-substance theory (immaterial mind and material body) is not a satisfactory explanation [3]. And Descartes fails to reflect critically on speech and language. Although he asserts "*I* think" as a pivotal premise, "he nowhere proves that thoughts need a thinker", nor that the first-person pronoun is anything more than "grammatically convenient" [4, p. 567]. To be able to say "I think" (or even, "that hurts"), one must first have learned a language, and no one learns to speak in isolation from others.

Nonetheless, pain impresses upon us our embodiment, supporting a phenomenological view that "the perceiving mind is an incarnated mind" *and* that "the body is wholly animated" [5]. Pain also evokes fears about our mortality and doubts about our well-being. We confront further doubts, however, when we communicate our pain and seek the empathy of others. And it is the verbal expression of pain and the responses thereto that preoccupy this chapter.

Ludwig Wittgenstein (1889–1951) traversed questions of language-use, pain and doubt, but averted the Cartesian problem. Early in life, he observed, we learn to use words such as *ouch*, *hurt* and *pain*. A child gets hurt and cries and so learns from adults how to use *pain* (or in German, *Schmerz*), thus learning "new behaviour", such that "the verbal expression of pain *replaces* crying, it does not describe [pain]" [6, p. 95]. We learn the meaning (or use) of the *word*. We don't learn *about* our pains; we simply *have* them. If I say, "I know that I'm in pain", then the phrase "I know" is nonsensical or redundant. It suffices to say, "I'm in pain". The pain is *expressed* verbally and non-verbally; it is not an incorrigible mental state that I "know about" and report on in a fashion unmediated by a particular language and socially acquired norms.

Similarly, "it makes sense to say about other people that they doubt whether I am in pain; but not to say it about myself" [6, p. 96]. So, when I talk about pain that I feel, the verbs *know* and *doubt* are redundant. It either hurts or it doesn't. In contrast, other people cannot directly know or experience the pain that I express openly, and it does make sense to say that they may doubt whether I am truly suffering pain. It is not easy "to imagine someone else's pain [...] for I have to imagine pain which I *don't feel* on the model of pain which I *do feel*" [6, p. 108]. But, when we are convinced that another person does feel pain, then we pity the person, the *sufferer*, not the body-part that hurts—"one looks into his [or her] eyes" [6, p. 105].

Certainty *and* doubt are both associated with pain. At one extreme, the torturer is presumably certain that what he plans to do will be excruciatingly painful, and the victim's cries will attest to that. A compassionate person, on hearing the victim's testimony, will not question whether it hurt. But, in many circumstances in which we care for others, in the home or in the clinic, it may require patience to hear and understand, with confidence, a person's complaint of pain. Empathy may be strained, and doubts may arise if complaints become repetitive, if there is no sign of illness or injury, or if we believe there are reasons for faking or hiding it.

Human pain-expression ranges from the spontaneous cry or yelp, not unlike an animal, through to poetic description (such as Keats's "My heart aches, and a drowsy numbness pains/My sense") and to elaborate pretending or even pretending to pretend. We are able to express, to hide, to minimise, to exaggerate and to fake experiences of pain [7]. A skilled actor on the stage may convincingly simulate being in great pain—but no one offers or calls for assistance, as the activity itself informs the audience that they are witnessing "pain behaviour without pain." We judge the authenticity and aims of the pain-complaint and the complainant in specific contexts. And we are predisposed to believe and to trust some people and their "performances" of pain more than others. So, a person complains; a hearer classifies the complaint. Does the complaint signify that something is wrong with the body and that a caring response or medical assistance is called for? Is this only a cry for sympathy? Is it a ruse to get time off work? How often have I heard this complaint?

An ordinary-language approach can help to describe the subtle things that happen in conversations about pain, if we *abandon* the idea that the only function of language is "to convey thoughts—which may be about houses, pains, good and evil, or whatever" [6, p. 109]. There is no grammar or rule (or ruler) by which one learns to judge "accurately" the authenticity of pain-complaints. A person's expression of pain and the hearer's interpretation of that expression are inter-woven with activities and shaped by their contexts.

2 Meanings of Pain

A verbal expression of pain may be simply *ouch* (or in French, *aie*, etc.) But this could mean various things. Imagine a couple engaged in private intimate play. One of them exclaims, "Ouch!" Depending on what they are doing, the tone of voice and the facial expressions, it could be an accusation ("you're hurting me") or a command ("stop that") or a permission ("you can keep doing that"). Such expressions of pain are *performative*: they are speech-acts, or utterances that do things, such as to change the interaction or situation between the parties. As such, they have "success-conditions," which may include responses of empathy, care and relief that a person in pain receives from others. Among those who trust one another, moreover, it is implicitly understood that complaints of pain should be heeded, and desired action should be taken. From countless experiences of pain, we learn many things, including the limits of our endurance, the capabilities of our bodies, and the things to avoid. We also learn about who cares, and how much they may be trusted to care, if we get hurt. This learning began before our earliest memories, before we even learned to speak and to say, "it hurts." It has also been argued that the experiences, expressions and sharing of emotions, and of nurturing and empathy, which begin during infancy, are necessary steps in the evolution, and in each individual's development, of symbolic and linguistic capabilities [8]. Pain, along with the other emotions, may indeed have made a critical contribution to our development as symbolically interactive beings.

To express pain verbally, one does not have to use the word *pain*. *Ouch* or *it hurts*, for example, may be easier, and more readily adopted by children. Conversely, common uses of *pain* do not always fit the definition provided by the IASP: "An unpleasant sensory and emotional experience associated with actual or potential tissue damage, or described in terms of such damage". To say, for example, "Sorry to be a pain," or, "Here's twenty dollars for your pains," is to recognise norms and obligations between people that necessitate speech-acts of apology, making amends or discharging debts. Such reciprocation maintains or restores balance in our relationships. There need not be unpleasant sensation or emotion associated with, nor described in terms of, "tissue damage" on such occasions. To quote a rather more "imperious" example, Queen Victoria's 1858 Proclamation to the Princes, Chiefs and People of India instructed the British colonial authorities to "abstain from all interference with the Religious Belief or Worship of any of Our Subjects, on pain of Our highest Displeasure." She was not describing a sensation; she was bringing into existence a lawful obligation.

These observations are not meant to challenge the IASP definition, which was devised for clinical purposes. They only exemplify the uses of *pain* in the "messy" linguistic world in which we live. For the present purposes, no recognisable usage of a word is regarded as better or more accurate; they are simply differing uses for diverse activities and contexts.

The peculiar etymology of *pain* adds another dimension. Along with *penalty* and *punishment*, *pain* derives from the Latin *poena*. Romance languages have descendants of the Latin *dolor* (pain, anguish, grief, sorrow, suffering), such as the French *douleur*; while, in English, *dolour* is rare and archaic. *Pain* supplanted *dolour*, even while synonymous with *punishment*. For example, Thomas Hobbes (1588–1679), in describing punishments, mentions non-fatal "corporal pains," including "stripes, wounds, [and] chains" [9, p. 208]. The pain is the physical lashing, in this case, not the noxious stimulus or the agonising experience that it aimed to induce. And Hobbes was aware that humans are, by nature, motivated to avoid such pains.

Prior to modern bureaucratic states with impersonal rule of law, sovereignty was founded upon the power to grant or to end a subject's life. The direct infliction of bodily pain and harm was normative in judicial and penal processes for offences against the sovereign or against the sovereign's law. The ultimate sanction was death, carried out in public ceremonies, preceded by the infliction of excruciating pain [10, 11]. We still occasionally use the phrase "on pain of death," as the word "pain" encompassed the public duties of the executioner. American states that retain the death penalty, in contrast, are nowadays obliged to ensure that the execution occurs with as little bodily injury and pain as possible; although there are "botched executions" [12, 13]. Today, if one violates a prohibition imposed on pain of death, one's final hour ought to be as "painless" as the executioner can make it.

Punishment of any kind, however, exacts retribution and inspires deterrence. This is based on a belief (or a suspension of disbelief) that there are equivalences between calculated, inflicted suffering and unpaid debts or unforgiveable harms [14]. The weakness in this belief is revealed by oft-heard statements that no punishment or compensation suffices to bring back a deceased loved one or restore a severely disabled person to a former status.

The state-sanctioned and deliberate infliction of pain—for investigation, punishment or intelligence-gathering—is now unlawful, or "taboo," although it does still occur [15]. Since the eighteenth century, incarceration has replaced the lash and the gallows as the primary form of penalty. The utilitarian philosopher Jeremy Bentham (1748–1832) designed an ideal prison called "the Panopticon," claiming that it would eliminate the painful and punishing use of irons, and introduce instead a "weightless" form of correction through confinement and surveillance [16]. There would be little bodily pain, but more penitence and correction. Bentham's calculative methodology included a classification of "pains": pains of privation, of the senses, of awkwardness, of enmity, of an ill name, of piety, of benevolence, of malevolence, of the memory, of the imagination, of expectation, and pains dependent on association [17]. This encompassed a generous range of sensory, cognitive, emotional and social phenomena, including, but far exceeding, the IASP definition. For example: "The pains of benevolence are the pains resulting from the view of any pains supposed to be endured by other beings. These may also be called the pains of good-will, of sympathy, or the pains of the benevolent or social affections" [17, p. 48]. Bentham recognised that it hurts to witness someone else's suffering. His general aim was to reform law and government in order to maximise the happiness of those who are governed, and so he viewed pains as "negatives" to be calculated and reduced, according to the principle of utility. Since his time, however, advances in anesthesia and pharmaceuticals have meant that pain is regarded more specifically as a sensory or emotional experience, largely falling under the authority of medical knowledge, and not often as a direct concern of government. And the pain-punishment nexus has been largely forgotten in contemporary understandings [18]. Hence, pain has a rich reservoir of historical and semantic associations, and the "arrival" of pain in the clinic, as a motive to seek help and as a symptom to aid diagnosis, is an important part of that history. But the messy ordinary-language uses of pain may lead to misunderstandings between patients and physicians.

3 Pain in Clinical Contexts

Pain has been an important symptom in medical practice since ancient times [19], but it is subjective and cannot be independently verified or inter-subjectively shared. There is no objective test for pain *per se*; the patient is the sole authority on his or her pains [20, 21]. This makes clinical assessment of pain reliant upon patient-clinician communication, along with relevant objective tests. The "epistemic asymmetries between the expert healthcare provider and the patient" [22, p. 31] are partially or temporarily reversed in favour of the patient. The clinician takes on trust the patient's verbal expressions of pain, and yet this may not eliminate doubts and biases.

Clinicians' assessments of patients' pain reports, and hence treatment decisions, are influenced by superficial characteristics, including age, gender and ethnicity. Judgements about the patient's authenticity when complaining of pain are prone to social biases, leading often to underestimation of pain and less analgesia. Clinicians' evaluations may also be influenced by their implicit judgements about patients' trustworthiness [23–26]. Not all patients' expressions of pain are regarded as equally believable. And individual and cultural differences mean that there are diverse norms and etiquettes about how much or to whom one may express pain. If the only noticeable difference between two patients in pain is their skin colour, and different decisions are made about pain medication, then there is an injustice. A wider politics of difference, recognition and discrimination surrounds pain assessment and treatment.

As a norm of clinical practice, "pain is what the patient says it is" [27]. But people's customs of language, appearance and action vary, and their self-expression is subject to variable interpretations. Observations of pain-expression are subject to personal and social biases, and sometimes hearers doubt a person's authenticity. On occasion, doubt is appropriate. Surely very few people can honestly say they have never hidden pain in order to save face, nor faked or exaggerated pain in order to evade an unwanted obligation. We doubt others' expressions of pain, knowing that we can simulate it ourselves, and this is a peculiarly human ability. Wittgenstein asks provocatively:

> Why can't a dog simulate pain? Is it too honest? Could one teach a dog to simulate pain? Perhaps it is possible to teach it to howl on particular occasions as if it were in pain, even when it isn't. But the right surroundings for this behaviour to be real simulation would be missing [6, p. 96].

Without doubting that dogs do feel and express pain, and that they could be trained to perform "pain behaviours," what are the "right surroundings" that Wittgenstein claims are missing here? A dog cannot *deliberately* simulate or lie about pain because "lying is a language-game that needs to be learned like any other one" [6, p. 96], and a language-game consists of "language and the activities into which it is woven" [6, p. 8]. A dog cannot speak and cannot tell lies, and hence it does not do things that involve "real" (or intentional) and not "taught" simulation.

So, our doubts about others' pain-complaints often reflect discriminatory biases or failures of empathy; nonetheless, any speaker of a language is able to simulate, and to disguise, pain. The best one can do, if in doubt, is to observe and listen carefully and to elicit more meaningful expressions—and hence to build trust.

4 Pain-Talk as Performative

Clinicians hear pain-talk in a constative (or descriptive) register, as symptom or sign[1] of pathology. The patient should accurately describe the location, severity, duration and quality of the pain to aid in diagnosis, alongside objective tests. The clinician should listen carefully and make an assessment. This contrasts with the performative interpretation of pain-talk. A performative utterance is one that *does* something, such as closing a meeting, launching a ship, or making amends [28]. An utterance may be both constative and performative, however, so there is no contest here over which is "correct." For instance, if I say, "Yesterday I promised to buy you a cake," I am *describing* or reporting something that happened; I am also confirming the obligation that was actively created by the original *performative* utterance ("I promise"). From a performative point of view, the basic genre of pain-talk is the complaint: an illocution that testifies that something is wrong, that influences others, and that seeks the perlocutionary effect of remedial assistance. If I say, "My head is killing me," I am using a metaphorical description; it may also be an oblique way of asking you for a pain-killer.

Viewing pain-expressive utterances as speech-acts, we set aside the idea that the speaker is trying to communicate a description of certain unpleasant sensations and feelings to a hearer. Instead of transmission of knowledge, we think of it as aiming to achieve some end or to shape a relationship or situation. We may express pain, rather than hide it, for instance, in order to elicit sympathy and relief. Most of the time, we are predisposed to believe others' expressions of pain as valid testimony based on a trust that complaints are genuine.

Models of trust in the patient-physician relation tend to be one-directional: that is, the patient knows less, is vulnerable, and hence has to trust the physician [22]. In the traditional Parsonian sick-role model, physicians have a fiduciary duty to act in the best interests of patients, and, in adopting the sick-role, patients are obliged to accept the authority of, and to cooperate with, their physicians [29]. Since Parsons' time, this relation of trust is regarded ideally as less hierarchical and more collaborative. But, the patient is typically represented as the trustor and the physician as the trustee due to asymmetry of expertise [22]. Trusting is a matter of reciprocity, however, and so we should ask, "In what ways do physicians trust patients?" Physicians trust patients to report symptoms accurately, follow advice, take prescriptions, pay their bills, etc. If we think of pain-expression as a kind of *testimony* (or, an invitation to the hearer to understand and believe the speaker), then it is a matter of the speaker's

[1]Pain is often referred to as "the fifth vital sign" in order to focus attention on unrelieved pain.

credibility [30]. How much do we "give the speaker credit"? Or, to what extent do we trust a person's testimony about pain?

The dictum "pain is what the patient says it is" enjoins us to trust, or to give the speaker full credit. But conversations about pain-relief and chronic pain can be challenging and complicated, sometimes questioning the patient's motives and credibility [31, 32]. If one party is the gatekeeper to prescription pain-relief, or if financial benefits may be at stake, surrounding factors come into play. Decisions about whom is entitled to drugs, health-care, compensation or subsidisation are ethical as well as clinical, and they are governed by statutory or common law and insurance policies. This means that pain and pain-relief are political concerns. The grounds on which a society allocates resources to relieving and compensating pain and suffering are "uneven" and open to accusations of bias or injustice. Advocacy groups advance the claims of people living with pain and disability, while conservative politicians and insurance companies seek to limit public and private liabilities [33].

Mutual recognition entails the awareness that others suffer pain, just as I do, and that others are aware that I am aware of that fact, and vice versa. Our full self-consciousness and belonging as *social* beings—who can understand "what we mean" and "how to act together"—depends upon inter-subjectivity of this kind. Such mutual recognition, however, is never a *fait accompli*; it is a process of continual dialogue, cooperation and struggle. Certain kinds of pain due to certain kinds of causes and the pain-complaints of certain kinds of people get greater credence, recognition and care than others. When this is due to non-medical factors such as "race," it is a form of social injustice and a cause for political contestation. In a democracy, liberal or socialist, one may reasonably demand, in principle, an equality of recognition of the "pains" of all citizens (or all sentient creatures), but the achievement, in practice, of an ideal "equality of treatment" is beyond reach, due to the lack of any direct neuro-imaging or other inter-subjectively comparable "measurement" of pain and suffering [20], and due to ongoing competition for social and economic resources.

5 Power-Relations and Pain

> If you prick us, do we not bleed? if you tickle us, do we not laugh? if you poison us, do we not die? and if you wrong us, shall we not revenge?
>
> *The Merchant of Venice*, III, i

Shakespeare's Shylock reminds us that the bodily experiences of bleeding and laughing, or generally of pain and pleasure, are much the same for all humans, regardless of our kind. His predicament raises critical questions about ethno-religious identification, discrimination and ghettoisation, in the face of which he demands recognition as a person who enjoys and who suffers. Even though we can neither experience nor inter-subjectively compare other people's pleasures and pains, we should recognise one another as similarly sentient-emotive beings.

Furthermore, Shylock's rhetorical questions implicate justice: "if you do us *wrong*, shall we not revenge?" From our basic *similarity* as bodies that suffer, the question of justice as *equivalence* arises to rebalance wrongs: three thousand ducats "for an equal pound of your fair flesh." But why?

The problem of revenge and equivalence was addressed by Friedrich Nietzsche (1844–1900), who asked: "How can inflicting pain provide satisfaction?" [14, p. 47] The human mind forgets much more than it remembers. And when one has made a promise, borrowed something, or acquired a debt, it is only too easy to forget about meeting one's obligations. Pain, however, serves as the most effective mnemonic. Debt was coupled with the pledge to allow the creditor a moment of cruelty—to inflict pain or injury—in case of default. A primitive means of ensuring repayment is a beating, or the threat thereof. It is from such relations of economic exchange that the legal subject *per se* emerged, Nietzsche argues. Pain *is* punishment and hence integral to the formation of legal systems. And the language of pain is eminently commercial. We say "ouch" when we receive an unexpectedly large bill, and payment is made "for your pains."

Nietzsche's genealogy addresses "the task of breeding an animal which is entitled to make promises" [14, p. 40]. He shows that there is no law of nature that links pain with justice, or with injustice. Promises are speech-acts, and systems of justice arise from our spoken and written claims upon one another. We all feel pain, but there is nothing pre-determined or "self-evident" about the *moral* significance of freedom from or infliction of pain. The relations between pain and justice, often taken for granted or considered "natural," have had to be bred into us and learned.

Our abilities to treat pain and suffering have expanded dramatically over time, moreover, and so our self-awareness as *deserving of relief* from pain has grown accordingly [34]. Nietzsche highlighted this historicity when he proposed that, in past ages when the infliction of pain was a *positive* tool for the advancement of life, "pain did not hurt as much as it does today" [14, p. 49]—although this need *not* encourage those stereotypes that held that "uncivilised races" were less sensitive to pain [34]. What outrages the modern person about suffering, Nietzsche asserts, is its meaninglessness. It no longer serves socially useful or redemptive purposes. Now we can effectively prevent and relieve pain; it is no longer a just punishment, divine or temporal; it is something that one "should not have to suffer." A failure to relieve "needless" pain is considered a "tragedy" [35].

Our standpoint in relation to the question of pain, equivalence, equality and justice has undergone historical changes. Not only do we now reject infliction of pain in the delivery of justice, we have also proposed a "self-evident" notion of a universal pursuit of, if not right to, happiness, or a pain-free existence. We now have a self-proclaimed "science" to guide us in this cause [36]. And much political theory today is premised upon the demand that everyone's pain and happiness be valued and treated equally, as a matter of justice.

For utilitarianism, equalisation is straightforward and exacting. Peter Singer argues that the child we can see about to drown in a duck-pond should excite neither more nor less sympathy and assistance than the starving child, whom we cannot see or hear, in a distant continent. And hence we owe to each endangered child *equally* a share of our resources if that's what it takes to save their lives. "Effective altruism" holds that a calculus based upon quality-adjusted life-years can advise us in which charities to invest. Applied ethics requires that we put ourselves in the position of others in order to prevent suffering [37].

According to Martha Nussbaum, the achievement of social justice requires compassion, or the recognition that other people's pain and suffering could be our own. And compassion requires "the belief in a common humanity." Hence, we need an account of human "powers and capabilities" that can be applied across time and place. "Otherwise, we will have no moral motive for other-regarding action toward people at a distance or toward people of other races and genders" [38, p. 239].

When Judith Butler talks of racism producing perceptions of some populations "who are eminently grievable" as compared with "others whose loss is no loss, and who remain ungrievable," then the implied call is for "frameworks of recognition" that would make "grievable beings" of us all [39, p. 24].

Underlying these three quite different theoretical approaches lies a common ethical dissatisfaction with unjustifiable inequalities of preventable pain and suffering, and with unequal distribution of the practical applications of care and relief, especially when poverty or discrimination has blocked our compassion or lessened entitlement. We are all equally worthy of care; we all live a life worth living; too many people are unjustly deprived of the resources needed to live with dignity. Pain and suffering unite us sympathetically as humans. With sufficient effort, we could extend this further to "show that we truly care about animal pain" as well. Morris argues that this would bring us closer to "a state in which we effectively address the pain of war refugees, migrant workers, rape victims, homeless veterans, and untold millions of sick and hungry children" [40, p. 391]. There is certainly too much preventable human and animal suffering, but should human victims' claims be in any way contingent upon a collective recognition of the suffering of animals? Are we to elevate animal suffering to the status of human suffering, and hence to raise claims for justice and equality for animals, or are we to reduce human suffering to that which any other mammal might feel?

The most affluent and civilised countries have taken great pains to prevent refugees and migrants from seeking safety and prosperity within their borders—from the Calais "Jungle" to Manus Island. And their domestic policies individualise and privatise responsibilities for the health-care and economic security of their own citizens, leading to gross inequalities of well-being. An otherwise well-intended respect for cultural differences and alterity, moreover, could deter us from venturing to understand or differentiate the experiences and suffering of others. Language and cultural norms do mediate the experiences and expressions of (that which we call in English) "pain" and "happiness" [41, 42]. And hence one may argue that, no matter how compassionate we are, there is no way to compare pain and suffering across cultures or between individuals. In some circumstances, our attempts to do so may even violate personal or cultural autonomy as they may be perceived as patronising or presumptuous. There is no certain foundation on which to judge whether one person's pain "hurts as much as" or "more or less than" another person's, or whether one person's pain deserves more care than another's, or how to resolve culturally diverse norms of responsibilities towards the sufferer. One cannot disagree *ethically* with Nussbaum's insistence on compassion for the suffering of all people of diverse ethnicity or gender. But the inability to weigh or compare—impartially, with justice—different people's pain and suffering provokes a logical problem. Does it lead us to accept each person's testimony of pain unquestioningly—even though we know that some may exaggerate or fake pain, while others may hide or minimise it? Or does inter-subjective non-comparability imply that one may pay *less* attention to the pain-complaints of others on the grounds that they express, in highly variable ways, "subjective" experiences of autonomous individuals? There is no *logically* compelling reason for following either path. We are still left with uncertainty as to the degree or quality of attention that the pain-complaints of others morally demand of us.

6 Pain, Law and Politics

The recognition, relief and infliction of pain is (and has always been) political, from the power to inflict pain as punishment through to the power to compensate for "pain and suffering" in litigation or legislation. I will outline two well-known matters of public policy that illustrate practically how pain is politicised. The first example involves formalised systems of compensation for "pain and suffering" arising from personal injury, based originally on the finding of fault, delivered either through royal judgement, or by magistrates and juries in courts, or more recently through social or private insurance on a no-fault basis.

Medieval civilisations developed systems such as *wergild*, blood-money or (in Islamic law) *diya* for settling disputes following an injury or homicide, to prevent the escalation of feuds and to recognise social status. In the ancient Germanic *wergild*, the "price" of a man's life, differentiated according to rank, was divided to recognise lesser damages such as injury, insult or other harm or loss, sometimes denominated in commodities such as cattle or furs in the absence of coin [43, 44]. This transformed private feuds, which risk ongoing retaliation, into enforceable obligations, creating "a unit of account to measure debts for the purpose of paying fines in compensation" [45, p. 84]. Increasingly, such fines or penalties became payable to the State. But common-law jurisdictions preserve the rights of injured parties to seek compensation directly from tortfeasors, for example in cases of negligence. In addition, private and social insurance systems have developed to cover the costs of personal injury. Workers' compensation schemes exist now in all industrial economies. In the past, they often applied a "table of maims" to determine lump-sum or ongoing payments to the victims of accidents; while today, they are more likely to use individualised assessments of impairment and functional incapacity. Naturally, bodily injuries cause pain, and functional abilities are limited by pain, but different jurisdictions recognise or directly compensate for pain *per se* to differing degrees and in different ways. Pensions for permanent partial disability, for instance, are in some jurisdictions awarded for, or adjusted through assessments of, chronic pain—but, in other jurisdictions, not.

Low back pain is now "the leading worldwide cause of years lost to disability," a burden that is growing due to population ageing, affecting especially the low- and middle-income economies [46]. Back pain was found to be "the leading cause" of disability benefits and pensions in Brazil, for example, and mostly this was pain due to "idiopathic" rather than "defined" cause [47]. This leads to political challenges, and governments have been encouraged by leading researchers to "tackle entrenched and counterproductive reimbursement strategies, vested interests, and financial and professional incentives that maintain the status quo" [46, p. 3].

What used to be called, pejoratively, "compensation neurosis" implies that the mere availability of financial compensation (from litigation, insurance or social security) changes the private perception of "pains" into publicly recognisable "injuries," and that efforts to prove to courts or administrators that one is genuinely incapacitated for work only delay recovery and rehabilitation. These concerns began with the earliest forms of industrial accident compensation and persist in contemporary times [48–50]. The pain-complaint may be conditioned by factors well beyond the unpleasant sensations and emotions, to include regulated medico-legal institutions that provide care, relief and/or financial support—and that sometimes decline or fail to do so. Such thoughts were advanced by the biopsychosocial model of pain. The psychological theory adopted by Fordyce and Loeser was behaviourism, assuming the contingent reinforcement effects of compensation payments; while their social theory was an economics of incentives [51, 52]. If pain-behaviour persists, it was argued, then some form of positive reinforcement is rewarding its persistence, and health-care, time off work, income-replacement insurance and social-security entitlements are the obvious candidates.

States permit or mandate combinations of common-law actions, private insurance, workers' compensation, veterans' administration and disability pensions. These are prone to "compensation neurosis" and "malingering"—and fraud—and hence there is political and clinical distrust in the testimonies of many patients who express pain, especially chronic pain. "Disability compensation is associated most consistently with distortions in self-reported pain, depression, and disability, both before and after rehabilitation has been instituted" [53]. The hearer may doubt the speaker's word about intrinsic pain due to such extrinsic factors.

The second example concerns the claim to a "right to pain management." There are only limited circumstances where people today may accept that pain serves a positive role: when it is autonomously chosen and endured, or sought for pleasure, in pursuit of valued religious, athletic, sexual or cultural objectives [54–56]. So, there are moments at which one could appropriately say, "If one can *endure* pain, then one ought to endure pain." Otherwise, it is more common to say, "one should not have to suffer pain." Access to effective pain-relief should be recognised as a human right enforceable through statute and even recognised in constitutions, it is said [57, 58]. And a serious governmental failure to provide health-care, such that many people suffer unnecessarily from pain, could, it is argued, breach the international-legal prohibition of torture and ill-treatment [59].

We have come a long way from those times when the infliction of pain was regarded as a normal and necessary legal tool; when pain and suffering were welcome signs of imminent salvation, and when many people underwent extreme pain during surgery. Today, in contrast, because one *can* relieve pain, then, as a moral and political duty, one *ought* to relieve pain—in as much as possible. This is reinforced by the IASP's Declaration of Montreal of 2010, upholding a universal human right to "access to pain management without discrimination."

Human rights are neither absolute nor perpetual, however. They have a history, and the history of human-rights claims is not only a history of collective responses to totalitarianism and mass atrocities, it is also a history of technical innovation. People used to live quite happily without the internet, for instance, and the idea of a human right to internet-access was inconceivable. But such a right is now claimed and advanced, as an extension of rights to freedom of expression, association and cultural development. This is a consequence of technological inventions, not a consequence of "the essentially human." Declarations of rights move with the times, as technological capabilities and social norms change, enlarging our political obligations to one another. So, it is not simply that "being human" determines what our rights are; innovation and the realisation of "new" human rights transform our perceptions of what it is to be "fully human."

The convergence of pain management and human rights is a product of the effectiveness and availability of pain-relieving substances. This invokes the Kantian principle that "ought implies can": if I am morally obliged to do something, then logically I am able to do it. Without the pharmaceutical "can-do," there'd be no justification for saying "should do." But there are unjustifiable global inequalities of access to pain management, especially in palliative care, and especially affecting poorer countries [60]. This inequality of supply can be overcome and is morally unacceptable. At the other extreme, however, many thousands of Americans die each year from overdoses of pharmaceutical opioid analgesics [61, 62]. And so, political responses are called for to address the shortage of pharmaceuticals in some places, and the over-supply of them in others [63].

7 Pain and Human Rights

As it *threatens* to reduce any human being, no matter how civilised or elevated, to the status of an animal in misery that welcomes death, pain equalises us. At its worst, it can rob us of the dignity on which the abstract equality of universal human rights was conceptually founded. Political debates in many countries about the legalisation of voluntary medically-assisted dying for those who suffer severe and intractable pain, especially during incurable and terminal illness, illustrate this point. But the insurmountable privacy of pain means that, publicly, it can be neither equally evaluated nor equally relieved. There is no consistent method that can be regarded as objective or impartial. Verbal and non-verbal expressions of pain, and our interpretations thereof, are subtly and continuously variable, from moment to moment, and from person to person. They do not fall into fixed categories of meanings; they do not signify determinate qualities or intensities. The uncertainty that this entails is thus constitutive of the very means of mutual recognition of our pains [64]. Hence, the uncertainty need not be regarded as a shortcoming in our perception that must be overcome by "more science." A fair comparability of our different needs when in pain is lacking because there is neither a common measure of pain nor a "common-wealth" of pain-relief. In pain, we are thrown back upon a "non-justiciable" subjectivity. So, how may human-rights discourse assist?

In the twentieth century, two world wars and other conflicts reshaped borders, creating new nations and new minority-groups, and forced millions of people to migrate, rendering them homeless, stateless or without protections under a rule of law. At worst, millions were stripped of citizenship and exterminated. The ideals of universal human rights that had evolved since the eighteenth century found their highest expression in the United Nations, in the aftermath of the gravest of "crimes against humanity;" although one should not assume a linear historical progress [15]. Today, the sight of millions of refugees who suffer the pains and humiliations of beatings, torture, rape, hunger, and exposure to heat and cold show we are as far as ever from designing (let alone constructing) frameworks of mutual political and civil recognition that could comprehend each individual's pain, equally weighted, and with full effect within a regime of justice. The idea of universal human rights gathers its force, however, from the appearance around the globe of large multitudes rendered stateless or displaced by war and ethnic cleansing.

In discussing the paradoxes that have arisen from declarations of universal human rights, Hannah Arendt referred to "the disturbing miracle" that each person is "single, unique, unchangeable," given an ineffable and unencodable inner privacy that radically differentiates each one of us from every other. This privacy presents a permanent challenge, or even resistance, to a public life in which we appear as persons with rights. No *equality* of rights is guaranteed to us by virtue of merely existing. Such equality as we can claim "is the result of human organisation insofar as it is guided by the principle of justice" [65, p. 301].

An appeal to universal rights encompasses every *person* by virtue of being *human*, neither more nor less. As Aristotle noted, however, animals give voice to pain, much like humans. But the feature that distinguishes a human association from a herd is its language, which enables reasoned deliberation over right and wrong, just and unjust. And this implies belonging to a particular community or state with its particular language. The idea that the human being as such (without particular association or shared language) has a right to recognition of his or her pain springs from compassion, of course. Paradoxically, it also reduces the person to a member of "humanity" in general, as a species. The central figure of rights discourse, after all, is the abandoned migrant, denied full recognition and protection as a citizen of a state. And it is by means of the state that rights and duties have historically been allocated and most effectively upheld [66]. Talk of universal human rights, rather than the particularity of citizenship, risks the reduction of the sufferer to *nothing but human*, in the abstract, detached from community or nation. The pain-sufferer would present as a living body, partaking of a universal "humanity," with an affliction that cannot be publicly verified, detached from the shared actions and experiences and the warm compassion of family, home and community.

So, what does this mean for the ideal of a human right to pain management? Declarations of human rights, as speech-acts, have illocutionary force (on saying "we declare," we have indeed declared something), but their perlocutionary effect is complex. That is to say, on declaring a right, we have not (yet) increased the enjoyment thereof, nor reduced the correlative evils or pains. Declaration of a right of all people to pain management neither increases the availability of pain-killers nor reduces the suffering in the world. To have effect, there must be the practical economic means to provide, the cultural acceptance of moral obligations to care, and lawful ways to enforce the right. Rights are ineffective without enforcement. Case law from Anglophone jurisdictions shows that the right to relief from pain can been recognized by courts, even though there remain some practical concerns about resource-allocation and about the weight given to symptoms or underlying causes [67]. It is on domestic law and public policy, rather than universal human rights, that enforcement of any legal right to pain management will depend.

Furthermore, one of the contradictory usages of human rights is to conflate "I have a right to X" with "I want X, and it should be given to me" [68]. The acceptance by clinicians of a universal right to pain management, followed by the growth in consumption and abuse of opiates in the USA and hence by dramatic rises in overdose rates, drug diversion and supply to addicts, led to questions about the unintended consequences of pain-screening practices and over-treatment [69–71]. The excessive growth in the prescription and consumption of opiates was facilitated, moreover, by educational programmes and financial contributions for medical practitioners and associations from the major pharmaceutical manufacturer [72, 73]. It was as if, for many thousands of people, "the right to kill pain" should mean implicitly "the right to kill myself." An over-supply of opiates may have created an unintended form of medically-assisted dying for those who cannot endure the pain of being human, mirroring the fact that they are reduced to "nothing but human." One must ask, "When did their communities cease caring for them as fellow citizens?"

Pain was once a prerogative of rulers, inflicted as revenge for and deterrence against violations of civil law. For Christians, it has also been a source of spiritual instruction, correction and redemption [34]. But rights-discourse has taken the place of the sovereign decision and religious orthodoxy. It has over-stepped the boundaries of modern nationhood and citizenship, within which humane redistribution policies, public-health measures and civil liberties made political sense, to encompass international relations, warfare and "humanitarian intervention" [15]. Furthermore, the language of rights, intended as a collective response to mass atrocities, tyrants and torturers, is now extended to ordinary forms of suffering to which everyone is prone, due to age, disease or accident, or due to anxiety and depression. A universal right to pain-relief would turn the care of the injured and the dying into a technical medico-legal matter, affecting a body in pain, no different from any other body, deracinated from pre-existing relationships, obligations, rituals and beliefs; symbolically removed from the social world in which the sufferer lived, one becomes a "consumer" in an economy over-supplied with addictive and lethal pain-killers.

8 Conclusion

Pain is political. The politics of pain tells a complex history. The uses of and responses to pain contribute to the formation and transformation of politico-legal subjectivities. The multiplication of the meanings and practices concerning pain in everyday private and public life should be acknowledged, moreover, in their performative effects. All of this is relevant to understanding the politics of the clinician—patient relationship and the policy environment in which medicine is practised in particular jurisdictions. It helps us to comprehend the politics of health-care, compensation, disability and pain management.

Torture has been said to "unmake" meaning, but pain is also an integral effect of routine productive labour [74]. In general, humans *make* meaning through their exposure to and expressions of pain. Pain is expressed *meaningfully* and *productively*. It engenders words with socially-shared memory and moral encoding, invokes interpersonal and institutional obligations, and generates law. Pain necessitates expression, using a performative language that initiates activities and reinforces obligations; often through highly regulated industries of health-care and pharmacy. As punishment, or as a call-for-relief, pain forms the legal-political subject. Pain is more than "an unpleasant sensation or emotion;" it is also a catalyst of the meanings of justice and equality.

The expression of pain and the kinds of attention it is given rely upon, and sometimes undermine, the trust that is reciprocated in relationships. The pain-complaint is taken as a kind of testimony, the authenticity of which may be judged by anyone who hears it. The hearer is often (but by no means always) a physician who is the lawful "prescriber" of care and relief.

Acknowledgement I am indebted to Dr. John Matthewson for his thoughtful peer-review of an earlier version and to Dr. Simon van Rysewyk for his helpful suggestions and careful editorship.

References

1. Melzack R, Wall P. Pain mechanisms: a new theory. Science. 1965;150(3699):971–9.
2. Morris DB. The culture of pain. Berkeley: University of California Press; 1991. 354 p.
3. Duncan G. Mind-body dualism and the biopsychosocial model of pain: what did Descartes really say? J Med Philos. 2000;25(4):485–513.
4. Russell B. A history of western philosophy. New York: Simon and Schuster; 1945. 955 p.
5. Merleau-Ponty M. Prospectus (a report to the Collège de France). In: Solomon R, editor. Existentialism. New York: Modern Library; 1974. p. 250–3.
6. Wittgenstein L. Philosophical investigations. Chichester: Wiley-Blackwell; 2009. 256 p.

7. Hadjistavropoulos T, Craig KD, Duck S, Cano A, Goubert L, Jackson PL, et al. A biopsychosocial formulation of pain communication. Psych Bull. 2011;137(6):910–39.
8. Greenspan SI, Shanker SG. The first idea: how symbols, language and intelligence evolved from our primate ancestors to modern humans. Cambridge, MA: Da Capo Press; 2004. 366 p.
9. Hobbes T. Leviathan. Oxford: Oxford University Press; 1998. 576 p.
10. Foucault M. Discipline and punish: the birth of the prison. New York: Vintage Books; 1995. 333 p.
11. Spierenburg P. The body and the state: early modern Europe. In: Morris N, Rothman DJ, editors. The Oxford history of the prison. Oxford: Oxford University Press; 1995. p. 49–77.
12. Sarat A. Gruesome spectacles: botched executions and America's death penalty. Stanford CA: Stanford University Press; 2014. 288 p.
13. Steiker JM. The American death penalty: constitutional regulation as the distinctive feature of American exceptionalism. Univ Miami Law Rev. 2013;67(2):329–55.
14. Nietzsche F. On the genealogy of morals. Oxford: Oxford University Press; 1998. 195 p.
15. Moyn S. Human rights and the uses of history. London: Verso; 2014. 160 p.
16. Bentham J. The Panopticon writings. London: Verso; 1995. 169 p.
17. Bentham J. An introduction to the principles of morals and legislation. Oxford: Clarendon Press; 1996. 416 p.
18. Duncan G. The meanings of 'pain' in historical, social, and political context. Monist. 2017;100 (4):514–31.
19. Rey R. The history of pain. Cambridge, MA: Harvard University Press; 1995. 398 p.
20. Robinson ME, Staud R, Price DD. Pain measurement and brain activity: will neuroimages replace pain ratings? J Pain. 2013;14(4):323–7.
21. Seers T, Derry S, Seers K, Moore RA. Professionals underestimate patients' pain: a comprehensive review. Pain. 2018;159(5):811–8.
22. Buchman DZ, Ho A, Goldberg DS. Investigating trust, expertise, and epistemic injustice in chronic pain. J Bioeth Inq. 2017;14(1):31–42.
23. Ashton-James CE, Nicholas MK. Appearance of trustworthiness: an implicit source of bias in judgments of patients' pain. Pain. 2016;157(8):1583–5.
24. Hoffman KM, Trawaltera S, Axta JR, Oliver MN. Racial bias in pain assessment and treatment recommendations, and false beliefs about biological differences between blacks and whites. Proc Natl Acad Sci USA. 2016;113(16):4296–301.
25. Schäfer G, Prkachin KM, Kaseweter KA, Williams AC. Health care providers' judgments in chronic pain: the influence of gender and trustworthiness. Pain. 2016;157(8):1618–25.
26. Goyal MK, Kuppermann N, Cleary SD, Teach SJ, Chamberlain JM. Racial disparities in pain management of children with appendicitis in emergency departments. JAMA Pediatr. 2015;169 (11):996–1002.
27. McCaffery M. Nursing practice theories related to cognition, bodily pain, and man-environment interactions. University of California Print Office; 1968. 207 p.
28. Austin J. How to do things with words. Cambidge, MA: Harvard University Press; 1975. 178 p.
29. Parsons T. Definitions of health and illness in the light of American values and social structure. In: Jaco E, editor. Patients, physicians and illness. New York: Free Press; 1958. p. 165–187.
30. Adler J. Epistemological problems of testimony [Internet]. Stanford CA: The Metaphysics Research Lab; 2017 [cited 20/09/2018]. Available from https://plato.stanford.edu/archives/win2017/entries/testimony-episprob/.
31. Miller L, Eldredge S, Dalton E. "Pain is what the patient says it is": nurse-patient communication, information seeking, and pain management. Am J Hosp Palliat Care. 2017;34 (10):966–76.
32. Wernera A, Malterud K. It is hard work behaving as a credible patient: encounters between women with chronic pain and their doctors. Soc Sci Med. 2003;57(8):1409–19.
33. Wailoo K. Pain: a political history. Baltimore, MD: Johns Hopkins University Press; 2014. 296 p.

34. Bourke J. The story of pain: from prayers to painkillers. Oxford: Oxford University Press; 2014. 411 p.
35. Melzack R. The tragedy of needless pain. Sci Am. 1990;262(2):27–33.
36. Layard R. Happiness: lessons from a new science. New York: Penguin; 2005. 401 p.
37. Taylor A, director. Examined life [Motion Picture]; 2009.
38. Nussbaum M. Human functioning and social justice: in defence of Aristotelian essentialism. Political Theory. 1992;20(2):202–46.
39. Butler J. Frames of war: when is life grievable? London: Verso; 2009. 224 p.
40. Morris DB. Animal pain: the limits of meaning. In: van Rysewyk S, editor. Meanings of pain. Cham: Springer; 2016. p. 389–401.
41. Goddard C, Ye Z. Exploring "happiness" and "pain" across languages and cultures. Int J Lang Cult. 2014;1(2):131–48.
42. Wierzbicka A. Is pain a human universal?: a cross-linguistic and cross-cultural perspective on pain. Emot Rev. 2012;4(3):307–17.
43. Grierson P. The origins of money. Res Eco Anthropol. 1978;1(1):1–35.
44. Hough C. Wergild. In: Schaus M, editor. Women and gender in medieval Europe: an encyclopedia. New York: Routledge; 2006. p. 831–2.
45. Wray LR. A new meme for money. In: Pixley J, Harcourt G, editors. Financial crises and the nature of capitalist money: mutual developments from the work of Geoffrey Ingham. Houndmills: Palgrave Macmillan; 2013. p. 79–100.
46. Buchbinder R, van Tulder M, Öberg B, Costa LM, Woolf A, Schoene M, et al. Low back pain: a call for action. Lancet. 2018;391(10137):2384–8.
47. Meziat Filho N, Silva GA. Disability pension from back pain among social security beneficiaries. Brazil. Rev Saúde Pública. 2011;45(3):494–502.
48. Deyo RA. Pain and public policy. N Engl J Med. 2000;342(16):1211–3.
49. Duncan G. Workers' compensation and the governance of pain. Econ Soc. 2003;32(3):449–77.
50. Figlio K. How does illness mediate social relations? workmen's compensation and medico-legal practices, 1890–1940. In: Wright P, Treacher A, editors. The problem of medical knowledge: examining the social construction of medicine. Edinburgh: Edinburgh University Press; 1982. p. 174–224.
51. Fordyce W. Back pain in the workplace: management of disability in nonspecific conditions. Seattle: International Association for the Study of Pain Press; 1995. 75 p.
52. Loeser J, Fordyce W. Chronic pain. In: Carr J, Dengerink H, editors. Behavioral science in the practice of medicine. New York: Elsevier; 1983. p. 331–46.
53. Hayes BJ, Craig KD, Wing PC. Diagnostic judgment: chronic pain syndrome, pain disorder, and malingering. BCMJ. 2002;44(6):312–6.
54. Güzel H. Pain as performance: re-virginisation in Turkey. Med Humanit. 2018;44(2) https://doi.org/10.1136/medhum-2017-011414.
55. Newmahr S. Power struggles: pain and authenticity in SM play. Symb Interact. 2010;33(3):389–411.
56. Rivera DC, Alex JL, Nehrenz GM, Fields BJ. Dancers' perceptions of injuries. J Music Dance. 2012;2(1):9–12.
57. Brennan F, Carr DB, Cousins M. Pain management: a fundamental human right. Anesth Analg. 2007;105(1):205–21.
58. Phillips CJ. The cost and burden of chronic pain. Rev Pain. 2009;3(1):2–5.
59. Amon J, Lohman D. Denial of pain treatment and the prohibition of torture, cruel, inhuman or degrading treatment or punishment. Interights Bull. 2011;16:172–84.
60. Manjiani D, Paul DB, Kunnumpurath S, Kaye AD, Vadivelu N. Availability and utilization of opioids for pain management: global issues. Ochsner J. 2014;14(2):208–15.
61. Division of Unintentional Injury Prevention. Understanding the epidemic [Internet]. Atlanta GA: Centers for Disease Control and Prevention; 2017 [updated 30/08/2017; cited 21/09/2018]. Available from https://www.cdc.gov/drugoverdose/epidemic/index.html.

62. Jones CM, Mack KA, Paulozzi LJ. Pharmaceutical overdose deaths, United States, 2010. JAMA. 2013;309(7):657–9.
63. Goldberg DS. Intervening at the right point in the causal pathways: law, policy, and the devastating impact of pain across the globe. Ann Health Law. 2013;22:198.
64. Wittgenstein L. Remarks on the philosophy of psychology, vol. II. Oxford: Basil Blackwell; 1980. 270 p.
65. Arendt H. The origins of totalitarianism. New York: Harvest; 1968. 548 p.
66. Agamben G. Homo sacer: sovereign power and bare life. Stanford: Stanford University Press. 213 p.
67. Arnold P. Is there a legal right to pain relief in New Zealand? Pub Int LJNZ. 2013;3:15–38.
68. Douzinas C. Human rights and empire: the political philosophy of cosmopolitanism. Abingdon: Routledge-Cavendish; 2007. 408 p.
69. Manchikanti L, Fellows B, Ailinani H, Pampati V. Therapeutic use, abuse, and nonmedical use of opioids: a ten-year perspective. Pain Physician. 2010;13(5):401–35.
70. Volkow ND, McLellan AT. Opioid abuse in chronic pain—misconceptions and mitigation strategies. N Engl J Med. 2016;374(13):1253–63.
71. Ahluwalia SC, Giannitrapani KF, Dobscha SK, Cromer R, Lorenz KA. "It encourages them to complain": a qualitative study of the unintended consequences of assessing patient-reported pain. J Pain. 2018;19(5):562–8.
72. Kelvey J. How advertising shaped the first opioid epidemic and what it can teach us about the second [Internet]. Smithsonian Magazine. 2018 [cited 21/09/2018]. Available from https://www.smithsonianmag.com/science-nature/how-advertising-shaped-first-opioid-epidemic-180968444/
73. Kolodny A, Courtwright DT, Hwang CS, Kreiner P, Eadie JL, Clark TW, et al. The prescription opioid and heroin crisis: a public health approach to an epidemic of addiction. Annu Rev Public Health. 2015;36:559–74.
74. Scarry E. The body in pain: the making and unmaking of the world. New York: Oxford University Press; 1985. 400 p.

The manufacturer's authorised representative in the EU is Springer Nature Customer Service Centre GmbH, Europaplatz 3, 69115 Heidelberg, Germany. If you have any concerns regarding our products, please contact ProductSafety@springernature.com

Printed and bound by CPI Group (UK) Ltd, Croydon, CR0 4YY
30/09/2026
02225434-0012